W9-BSG-418

About Island Press

Island Press, a nonprofit organization, publishes, markets, distributes the most advanced thinking on the conservation of our natural resources – books about soil, land, water, forests, wildlife, and hazardous and toxic wastes. These books are practical tools used by public officials, business and industry leaders, natural resource managers, and concerned citizens working to solve both local and global resource problems.

Founded in 1978, Island Press reorganized in 1984 to meet the increasing demand for substantive books on all resource-related issues. Island Press publishes and distributes under its own imprint and offers these services to other nonprofit organizations.

Support for Island Press is provided by Apple Computers, Inc., Mary Reynolds Babcock Foundation, Geraldine R. Dodge Foundation, The Educational Foundation of America, The Charles Engelhard Foundation, The Ford Foundation, Glen Eagles Foundation, The George Gund Foundation, The William and Flora Hewlett Foundation, The Joyce Foundation, The J. M. Kaplan Fund, and John D. and Catherine T. MacArthur Foundation, The Andrew W. Mellon Foundation, The Joyce Mertz-Gilmore Foundation, The New-Land Foundation, The Jessie Smith Noyes Foundation, The J. N. Pew, Jr., Charitable Trust, Alida Rockefeller, The Rockefeller Brothers Fund, The Florence and John Schumann Foundation, The Tides Foundation, and individual donors.

ENVIRONMENTAL RESTORATION

Environmental Restoration

Science and Strategies for Restoring the Earth

Edited by
John J. Berger
Restoring the Earth

Island Press
Washington, D.C. □ Covelo, California

Library of Congress Cataloging-in-Publication Data

Restoring the Earth Conference (1988: University of California, Berkeley)
Environmental restoration: science and strategies for restoring the Earth: selected papers from the Restoring the Earth Conference / John J. Berger, editor.
p. cm.
Includes bibliographical references.
ISBN 0-933280-94-7. — ISBN 0-933280-93-9 (pbk.)
1. Restoration ecology—Congresses. 2. Reclamation of land—Congresses. I. Berger, John J. II. Restoring the Earth (Organization) III. Title.
QH541.15.R45R49 1988
333.7′153—dc20 89-26719
CIP

Printed on recycled, acid-free paper

Manufactured in the United States of America
10 9 8 7 6 5 4 3 2 1

Contents

PREFACE

This volume is a result of a four-day national conference on ecological restoration held at the University of California, Berkeley in January 1988. The conference was convened by the staff of the Restoring the Earth organization of Berkeley, California, assisted by more than 100 volunteers, three cosponsors, and several nonprofit foundations. Restoring the Earth is a project of The Tides Foundation of San Francisco.

The meeting was held to consider the restoration of all major natural resource systems and the planning of environmentally sustainable urban areas. The intent was to focus public attention on restoration accomplishments and capabilities and to formulate new solutions to environmental problems through restoration. The aim was also to provide an opportunity for exchanging knowledge and information about restoration and its relationship to the environmental movement as a whole.

More than 1,000 participants, including more than 200 speakers, attended the conference, and many other interested people could not be accommodated for lack of space in our hall. Judging from the excitement and enthusiasm apparent at the meeting and the international publicity it generated, the gathering met its objectives.

The conference consisted of scientific and technical sessions as well as sessions for the general public and discussions of environmental policy issues. Keynote panels and plenary sessions provided participants with a forum for moving toward consensus on a restoration agenda for the future.

This volume consists mainly of the scientific and technical portion of the conference. It is divided in three principal parts treating aspects of (1) terrestrial restoration, (2) aquatic restoration, and (3) law, planning, land acquisition, and conflict resolution, all as related to restoration. A list of general references is provided at the end of the book, and an appendix contains information about obtaining portions of the conference not included in this book (because of space limitations, some papers are represented as abstracts in the pages that follow).

The publication of this book is but one of many positive developments stemming from the conference. Another direct consequence is the decision by the National Research Council's Water Science and Technology Board to proceed with a national study titled *Restoring Aquatic Systems: Science, Technology, and Public Policy.* The study is expected to begin in 1989 and to focus on the scientific, technological, and institutional causes for success and failure in aquatic restoration. The conference also led directly to the establishment of a new local organization in our community called Berkeley Citizens for Creek Restoration, which is currently working to build a constituency for restoration of now buried and channelized creeks of Berkeley.

On a personal level, the conference was a deeply rewarding experience for us at Restoring the Earth. It seemed to us that the conference was an important event for participants and that, through the media coverage that ensued, many people and organizations for the first time understood that our worsening environmental conditions make it imperative to restore our damaged natural resources, not just to conserve them.

This volume would not have been possible without the help of those who made the Restoring the Earth Conference possible, namely our staff, our volunteers, our advisers, and our sponsors. We are deeply grateful to them all for their generous help. We hope the information in the pages that follow will be of value and interest to you.

Acknowledgments

The dedication of many hard-working people whose contributions may not at first glance be evident to the reader actually made this volume possible. Kathleen Karn, Restoring the Earth's Program Manager, cheerfully contributed her extraordinary management talents, organizational skills, and many extra hours, despite having to contend with the needs of a toddler at home. Her diligence, interpersonal skills, and professionalism were crucial to the team effort that made this project successful.

Sally Smith, Project Coordinator, donated hundreds of hours of her time to Restoring the Earth for this proceedings volume. Ms. Smith served as our liaison with the volume's many authors and also was a principal member of the word processing team.

Restoring the Earth also is indebted to the late Professor Tom Dickert, Associate Dean and Director of the Center for Environmental Design Research. Dr. Dickert was a crucial early supporter of the Restoring the Earth Conference and took a special interest in its Proceedings volume. Until his untimely death, the Center was to have been the publisher of the Proceedings. Ms. Nora Watanabe, Dr. Dickert's Administrative Assistant at the Center, provided us with excellent administrative support by corresponding with authors, collecting the papers for the volume, accepting orders for the proceedings, and performing countless other important tasks.

Ms. Arlene Magarian contributed her time to assembling the Selected Bibliography section of this volume from my rough notes and files. Dr. Claire Chapin, another volunteer, generously assisted with the copyediting of the references. Our word processors were Tom Atlee, Ann Malamud, Gar Smith, Sally Smith, Beth Weinberger, and Stefanie Wasserman. Linda Healy, Betty

Lyman, and Jane Witkin helped with proofreading. Joan Rummelsberg contributed research to help us identify conference speakers, and David Patton provided valuable administrative assistance. Mary Dee Bowers was kind enough to lend us the computer and laser printer on which this manuscript was produced.

Special thanks to members of the Conference Planning Committee for giving us advice on conference content and to chairs of the scientific and technical sessions for assistance in screening papers in their fields of expertise. We also appreciate the advice on the manuscript we received through editor Ernest Callenbach at the University of California Press in Berkeley.

Last but not least, we thank Island Press Executive Editor Barbara Dean for her continuing enthusiasm for this project, her patience with the many delays, and her moral support.

Sponsors

Restoring the Earth Conference • 88 was cosponsored by the College of Natural Resources and the Center for Environmental Design Research of the University of California, Berkeley and by the San Francisco Bay Conservation and Development Commission. In addition, conference support was provided by: The Beldon Fund, The Bread and Roses Fund, Chevron U.S.A. Inc., Eschaton Foundation, The Evergreen Fund, The Wallace Alexander Gerbode Foundation, Jasmine Technologies, The Joyce Foundation, The Max and Anna Levinson Foundation, The New York Community Trust, Nu Lambda Trust, The San Francisco Foundation, The Sigma Xi chapter of the University of California, Berkeley, Smith & Hawken, The Threshold Foundation, The Tides Foundation, The Weeden Trust, and individual donors. Conference endorsers included Earth Island Institute and The Trust for Public Land.

Staff

Dianne Ayres, Conference Associate
Jerry Bass, Development Director
John Berger, Executive Director
John Cloud, Conference Field Trip Coordinator
Diane Delany, Database Manager
Kathleen Karn, Conference Coordinator and Program Manager
David Patton, Conference Associate
Marianne van Zeeland, Conference Coordinator

—J. J. B.

Introduction

John J. Berger

We now live in a critical time in human history. The air, land, water, and wildlife resources of this planet are being decimated—with astonishing speed. Rapid industrialization, militarization, and rampaging population growth throughout much of the world is destroying not only the quality of life but the earth's very capacity to support life. In a twinkling, through extinctions and habitat loss, the results of millions of years of evolution are being wiped out.

We are thus in the midst of an environmental emergency. And emergencies require extraordinary responses. Not to respond effectively can turn an environmental emergency into an environmental nightmare. Misguided attempts to conduct business as usual in an emergency can be suicidal.

Environmental problems today are unprecedented in nature and magnitude. This is the essence of the emergency: life as we know it today is being forever altered.

Major climatic disturbances are in prospect. Profligate fossil fuel burning and the simultaneous incineration of our tropical forests has raised atmospheric CO_2 concentration. In only a few decades, global temperature may increase several degrees. Swollen oceans may flood heavily settled coastlines throughout the world. World agriculture and food production will suffer. The half billion people already hungry in the world may find food even less affordable. And property losses alone will be in the billions—or more.

The protective ozone layer that shields us from ultraviolet rays, skin cancers, and cataracts is disintegrating. A continent-sized hole has appeared over the Antarctic, and that hole is growing larger as chlorofluorocarbons we release destroy stratospheric ozone. The ozone over the Arctic is also being attacked.

By cutting and burning tropical forests at an alarming rate, we now threaten to annihilate up to a fifth of all species on earth. The extinction rate is increasing globally. In Africa, massive desertification is scarring millions of acres, turning it to wasteland; African wilderness is vanishing; and much of the continent's wildlife is being slaughtered.

Here in the United States, fulfillment of clean air and water goals are postponed, and billions of pounds of toxic chemicals are being released into air and water.

While more than 50 percent of our wetlands have already been destroyed, the United States is still losing over 400,000 acres every year. In California, more than 94 percent of the wetlands in the Central Valley are now gone, along with about 90 percent of San Francisco Bay's salt marshes.

On dry land, billions of tons of topsoil above replacement rates are eroding from abused and pesticide-ridden lands. More than 20,000 toxic waste dumps now decorate the landscape, many slowly poisoning our groundwater. Measurable quantities of dioxin and DDT can now be found in human breast milk and body fat throughout the country. Meanwhile, our nation is running out of landfill space, and so much sewage and filthy runoff has flowed into our offshore areas that, from the Bay of Fundy to the Gulf of Mexico, zones of death are spreading where rotting waste has consumed all oxygen and no sea life can survive.

In the western United States, the scourge of air pollution, once confined only to large cities like Los Angeles, has become a pervasive problem, and the entire state of California faces an air pollution crisis by the end of this century if bold measures are not taken to avert it.

In the Northeast, thousands of lakes are dead or near death, and forests, too, are sick and dying from acid rain and other air pollution. Losses to agriculture from acid precipitation cost us billions. Meanwhile, the last refuges for embattled nature, our national parks, are losing species and are being severely degraded through overuse.

When the rain in the Northeast is 30–40 times more acid than normal; when we're losing billions of tons of topsoil a year; when we're tampering with the world's climate and may even melt the polar ice caps; when estuarine and offshore marine ecosystems are breaking down; when we may be killing a fifth of all the species on earth, that's my idea of an emergency.

In response to this crisis, we must mobilize the broadest possible societal coalition to stop further harm to our life-support systems. The damage is occurring so fast—scarcely an instant on a geological time scale—that we cannot act too quickly to stop it. Yet stopping it will not be enough: we must also repair the wounds made, both to prevent further deterioration and to recreate the living natural heritage that connects us biologically and historically to the past.

In caring for what was here before us, we are simultaneously looking toward the future, for we hold the earth in trust, and we have a moral responsibility to pass on a productive, resplendent Earth to its future stewards.

But how to do it?

I have been told that we should not publicize or even propose restoration because society will use it to excuse new assaults on the environment. Yet the technology of resource restoration is developing rapidly and cannot be wished away. Just as our understanding of atomic energy cannot be unlearned, so, too, the genie of restoration is "out of the bottle." Our task is to see that restoration is used properly—to repair past damage, not to legitimize new disruption.

An epochal development has clearly begun: For the first time in human history, masses of people now realize not only that we must stop abusing the earth, but that we also must restore it to ecological health. We must all work cooperatively toward that goal, with the help of restoration science and technology.

The science of restoration ecology is young and rapidly evolving. Restoration ecologists are currently developing and refining an ever-expanding array of technologies suitable for responding to a broad and growing range of environmental problems. Their efforts are also contributing fundamental insights to the science of ecology (Jordan et al. 1987). The papers in this book are but one example of the new and burgeoning literature on restoration (see Selected Bibliography, this volume).

Ecologists are not alone in contributing to the development of restoration ecology. The restoration movement is unusual in that it includes not only scientists but consultants, students, environmentalists, corporate officials, members of sporting and youth organizations, and ordinary citizens. These people have heard the earth's cry for help, and they have responded.

Few would dispute that the damage done to the earth is reprehensible and should be corrected. But debate often arises between restorationists and those who contend that the repair of ecological damage can and must wait until the more pressing task of conserving natural resources, especially biological diversity, has been accomplished.

Certainly we as a society must make heroic efforts at this time in our history to protect all remaining relatively pristine resources and to prevent any further loss of species. Threatened and endangered species unquestionably should not have to wait any longer for effective protection. But I am convinced that restoration also cannot wait and must be accorded very high priority.

The problem with the argument that restoration should be put "on hold" until biological diversity has been conserved is based partly on the fallacy that energy directed to resource restoration would otherwise have been focused on conservation of diversity or on the protection of pristine resources. Whereas this may at times be true, some of the people and resources engaged in restoration might otherwise not be active in conservation at all.

Moreover, even though the need to conserve biological diversity is absolutely critical today with vast numbers of species threatened by tropical deforestation, serious threats to diversity will be present for the foreseeable future in a world of exploding human population. We must not in the meantime freeze all other environmental work. And we must not be prevented from opening another much-needed front of environmental work. As the great conservation leader and restoration advocate David Brower once pointed out, a resource protected can always be threatened again by future proposed development, so eternal vigilance is required from the conservation community. In other words, the job of resource protection is an ongoing process that may never be complete. Therefore we cannot wait for its completion before setting to work on the tremendous backlog of environmental damage that awaits our attention.

Restoration and conservation are in general interrelated and complementary. Failure to restore damaged resources may damage pristine resources. For example, failure to restore a site contaminated with toxics may pollute a pristine aquifer with long-lived PCBs or permit difficult-to-remove heavy metals to find their way into stream and river sediments. Unreclaimed mines pose another serious hazard: rainwater or underground water from disrupted aquifers can leach sulfates or toxic metals, such as arsenic, chromium, or mercury from ore or tailings into surface water, killing aquatic life.

Damaged resources are often unstable and actively deteriorating. Failure to restore clearcut or otherwise disturbed hillsides may result in damage to streams, rivers, lakes, and fisheries as well as to engineered waterworks (dams and canals). In general if deterioration is not arrested, repair becomes progressively more expensive and difficult. Eventually the damage may become irreparable, as when the erosion of a steep slope causes the loss of topsoil down to bedrock. Successional processes may be set back a thousand years, and revegetation may become virtually impossible.

Failure to arrest deteriorating conditions through restoration can also lead to the loss of biological diversity by the extinction of endangered species dependent on the habitat being lost. By contrast, in his inspiring and unprecedented effort to restore hundreds of square kilometers of the very scarce dry tropical forest ecosystem in Costa Rica, University of Pennsylvania biologist Daniel H. Janzen is recreating habitat and providing seasonal refuges for numerous threatened species.

To cite another example, along western U.S. rivers, such as the Kern and the Colorado, native songbirds are threatened by loss of riparian gallery forests. In California, only a tiny percent of the native riparian forests remain today, but some of them are being restored by The Nature Conservancy along the Kern River and elsewhere. And restorations of native prairie may be important to the survival of endangered prairie plants, prairie-linked insects, and birds.

Restoration and conservation are related in another way, too. Degraded properties left in ugly and unappealing condition are sometimes more susceptible to being developed and irretrievably lost to conservation since their natural resource values are less evident, and hence they have fewer defenders. Neglected, abused, and derelict, these lands may be relatively inexpensive for the developer to acquire. But this is also an opportunity for the conservation and restoration forces to join, acquire the land, and see to its repair.

Another argument put forward by conservationist Brower is that by restoring resources, we can offer society alternatives to overusing or consuming the few unspoiled and truly natural areas left. A beautiful restored forest or prairie near a population center can provide a readily accessible, partial substitute for the increasingly overused and shrinking wilderness.

Looking ahead toward the future, we need to develop, test, and refine the science and technology of restoration ecology now, so it is capable of meeting the challenge of global ecosystem repair in the tropics and temperate zones tomorrow. Adequate technology is not likely to be available unless we support and nurture it today. Society shouldn't be forced to choose between restoration and conservation. We need and must do both.

Although restoration activity itself is far from new, it was uncommon earlier in this century.[1] However, a tremendous increase in restoration activity has been occurring in the United States, especially in the past 15 years. This is in part a response to the nation's increasingly grave environmental problems.[2] Many of the new restoration projects have been chronicled by Jordan (1983) and Berger (1985). The Restoring the Earth organization recently conducted a survey of restoration work just in the San Francisco Bay Area (Berger, Karn, and Witkin 1989) and discovered hundreds of restoration efforts just in that nine-county region; one consulting firm alone reported 200 projects.

Federal and state agencies are engaged in numerous kinds of restoration work (Berger 1989a). In particular the U.S. Fish and Wildlife Service has undertaken a large number of wildlife restoration projects (U.S. Department of the Interior 1987). Corporations, ranging from small consulting firms to major companies, such as Chevron USA,[3] are also showing increased interest in restoration. All this fieldwork has been accompanied by intensified academic interest in restoration and by the advent of restoration ecology as a distinct scientific discipline. A new professional association, the Society for Ecological Restoration, was recently formed with headquarters at the University of Wisconsin–Madison Arboretum.

Despite the newness of this field, it is already apparent and encouraging to see that restoration technology, along with traditional conservation efforts (preservation and pollution control), can greatly ameliorate many of our environmental problems.

For this reason, the demand for restoration is likely to continue growing for the foreseeable future, and the business of restoration is likely to become

---

[1]Antecedents of current restoration activity go back at least to the early twentieth century school of naturalist landscape architecture, exemplified by Jens Jensen and others. Soil, forest, and range restoration were extensively conducted during the Civilian Conservation Corps era from 1933–1943. Stream restoration also was occurring at least from the early 1930s. Major ecosystem restorations with modern ecological goals began in the 1930s at the University of Wisconsin's Arboretum in Madison. Certain branches of restoration ecology, such as lake restoration, are new areas of science (Cooke et al. 1986). For early federal restoration statutes, see also Trefts (this volume) and Berger (1989a).

[2]As Cairns (1988) has observed, "It is probably not an exaggeration to say that much of the planet is occupied by partially or badly damaged ecosystems." And according to Bradshaw (1987), ". . . destruction continues to outstrip restoration or natural recovery processes at an alarming rate. All types of ecosystems are being destroyed."

[3]For example, Chevron has provided support for bighorn sheep (*Ovis canadensis*) and meadow restoration projects in Yosemite National Park.

a multibillion dollar global enterprise (Berger 1989b). (If one includes spending for wastewater treatment and airborne emissions control, then this threshold has already been passed.)

As the papers in this volume indicate, a broad spectrum of serious ecological problems are amenable to restoration solutions: deforestation, desertification, endangerment of species, soil erosion, surface mining, degradation of rivers, lakes, and streams, and damage to coastal marine resources.

Sophisticated wildlife biology techniques can now be used to breed and release numerous endangered and threatened species, including some that once defied captive breeding efforts, such as the peregrine falcon (*Falco peregrinus*) and the California condor (*Gymnogyps californianus*).

Advanced wastewater treatment plants can now remove almost all kinds of pollutants from waste streams. Capital and energy costs for these facilities are high, however, and federal funding is being reduced for plant construction. Once the causes of ecosystem perturbation are corrected, water quality in lakes, rivers, and streams can often be partially or fully restored by instream and watershed-wide measures, or by both (Cooke 1986; Gore 1985). Marine ecosystems can be revitalized by eliminating toxins and other disturbances and by reestablishing salt marshes, seagrasses, and kelp beds (Thorhaug, this volume).

In general, the papers presented here document the need for and feasibility of restoring a wide variety of natural resources, from deserts and barrens to wetlands; from forests and mined lands to lakes and streams. Bainbridge, for example, draws attention to the pervasiveness of dryland and agricultural land deterioration in the United States and to the interaction of cultural and ecological factors causing the degradation. Thus he sets the stage for later papers that discuss techniques and processes of restoration, such as those by Dixon on the inexpensive revegetation of desertified or overgrazed land; Virginia on restoration of disturbed desert using woody legumes; Pickart and Guinon on coastal dune revegetation; Stritch on the restoration of barrens using fire; and others.

Cairns (1988) has pointed out that ecosystem restoration can be a tool for protecting biological diversity on the planet, and that restoration of an ecosystem requires the recreation of prior structure and function (Cairns 1980), including ecosystem services (Cairns 1986). Merely reestablishing indigenous species is not sufficient; ecosystem structure includes "spatial relationships, recruitment rates, population and community dynamics, predator/prey relationships, trophic interrelations" (Cairns 1986). Shonman (this volume) echoes some of these concerns in his treatise on resolving coastal restoration conflicts.

Readers of this volume will encounter many examples of the subtle and complex biological interactions that must be understood for restoration to succeed. Weiss, for example, demonstrates how the suitability of habitat for endangered butterfly species depends on interrelationships of topography, climate, and butterfly/host plant life cycles.

The papers of Perry and St. John provide insights into the role of rhizosphere organisms – particularly mycorrhizal fungi – in plant nutrition, soil structure, and soil fertility. Perry suggests that clearcutting and overgrazing severely reduce mycorrhizae, and thus impair nutrient cycling, soil aggregation, and water retention. In his research, Perry found that inoculation of conifer planting sites with soil containing mycorrhizae seemed critical to seedling survival.

St. John discusses techniques for reintroducing mycorrhizae and explains their role in nutrient uptake and in the conferral of drought resistance. He finds that natural reinvasion by mycorrhizae after clearcutting may be slow and that artificial reintroduction of mycorrhizae can dramatically aid reforestation.

Horowitz presents a case for "restoration reforestation" and against such current forestry practices as clearcutting, creation of even-aged forest monocultures, and the "overplanting of seedlings and the overuse of herbicides." The papers of Horowitz, Perry, and St. John taken together suggest that clearcutting may do much more than denude the landscape and predispose it to erosion: It may also damage soil structure, impair nutrient cycling, and reduce moisture uptake, all through the impacts on mycorrhizae. And because of possible slow reinvasion of certain mycorrhizal organisms, the damage can be long lasting. Soil deterioration initiated by clearcutting, overgrazing, or removal of noncommercial native forest species can begin a cycle of soil impoverishment that can render a site progressively more difficult to revegetate.

Responding to her concerns about other forms of land degradation – including excessive erosion and sedimentation – Sotir describes "bioengineering" approaches to land instability problems and distinguishes bioengineering techniques from those of conventional engineering. Among other distinctions, she points out that properly designed bioengineered structures tend to be self-repairing and to grow stronger with age. Also on the subject of erosion, Harding compares the effectiveness of various erosion control products and mulches. Regarding mined land, Covert gives the results of a treatment program for revegetation of acid mine spoils; and Kleo reports on a successful mined land reclamation program under harsh climatic conditions. Other authors discuss land restoration in urban settings.

In Part II of this book, Restoration of Aquatic Systems, Williams's study of the riparian vegetation along the Carmel River in California highlights the importance of considering the physiology and phenology of willow (*Salix*) species in riparian zone rehabilitation and protection. Whereas selecting species for drought hardiness instead of reestablishing the original species mix may initially seem advisable, it may ultimately degrade riparian habitat by reducing food resources available to insects, fish, and birds.

Also writing on the restoration of streams and rivers, Kondolf focuses attention on the need to understand the causes of recent historical channel changes as a prerequisite for design of a new stable channel. He points out the need to consider stream-groundwater interactions in determining minimum flow requirements on regulated-flow rivers. Describing the case of Sugar Creek in Ohio, Magsig relates how a program of removing stream obstructions saved Sugar Creek from channelization.

Other authors in Part II describe techniques and case study experiences in salt marsh restoration (Coats and Williams); freshwater marsh establishment (Silverman and Meiorin; Buckner and Wheeler); and in aquatic weed removal as a remedy for eutrophic conditions in Lake Okeechobee, Florida (Mericas and Gremillion). In one paper, Thorhaug describes the progress of an ambitious long-term restoration program on Biscayne Bay, Florida, and in another paper she provides an authoritative international review of seagrass and mangrove restoration.

Although papers such as Thorhaug's on Biscayne Bay in Part II and Bainbridge's overview in Part I include some issues of environmental policy and planning, it is in Part III, Strategic Planning and Land Acquisition for Restoration, that policy, planning, law, and conflict resolution are more fully addressed.

Sowl analyzes existing plans providing for preservation, restoration, and management of middle Missouri River oxbow lakes and proposes a new alternative plan that scores higher on critical decision criteria. Brumback and Brumback discuss land acquisition for restoration and protection in one paper and, in another, Barbara Brumback explains the comprehensive strategic planning approach being used to protect and restore the Florida Everglades. The latter paper begins with a fascinating ecological history of the Everglades and its degradation.

In the only paper of this collection to deal explicitly with atmospheric issues, Mussen reviews existing air quality regulations and proposes air quality management plans and intergovernmental coordination on air quality issues to protect and restore clean air. Ford, Glatzel, and Piro also emphasize the need for interjurisdictional coordination for successful watershed planning and restoration. Shonman underscores the need for including strong enforcement mechanisms to insure that restoration work mandated in development agreements is properly performed. Performance bonds, fines, and liens on property are among the enforcement mechanisms discussed.

Trefts is not only concerned about the need to enforce restoration agreements, but that resource managers generally are neglecting to provide for the repair of past ecological damage. She contends in a legal review of the public trust doctrine that it could be expanded to create a legal obligation to restore previously damaged resources.

This is indeed an exciting idea with broad political and economic implications. The nation has an enormous backlog of damaged natural resources of all types—forests, mined lands, coastal lands, crop lands, wetlands, rivers, lakes, aquifers, and streams. The damaged areas extend over millions of hectares and thousands of kilometers. In some cases, such as California's riparian forests, less than 1% of the original habitat remains today. America's dwindling wildlife is imperiled by all these conditions, as is the quality of human life.

Restoration of some damaged sites will not be possible or may not be desirable (Cairns, this volume). But it is imperative that large-scale resource restoration begin. I hope that many of the restoration techniques and ecological insights presented in this volume will be of use in this awesomely challenging but necessary enterprise. The same intelligence, energy, and ingenuity with which humanity subdued the earth is needed now to heal it.

References

Note: All citations without dates refer to articles in this volume; information on all other citations is provided below.

Berger, J. J. 1985. *Restoring the earth: How Americans are working to renew our damaged environment.* New York: Alfred A. Knopf and Doubleday & Co. [1987].

Berger, J. J. 1989a. Doctoral dissertation (in progress). Davis, CA: Graduate Group in Ecology. University of California.

Berger, J. J. 1989b. Reflections on the environment (essay). Berkeley, CA: Restoring the Earth.

Berger, J. J., K. Karn, and J. Witkin. 1989. *Restoration database and directory for the San Francisco Bay Area.* Berkeley, CA: Restoring the Earth.

Bradshaw, A. 1987. Restoration: An acid test for ecology. In *Restoration ecology: A synthetic approach to ecological research,* ed. W. R. Jordan III, M. E. Gilpin, and J. D. Aber. Cambridge, England: Cambridge University Press.

Cairns, Jr., J. 1980. *The recovery process in damaged ecosystems.* Ann Arbor, MI: Ann Arbor Science Publishers, Inc.

Cairns, Jr., J. 1986. Restoration, reclamation, and regeneration of degraded or destroyed ecosystems. In *Conservation biology,* ed. M. E. Soulé. Sunderland, MA: Sinauer Associates, Inc.

Cairns, Jr., J. 1988. Increasing diversity by restoring damaged ecosystems. In *Biodiversity,* ed. E. O. Wilson. Washington, DC: National Academy Press.

Cooke, G. D., E. B. Welch, S. A. Peterson, P. R. Newroth. 1986. *Lake reservoir restoration.* Boston, MA: Butterworth Publishers.

Gore, J. A. 1985. *The restoration of rivers and streams: Theories and experiences.* Boston, MA: Butterworth Publishers.

Jordan, W. R. III, M. E. Gilpin, J. D. Aber. 1987. *Restoration ecology: A synthetic approach to ecological research.* Cambridge, England: Cambridge University Press.

Jordan, W. R. III, ed. 1983-. *Restoration and Management Notes,* vol. I–VII.

U.S. Department of the Interior, Fish and Wildlife Service. 1987. *Restoring America's wildlife.* Washington, DC: U.S. Department of the Interior.

John J. Berger, author and environmental specialist, is Executive Director of Restoring the Earth.

PART I

Restoration of Land

Agricultural Lands, Barrens, Coastal Ecosystems, Prairies, and Rangelands

The Restoration of Agricultural Lands and Drylands

David A. Bainbridge

ABSTRACT: *Throughout the world, drylands used for agriculture and grazing are deteriorating. In the United States, drylands are also experiencing moderate-to-severe desertification and declining productivity. Lands once productive and profitable are being abandoned as a result of declining fertility, increased sensitivity to drought, high water tables, salinization, and groundwater overdrafts. Farmland productivity in humid and subhumid areas of the world is also declining as a result of unsustainable management practices. These lands can be restored by reversing the social and ecological factors that led to their deterioration and by establishing new incentives for restoration.*

KEY WORDS: *drylands, farmland, rangeland, restoration, reforestation, revegetation.*

Introduction

THE WORLD'S AGRICULTURAL lands and drylands are deteriorating as a result of mismanagement. Fertility of agricultural lands is declining, erosion is widespread, and production increases are not matching population gains in many areas.

The drylands are in the worst condition, as a result of inappropriate agricultural practices, overgrazing, and tree cutting for fuel. Recent estimates place the worldwide area of land affected by moderate-to-severe desertification at 22.5 million km^2 (Dregne 1986), an area two and one half times the size of the United States. More than a fourth of the drylands of the United States are more than moderately desertified.

The deterioration of these lands can be measured by the reduced productivity of desirable plants, alterations in biomass and diversity of the micro- and macrofauna, accelerated soil erosion, and increased risk for human occupants. The causes for this decline in drylands productivity include overgrazing, inappropriate dryland cultivation, overpumping of groundwater, deforestation, and poor drainage of irrigated lands (Sheridan 1986).

The problems of farmland deterioration in the United States first received national attention during the Dust Bowl years of the 1930s. After the Second World War new forces emerged that encouraged further land degradation. Foremost among these were the massive irrigation schemes of the West; the development of farm chemicals which offered the illusion of fertility maintenance; and the short-term control of pests and diseases needed to grow extensive monocultures of the same crop year after year. These problems were compounded by government research and regulatory programs which subsidized chemicals, energy, water, and commodities without concern for environmental or social impact.

The area of agricultural land that has been abandoned in the United States is not known, but Pimentel et al. (1976) suggested that an estimated 80 million ha have been either totally ruined for crop production or so heavily damaged as to be only marginally productive.

Range deterioration in the United States was most catastrophic in the late 1800s as a result of serious overstocking. Within this period the carrying capacity of the California range was reduced by half (Burcham 1957).

Land deterioration results from complex interactions of cultural and ecological factors. Finding solutions that will both prevent further decline and restore degraded lands will require an approach that combines ecological, technical, and cultural understanding of these problems. Developing restoration programs that work in the United States will provide a sound base for addressing similar problems elsewhere in the world.

Restoring Agricultural Lands

Most of the land suited for continued agricultural production is already in production but much is in very poor condition. The elements of a restoration program for farmland will depend on the soil, climate, cropping system, market, and the farmer's experience and skill. In general, a restoration program will include a decreased emphasis on chemical inputs, an increased diversity of crops, and an attempt to mimic the structure and function of natural ecosystems. In many cases, trees and animals are included in the farm system to provide better utilization of the farm resources.

A restoration program will often include: subsoiling or deep chiseling to break up compacted soil and facilitate root growth (Sykes 1946); use of manure, mulch, compost, or green manures to restore soil organic matter and biological activity, and improve soil structure and tilth (Pieters 1927; Turner 1951); primary reliance on biological nitrogen fixation rather than commercial fertilizers (Subba Rao 1982); conversion from moldboard plowing to conservation tillage (Sprague and Triplett 1986); establishment of a rotation program; intercropping and/or multiple cropping (Francis 1986); development of integrated pest management programs that maximize use of biological controls and minimize use of chemical controls (Huffaker and Messenger 1976); and establishment of windbreaks, hedgerows, and drain channel vegetation to control erosion (Bennett 1939). In addition the farm program should include a monitoring program to track conditions in each field and to ensure that both macro- and micronutrients removed in harvested crops are replaced.

Many excellent farm restorations have been achieved by improved management (Howard 1943; Berry 1981). Agricultural land restoration will be aided by the development of perennial grains and tree crops which can be grown with limited inputs and beneficial environmental impacts in areas where production of conventional annual crops can be very destructive (Jackson 1980; Bainbridge 1986; Wagoner 1986).

Restoring abandoned agricultural land can be relatively easy and economical because conventional farm equipment can be used for cultivation and seeding. Thousands of hectares of abandoned agricultural land have been restored to productive use as rangeland in the western United States. In southeastern Oregon, for example, range managers estimate that the livestock carrying capacity was doubled by restoration efforts (Heady and Bartolome 1976).

The low value and limited economic potential of much of this abandoned land makes low-cost restoration essential. Although long-term fallow periods will lead to revegetation in some cases, the native seed stock is commonly exhausted, and the soil structure and fertility have deteriorated sufficiently to limit or prevent revegetation. Where funding is limited, treatment may have to be limited to pitting or imprinting to increase surface roughness and infiltration. If more money is available, direct seeding with mixes of forbs, grasses, shrubs, and trees can be added.

Restoration of agricultural land in subhumid and humid areas is much easier than in arid areas; extensive areas of the United States that were in poor condition have been reforested by natural processes. In New England, for example, where the boom years for farming in the mid-1800s led to extensive conversion of forest to agricultural use, most of the marginal land has now reverted to forest. In the southeastern United States, direct seeding of oaks has proved effective for reforesting abandoned farmland (Krinard and Francis 1983), and trees should be an integral element in most restoration schemes (Smith 1988).

Every year more than a million hectares of agricultural land is developed for housing, highways, and other uses. Much of this land could be kept in production if development were more wisely managed. Village Homes, a 200-unit residential development in Davis, California, kept more than 17% of the land area in agricultural use (Bainbridge et al. 1978).

Restoring Drylands

Some of the little known yet important factors that are involved in restoration of drylands are the living soil crust, soil structure, chemistry, microbiology, microsite differences, and fire.

Soil crusts include lichens, ferns, algae, and other cryptogams. Cryptogamic crusts have been ignored by range managers until recently, despite early suggestions that they might be important for plant establishment (Booth 1941). More recent studies have shown that grazing can degrade the cryptogamic crust (Anderson et al. 1982) and reduce infiltration (Loope and Gifford 1972).

Other recent work on the effects of grazing on soil properties and plant succession has demonstrated the importance of microsite changes and the role these play in determining which species increase and which decrease (Eckert et al. 1987). They found that moderate trampling encouraged the establishment of desirable plants on range in excellent condition, but led to further deterioration of land in poor condition.

Fire was used extensively by indigenous people to manage vegetation. Fires of natural origin also strongly influence the course of plant succession. Fire suppression may have many unintended and unwanted effects and should be more carefully considered in land management (Minnich 1987). Fire is an important tool for grassland management and restoration (Berger 1985).

Dryland restoration is made difficult by the limited potential for immediate return on investment. Sheridan (1986) estimated that the cost of dryland restoration ranges from $60-3,000/ha. Presentations at the Second Native Plant Revegetation Conference held in San Diego (Rieger and Steele 1987) made it clear that even larger expenditures can lead to poor results if ecological and soil factors are not considered. Attempts to reestablish native vegetation without restoring native soil conditions failed completely in many cases. Yet plant establishment and restoration have been successful when ecological considerations are properly addressed (Virginia and Bainbridge 1987; Khoshoo 1987).

Principles of Successful Restoration

The foundation of an economical and successful restoration program is a clear understanding of the environment and the plants, animals, and people involved. A restoration program should begin with a study of the history of the land, its native vegetation (and human influences), the soil characteristics of comparable undisturbed native soils, and as much information as possible on the interactions between plants, animals, and humans. When this information is available, a draft plan for restoration can be developed.

The second step should be a series of test plots to evaluate the strategies for restoration that appear promising. This is particularly important in areas where little information is available. While the test plots are underway, a seed-collection program should be initiated, and seed nurseries should be established if needed to increase seed stocks.

The essential elements of a minimum-cost restoration effort are the introduction of appropriate seeds and related symbionts to microsites that provide suitable soil and moisture conditions for rapid root growth and plant establishment. This may include preparation of the soil by deep ripping, chiseling, and/or discing; application of soil crust inoculum (St. Claire et al. 1986); seeding with a complex mix of species inoculated with appropriate symbionts (St. John 1985; Virginia and Bainbridge 1986); and imprinting (Dixon 1982). Weed control can help slow-growing native plants to compete. Controlled burning at the time weed species are most vulnerable (Haverkamp et al. 1987) or soil solarization (Horowitz et al. 1983) can provide weed control without chemicals. Solarization involves moistening the soil and covering it with transparent plastic, letting the sun heat the soil, thereby killing weed seeds and many pathogens.

Other low-cost techniques that have been successful include the use of discs that have been modified to create pits which provide a variety of microsites and collect water. Large pits were found to be generally more effective than small pits in arid areas (Medina and Garza 1987). Pits are most effective on slopes of less than 8% where natural infiltration is limited (Vallentine 1980).

When very little money is available, the best option is simply to roughen up the soil surface. A rough surface increases infiltration and traps blowing soil and seeds. On some soils an imprinter would be most appropriate for roughening; others would benefit from use of a plow or disc pitter. This can often be done for less than $50/ha.

More expensive treatments will provide more rapid revegetation. These treatments might include ridging, catchment basins, mulch, and pest control. Ridging provides many benefits, including water collection and development of a microsite gradient that should provide favorable conditions for seeds over a wide range of precipitation (Medina and Garza 1987). Ridging is also very effective in areas that may experience waterlogging or standing water (Bainbridge 1987). Micro- and macrocatchment basins may also be used to aid in establishing vegetation (Shanan et al 1970).

Mulching and composting can also provide many benefits. Native grasses with seed are excellent for mulching if they are available (Wenger 1941; Schiechtl 1980), but straw is also of value. High application rates, from 5-10 tons/ha, with crimping to retain the straw, are desirable, particularly on erosive slopes (Schiechtl 1980; Kay 1987). Compost is also of value.

Other restoration program elements that may be of value include pest control (cages or fencing to protect plants), rodent control, limited irrigation, and fertilizer. Fertilizer should be used with care because it may increase shoot rather than root growth, increase weed competition, depress microsymbiont development, and make plants more palatable for pests. Adams et al. (1987)

found that even a slow release fertilizer decreased transplant survival in all cases and by as much as 90% in the worst case. Similar problems might be expected with direct seeding, and few experienced revegetation groups in the California deserts used fertilizer of any kind except on cut slopes and exposed subsoil (Virginia and Bainbridge 1986).

Transplants are expensive but make it possible to establish plants that are not easily started from seed in the field. Containers and nursery management should enable plants to develop a root system (with symbionts) suited for survival in a difficult environment. Deep containers may provide substantial benefits in this regard (Virginia and Bainbridge 1986). Transplants will usually require cages or screens to reduce grazing pressure from insects, rodents, livestock, and deer.

Timing of transplanting can be critical for establishment. Transplanting in the desert may be feasible only after a flood event. Lovenstein (1988) has achieved 95% establishment in the Negev Desert at such times. Even transplants that die may provide some cover and increase establishment of seedlings.

It may be desirable to combine expensive treatments, i.e., transplants, on a very limited area (1–2%), with contour strip treatments, i.e., pitting and direct seeding, on a larger area (perhaps 10–20%). This approach can establish seed sources for subsequent natural revegetation of the remaining land. Heady and Bartolome (1976) found that revegetating 10% of the land in their study area had a very positive effect on the remaining 90% by reducing grazing pressure.

Supporting Restoration Efforts

The key to developing a large-scale restoration movement is improving the understanding of ecological principles and land management in the general population (Bainbridge 1985). This can be done by developing appropriate curricula for colleges and schools. The reestablishment of a Civilian Conservation and Restoration Corps could also be included to provide training in the methods of restoration and field research.

One of the most important and currently neglected areas is the development of an accurate understanding of the condition and trends in land use and condition. The establishment of a U.S. Ecological Survey (U.S.E.S.) with status and funding comparable to the U.S. Geological Survey would be appropriate to undertake baseline monitoring and restoration studies. The general outlines for a national biological survey have been developed (Kim and Knutson 1986) and would provide a good starting point for the U.S.E.S., although it should have a strong restoration research and demonstration program. Much of the work could be done under long-term, ten-year cycle, competitive grants.

The redirection of agricultural research at the university and college level could be accomplished by changing program emphasis and funding availability. Sustainability and restoration should be the cornerstones of this program. The foundation for this would be long-term sustainability and self-reliance and economic efficiency in the most conservative sense (Bainbridge and Mitchell 1988).

The University of California, for example, might set up a series of restoration reserves, comparable to the current natural reserves, but specifically established on degraded lands for hands-on study of ecology and restoration. This would provide valuable training for students and would aid in developing techniques for farmers and ranchers to use on their degraded lands.

On the national level, the establishment of several major restoration reserves would be desirable. Three candidates for this are the proposal for a 40,000 km^2 wildlife area in Montana (Anon. 1987), an Ogalla buffalo range of comparable size, and a 300 km^2 California Grassland in the western San Joaquin Valley. Much of this land could be kept in private ownership under easements, with farmers paid for restoration work under a modified federal conservation program. As land health is restored, these farms could become self-supporting from tourism, hunting, and fishing. In wilderness areas where these uses will not be allowed, the land might be owned and managed by the state or National Park Service. Management of state, federal, and private land in these reserves could be coordinated by a working group representing the various participants, as has been proposed for the 1 million ha Greater Yellowstone Ecosystem.

Restoring the American earth may provide the challenge that is needed to keep our society vital, the meaningful work essential to provide a feeling of self-worth and an escape from poverty, and a solid ecological basis for a future of sustained abundance.

Acknowledgments

With special thanks for the assistance of Hugh Smith, Ross Virginia, John Rieger, Arturo Gomez-Pompa, Andrea Kaus, and Wes Jarrell.

References

Adams, T. E., P. B. Sands, W. H. Weitkamp, N. K. McDougal, and J. Bartolome. 1987. Enemies of white oak regeneration in California. In *Proceedings of the symposium on multiple-use management of California's hardwood resources,* 459–462. General Technical Report PSW-100. Berkeley, CA: U.S. Department of Agriculture, Pacific SW Forest and Range Experiment Station.

Anon. 1987. Back to Lewis and Clark (converting farmland to game parks in Montana). *Time* 129 (March 23):29.

Anderson, D. C., K. T. Harper, and S. R. Rushworth. 1982. Recovery of cryptogamic soil crusts from grazing on Utah winter ranges. *Journal of Range Management* 35(3):355–359.

Bainbridge, D. A., J. Corbett, and J. Hofacre. 1978. *Village Homes solar house designs.* Emmaus, PA: Rodale Press.

Bainbridge, D. A. 1985. Ecological education: Time for a new approach. *Bulletin of the Ecological Society of America* 66(4):461–462.

Bainbridge, D. A. 1986. *The use of acorns in California, past, present and future.* In *Proceedings of the symposium on multiple-use management of California's hardwood resources,* General Technical Report PSW-100, 453–458. Berkeley, CA: U.S. Department of Agriculture, Pacific SW Forest and Range Experiment Station.

Bainbridge, D. A. 1987. Restoration of wetlands using methods of traditional agriculture. Presented at the Conference on Wetfield Agriculture. Riverside, CA: University of California at Riverside.

Bainbridge, D. A. and S. M. Mitchell. 1988. *Sustainable agriculture in California: A guide to information.* Davis, CA: University of California, Sustainable Agriculture Research and Education Program.

Bennett, H. H. 1939. *Soil Conservation.* New York: McGraw Hill.

Berger, J. J. 1985. The prairie makers. In *Restoring the earth.* New York: Alfred A. Knopf, pp. 106–122.

Berry, W. 1981. *The gift of good land.* Albany, CA: North Point Press, pp. 205–211.

Booth, W. E. 1941. Algae as pioneers in plant succession and their importance in erosion control. *Ecology* 22:38–46.

Burcham, L. T. 1957. *California range land: An historico-ecological study of the range resource in California.* Sacramento, CA: California State Division of Forestry.

Dixon, R. 1982. *Land imprinting activities.* In *36th Annual Report Vegetative Rehabilitation and Equipment Workshop,* 20–21. Missoula, MT: U.S. Department of Agriculture Forest Service Equipment Center.

Dregne, H. E. 1986. Magnitude and spread of the desertification process. In *Arid land development and the combat against desertification: An integrated approach,* 10–16. Moscow, U.S.S.R.: United Nations Environment Programme/GKNT.

Eckert, R. E. Jr., F. F. Peterson, and F. L. Emmerich. 1987. A study of factors influencing secondary succession in the sage brush type ecosystem. In *Proceedings of the symposium on seed and seedbed ecology of rangeland plants,* ed. G. W. Frasier and R. A. Evans, 149–168. Tucson, AZ: U.S. Department of Agriculture/Agricultural Research Services.

Francis, C. A. 1986. *Multiple Cropping Systems.* New York: Macmillan Publishing Company.

Haverkamp, M. R., P. C. Ganskopp, R. F. Miller, F. A. Sneva, K. L. Marietta, and D. Couche. 1987. Establishing grasses by imprinting in the Northwestern U.S. In *Proceedings of the Symposium on Seed and Seedbed Ecology of Rangeland*

Plants, ed. G. W. Frasier and R. A. Evans, 299–308. Tucson, AZ: U.S. Department of Agriculture/Agricultural Research Services.

Heady, A. F., and J. W. Bartolome. 1976. Desert repaired in Southeastern Oregon: A case study in range management. In *Desertification: Process, problems, perspective,* ed. P. Paylore and R. A. Haney, 107–117. Tucson, AZ: University of Arizona, Office of Arid Land Studies.

Horowitz, M., Y. Regev, and G. Herzlinger. 1983. Solarization and weed control. *Weed Science* 31:170–179.

Howard, A. 1943. *An agricultural testament.* New York: Oxford University Press.

Huffacker, C. B., and P. S. Messenger. 1976. *Theory and practice of biocontrol.* New York: Academic Press.

Jackson, W. 1980. *New roots for agriculture.* San Francisco, CA: Friends of the Earth.

Kay, B. 1987. Modifications of seedbeds with natural and artificial mulches. In *Proceedings of the Symposium on Seed and Seedbed Ecology of Rangeland Plants,* ed. G. W. Frasier and R. A. Evans, 221–224. Tucson, AZ: U.S. Department of Agriculture/Agricultural Research Services.

Khoshoo, T. N. 1987. *Ecodevelopment of alkaline land.* Lucknow, India: National Botanical Research Institute.

Kim, K. C., and L. Knutson. 1986. *Foundations for a national biological survey.* Lawrence, KS: University of Kansas, Association of Systematics Collections.

Krinard, R. M., and J. K. Francis. 1983. Twenty-year results of planted cherrybark oaks on old fields in brown loam bluffs. *Tree Planters Notes* 34(4):20–22.

Loope, W. L., and G. F. Gifford. 1972. Influence of a soil microfloral crust on select properties of soils under pinyon-juniper in southeastern Utah. *Journal of Soil and Water Conservation* 27:164–167.

Lovenstein, H. 1988. Personal communication, Desert Runoff Farms, Israel.

Medina, J. G., and H. Garza. 1987. Range seeding research in Northern Mexico. In *Proceedings of the symposium on seed and seedbed ecology of rangeland plants,* ed. G. W. Frasier and R. A. Evans, 246–259. Tucson, AZ: U.S. Department of Agriculture/Agricultural Research Services.

Minnich, R. A. 1987. Fire behavior in southern California chaparral before fire control: The Mount Wilson Burns at the turn of the century. *Annals of the Association of American Geographers* 77(4):599–618.

Pieters, A. J. 1927. *Green manuring: Principles and practice.* New York: John Wiley and Sons.

Pimentel, D., E. C. Terhune, R. Dyson-Hudson, S. Rochereau, R. Samis, E. A. Smith, D. Deman, D. Reifschneider, and M. Shepard. 1976. Land degradation: Effects of food and energy resources. *Science* 194(4261):149–155.

Rieger, J. and B. Steele. 1987. *Proceedings of the second native plant revegetation symposium.* Madison, WI: Society for Ecological Restoration and Management.

Schietchtl, H. 1980. *Bioengineering for land reclamation and conservation.* Edmonton, Alberta: University of Alberta Press.

Shanan, L., N. H. Tadmor, M. Evenari, and P. Reiniger. 1970. Microcatchments for improvement of desert range. *Agronomy Journal* 62:445–448.

Sheridan, D. A. 1986. Problems of desertification in the United States. In *Arid Land Development and the Combat Against Desertification: An Integrated Approach,* 96–100. Moscow, U.S.S.R.: United Nations Environment Programme/GKNT.

Smith, J. R. 1988 [1929]. *Tree crops: A permanent agriculture.* Covelo, CA: Island Press.

Sprague, M. A., and G. B. Triplett. 1986. *No-tillage and surface tillage agriculture.* New York: John Wiley and Sons.

St. Claire, L. L., J. R. Johansen, and B. L. Webb. 1986. Rapid stabilizaion of fire and disturbed sites using a soil crust slurry: Inoculation studies. *Reclamation and Revegetation Research* 4:261–269.

St. John, T. V. 1985. Mycorrhizal fungi and revegetation. In *Native Plant Revegetation Symposium,* ed. J. P. Reiger and B. A. Steele, 87–93. San Diego, CA: California Native Plant Society.

Subba Rao, N. S. 1982. *Biofertilizers in agriculture.* Rotterdam: A. A., Balkema.

Sykes, F. 1946. *Humus and the farmer.* London: Faber and Faber.

Turner, N. 1951. *Fertility farming.* London: Faber and Faber.

Vallentine, J. F. 1980. *Range development and improvement.* Provo, UT: Brigham Young University Press.

Vaughn, D., and R. E. Malcolm. 1985. *Soil organic matter and biological activity.* Boston, MA: Martinus Nijhoff.

Virginia, R. A., and D. A. Bainbridge. 1986. *Revegetation in the low desert region of California.* San Diego, CA: San Diego State University/Dry Lands Research Institute.

Virginia, R. A., and D. A. Bainbridge. 1987. Revegetation in the Colorado desert: Lessons from the study of natural systems. In *Proceedings of the second native plant revegetation conference,* 52–63. Madison, WI: Society for Ecological Restoration and Management.

Wagoner, P. 1986. Summary of research: development of perennial grain systems at the Rodale Research Center, NC-86/33. Kutztown, PA: Rodale Research Center.

Wenger, L. E. 1941. *Reestablishing native grasses by the hay method,* Kansas Experiment Station Circular 208.

David A. Bainbridge is a Restoration Ecologist, 535 W. 12th Street, Claremont, CA 91711.

LAND IMPRINTING FOR DRYLAND REVEGETATION AND RESTORATION

ROBERT M. DIXON

ABSTRACT: *Land imprinting directly counteracts desertification and appears to hold great promise for revegetation and restoration of the vast areas of the earth that have been severely degraded or completely denuded by human activities. Imprintation is a mechanical process that converts the smooth-sealed and compacted desertified soil surface into a rough-open surface with rapid infiltration and excellent rainwater retention.*

Imprinters force angular teeth into the desertified soil surface to form the fluid-exchange funnels needed for rapid rainwater infiltration. Imprinting has been proven effective in a variety of soils for both reducing runoff and erosion and aiding plant establishment.

KEY WORDS: *imprinting, imprintation, tillage, infiltration, revegetation, seeding, plant establishment.*

INTRODUCTION

THE MAGNITUDE OF global desertification was first clearly presented in the mid-seventies (United Nations 1977). It can be defined as climatic dryness induced by human disturbances of the topsoil and natural plant communities. Yet little progress has been made in halting desertification and the populations of the arid lands are increasing faster than the world population (Eyre 1985). These facts make the search for low-cost methods of controlling desertification and restoring dryland productivity increasingly important.

Land imprinting directly counteracts desertification and appears to hold great promise for revegetating the vast areas of the earth that have been severely degraded or completely denuded by human activities. A quarter century of research (1960-1985) by the author and his co-workers led to the Desertification-Revegetation Model shown in Figure 1. Most of this research was done to gain a better understanding of how rainwater infiltration works in natural systems; how land desertification alters this hydrologic process;

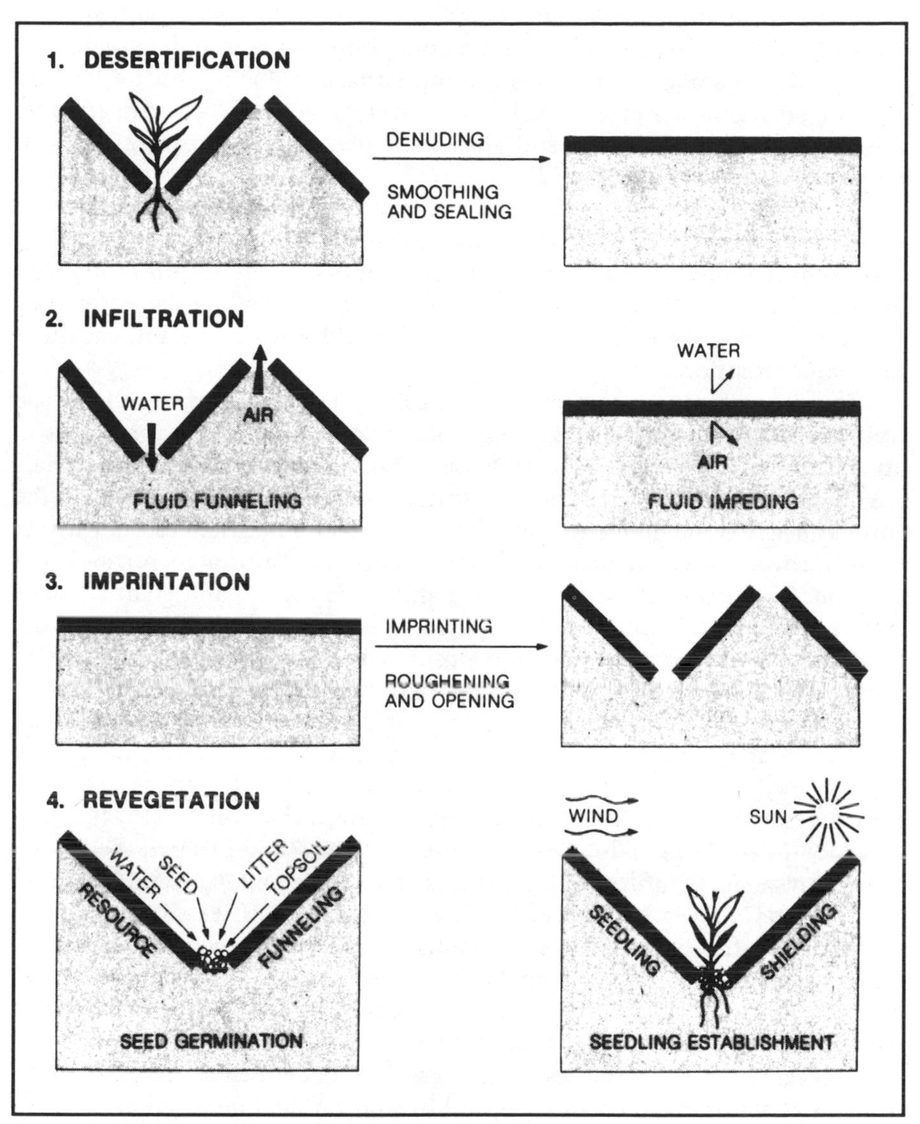

FIGURE 1. *The Desertification-Revegetation Model.*

and how rainwater infiltration, and thus desertification, can be controlled through cultural practices (Dixon 1983).

IMPRINTING

Land imprinting evolved from the search for an agricultural practice that would restore hydrologic function and reverse desertification. The Desertification-Revegetation Model encompasses four interdependent air-earth interface processes: desertification, infiltration, imprintation, and revegetation. Desertification smooths and seals the soil surface, reducing infiltration and retarding revegetation processes. Land imprinting reverses desertification by roughening and opening the soil surface, which improves infiltration and promotes revegetation (Figure 2).

Infiltration is defined as the exchange of rainwater and soil air across the air-earth interface (AEI) with the smooth-sealed AEI significantly impeding this exchange and the rough-open AEI greatly enhancing it. Imprintation may be defined as a mechanical process that converts the smooth-sealed desertified soil surface into a rough-open surface with rapid infiltration and excellent rainwater retention.

Imprinters force angular teeth into the desertified soil surface to form the fluid exchange funnels that are needed for rapid rainwater infiltration and soil air venting during intense storms (Dixon 1979). Above-ground plant materials are chopped into a moisture conserving mulch as the imprinter rolls over the soil surface. Unlike conventional tillage methods, imprinters do not invert topsoil, cover protective plant materials, and make continuous furrows that can concentrate and channel rainwater and erode topsoil and plant residues from even gently sloping hillsides. Imprinters range from hand-operated devices appropriate for areas where labor-intensive operations are desirable to massive machines requiring large tractors to operate, but little labor.

The imprinter forms interconnected water shedding and water absorbing imprints which function as a rain-fed microirrigation system. Down-slope furrows feed rainwater into cross-slope furrows where it collects and infiltrates. Tests with sprinkling infiltrometers showed that 1-hr infiltration volumes for the shedding and absorbing imprints were 3.7 and 1.0 cm respectively, with corresponding infiltration rates of 2.2 and 6.9 cm/hr (Dixon and Simanton 1981). From U.S. Weather Bureau data it was determined that this microfurrow system had sufficient capacity to control a 1-hr, 100-year maximum intensity storm without allowing any runoff or erosion at the test site near Tucson, Arizona (Dixon 1981). Control is accomplished by a combination of rainwater storage on the surface plus that which has infiltrated the soil.

Revegetation, for purposes of this research, is defined as the restoration of plant communities and their topsoil habitat on land denuded and eroded through desertification. The two initial phases of revegetation—seed germination and seedling establishment—are both improved by imprinting a desertified surface (Dixon 1987). Funnel-shaped imprints concentrate water, seed, litter, and topsoil together where these resources can improve seed germination and seedling establishment. The imprint also provides an improved and protected microsite to shield tender young plants from the desiccating effects of the hot sun and dry winds. Plant communities established in the imprints can maintain the rough-open air-earth interface and provide continued rainfall retention, runoff and erosion control, and improved soil fertility needed to rebuild topsoil.

FIGURE 2. *Land imprinter operating near Tombstone, Arizona, to restore perennial grasses in an overgrazed ecosystem.*

A deep absorptive topsoil is the best "insurance" against the devastation wrought by the twin desertification hazards of drought and deluge. That "insurance" could have protected southern Arizona from the October 1983 floods which caused damages estimated at almost half a billion dollars (Editorial Staff 1983). However, a century of overgrazing and overcropping has desertified much of this region (Dixon 1988).

Good Seeds

A good seed mix must be appropriate for existing soil and vegetative conditions; seasonal rainfall and temperature distribution; land treatment objectives; and post-treatment management plans. The cultivars and species in the seed mix must be adapted to the environment and capable of achieving the desired goals of the land manager.

The species in the mix must be compatible with each other and with species in the existing plant community. Compatibility includes such species interactions as competition, cooperation, and complementation. Where the land management objective is to improve the arid land ecosystem in overgrazed rangeland and abandoned cropland, an "ecomix" of several perennial and annual grasses, forbs, and shrubs is often appropriate. The annual grasses serve as cover, nurse, and mulch crops for the perennial plants.

Perennial grass is usually the most important component of these ecosystems because it permanently enhances infiltration and greatly increases the rainwater use efficiency by reducing runoff and surface evaporation. Native perennial grasses are preferable, but exotic species are often chosen because they may be better adapted to the degraded soil habitat, and their seeds are more readily available. Perennials have proven to be more responsive to imprinting than annuals and typically germinate only in the imprint troughs (Dixon 1987).

A good seed mix will accelerate the natural plant succession toward the major land-treatment objective; increase biomass production and topsoil reconstruction; and increase the diversity and stability of the ecosystem. (Examples of major land-treatment objectives include such aims as wildlife habitat enhancement, reduced erosion, improved flood control, etc.) A good seed mix will also include many species with seeds that will remain viable as long as the imprint lasts, because they may have to wait for months or even years until significant rainfall occurs (Dixon 1987).

Generally, seeds should be scattered uniformly on the ground ahead of the imprinter. Extremely small seeds can be scattered after imprinting but larger seeds should be imprinted over to ensure adequate covering and good seed-soil contact. Broadcast seeders and drill boxes mounted on either the tow tractor or the front of the imprinter have been used for seed distribution.

Good Imprints

Good imprints accelerate the revegetation process toward the major land treatment objective. The design of the imprinter and field operation should fit local soil, climate, seed characteristics, and management objectives.

Imprints are often made by the heel of steel angles having equal legs ranging in length from 10–20 cm which give maximum imprint depths of 7–14 cm, respectively. Imprinting teeth should be small where rains are gentle, and large where rain is more intense. Small teeth are also better for small seeds and large teeth for large seeds. Tooth spacing ranging from 15–60 cm can also be selected to fit local conditions. Spacing should be wide where rainfall is limited and water concentration is needed and narrow where rainfall is more plentiful and surface storage needs to be maximized.

The imprinter should be weighted enough to cause full-tooth penetration, which depends on tooth shape and spacing, soil characteristics, and existing vegetation. Narrow teeth-spacing requires more weight than wide spacing. Soil load bearing capacity typically increases as soils become finer and drier. A soft soil, such as wet sand, may require only 10 kg/cm of imprinter tooth-soil contact, whereas a dry clayey soil may require four times as much. Weight may have to be doubled if thick stands of shrubs are to be imprinted. In extremely hard soils, penetration can be increased by reducing the soil load bearing capacity by irrigating (or waiting for rain), shallow ripping (10–15 cm), and chopping shrubs where present.

A good imprint should be full-depth, smooth-walled, and stable enough to serve its dual function of funnelling resources to germinate seeds and protecting seedlings by moderating their microenvironment. A good imprint should also be stable enough to survive several seasons of post-treatment drought and still function properly when the rains finally occur.

The amount of rainfall required for good response of imprinted areas will depend on a number of factors, including the seed germination requirements, soil characteristics, climatic season (temperature), and existing vegetation. Usually one large rain (2.5–5 cm) or several small rains during consecutive days (when the soil is warm enough) will suffice.

Rain is the only factor over which the land manager has little control. However, if good imprints and good seeds are in place, time and patience will provide good rains.

Results: Revegetation and Restoration

Identification of seed, imprint, and rainfall as critical to the success of the Imprinting Revegetation System (IRS) came from field investigations of land imprinting, conducted in the southwestern United States since 1976 (Dixon 1986). A standard set of procedures was developed to aid in comparing test results from different areas (Dixon 1982). They may need to be modified for other applications of land imprinting and as new knowledge becomes available.

Funding was inadequate for thorough testing of land imprinting until 1982. However, some low-budget tests were conducted between 1976 and 1982 through cooperative arrangements with government agencies and private ranchers. Near Sierra Vista, Arizona, 200 ha of bulldozer-cleared land was imprinter seeded in July 1978 to control wind erosion and air pollution (Dean and Whorton 1980). Ten months after imprinter seeding of weeping and Lehmann lovegrasses (*Eragrostis curvula* and *E. lehmanniana*) grass (oven dry) accumulation was 3,250 kg/ha on imprinted land and only 56 kg/ha on adjacent unimprinted land. Eighteen months after seeding the imprinted acreage had 4,590 kg/ha while the unimprinted (but seeded) acreage had only 325 kg/ha. The cost was estimated at $25/ha for imprinting and $25/ha for seed. In this trial, the initial dozer clearing was completely unnecessary and probably handicapped IRS. The dozer removed much of the plant litter and above-ground growth which otherwise would have been pulverized by the imprinting teeth into a beneficial mulch.

Rancher Ralph Wilson used a homemade imprinter in a very successful revegetation effort on the Falcon Valley Ranch just north of Tucson, Arizona (Dixon 1981). Before imprinting in March 1980, the degraded rangeland was largely barren with a few scattered cacti and shrubs, and virtually no perennial grass. Lehmann lovegrass seed was seeded at 2 kg/ha. A good rain on July 30, 1980, initiated grass seed germination and in 6 weeks the grass was 46 cm inches high and forming seed heads. Grass forage was estimated at 1,120 kg/ha on the imprinted land and virtually zero on adjacent land.

Wilson has imprinted several times since 1980 with similar success. Don Martin, who operates the Rail X Ranch nearby, converted a sheepsfoot roller (used in compacting roadbeds) to an imprinter and seeded several thousand hectares with Lehmann lovegrass and buffelgrass (*Cenchrus ciliaris*). Excellent stands of grass resulted even when the "good" rain did not come until the second year after seeding.

Both ranchers have found that treating half to a tenth of the range in parallel strips is more economical than a 100% treatment. Lehmann lovegrass, when ungrazed during the growing season, produces about 10 kg of seed per ha per year—more than enough to self-seed the unimprinted strips. By strip treating at a 10% rate, the rancher can lower his treatment cost from $20 to just $5 per gross ha. Both ranchers have observed that the imprinter-seeded perennial grasses quickly crowd out the tumbleweed (*Salsola kali*) and burroweed (*Haplopappus tenuisectus*) that flourished on their ranches after the native stands of perennial grass were grazed out in the late 1800s and early 1900s. Imprinter grassed watersheds infiltrate water faster, thereby decreasing runoff filling of earthen ponds. Erosion is also decreased and gully headcuts recede downslope.

Imprinters, homemade from salvaged asphalt rollers, have been used to revegetate overgrazed rangeland in Utah since 1982. Don Larson, owner of the Broken Arrow Ranch near Gunnison, Utah, used one to successfully revegetate some of his degraded rangeland with crested wheatgrass (*Agropyron desertorum*) and Russian wildrye (*Elymus junceus*). He found that these grasses crowded out rabbitbrush (*Chrysothamnus* sp.) and cheatgrass (*Bromus tectorum*) within several years after imprinting.

Rancher Gerald Hall used an asphalt roller, converted to an imprinter, to compare imprinting with rangeland drilling for revegetation of mountain slopes after cheatgrass burns near Levan, Utah (Clary and Johnson 1983). Imprinting produced 3.4 times more seedlings per unit area than drilling in the first year. Imprinted areas had 4.5 times more crested wheatgrass (*Agropyron desertorum*), 2.2 times as much pubescent wheatgrass (*Agropyron trichophorum*), and 2.0 times as much Ladak alfalfa (*Medicago sativa*).

More comprehensive tests, begun in Arizona in 1982, compared seeded and imprinted (SI), ripped-seeded-imprinted (RSI), hand-seeded (HS), drill-seeded (DS), and untreated areas (U) on cactus-mesquite (*Opuntia-Prosopis*), burroweed (*Haplopappus tenuisectus*), creosote bush (*Larrea tridentata*), mesquite (*Prosopis juliflora*), and tumbleweed (*Salsola kali*) sites in the Upper and Lower Sonoran Desert (Dixon 1987; Dixon 1988).

Biomass production and grass plant population decreased in the order RSI>SI>DS>HS>U, with RSI slightly greater than SI; SI much greater than DS; DS greater than HS; and HS slightly greater than U. For all sites, the test results can be reduced to the equations:

Biomass production: RSI = 1.1SI = 2.3 DS = 2.8 HS = 3.3U
Plant population: RSI = 1.1SI = 4.1 DS = 9.8 HS = 16.6U

The most impressive figure is the increase in plant population, with more than 16 times as many perennial grass plants per m^2 in the imprinted plots compared to the untreated plots. On the most challenging sites, a creosote bush area and a barren area, the seeded and imprinted treatment averaged respectively 126 and 114 plants per m^2 versus 2 for both handseeded treatments, and 0 on each of the untreated areas. The imprinter established nearly twice as many plants as the rangeland drill with 35 cm rain and nearly 12 times as many at 25 cm (Dixon 1988). Imprinting has also been effective in areas affected by wind erosion (Clary and Johnson 1983; Haverkamp et al. 1987).

Conclusion

The Imprinting Revegetation System has proven itself in the field under a variety of conditions. It is particularly effective in areas with low rainfall and difficult soil conditions including brushy, rocky, sandy, and clayey land areas. It is an excellent method for reestablishing plant cover on degraded land and accelerating natural plant succession. It can be applied to large-scale rehabilitation of overgrazed range and abandoned agricultural land in the western United States and Mexico and should also be a valuable method for halting and reversing desertification of other drylands.

References

Clary, W. P., and R. J. Johnson. 1983. Land imprinting results in Utah. Vegetative rehabilitation and equipment workshop, *37th Annual Report*, 23–24. Missoula, MT: U.S. Department of Agriculture, Forest Service.

Dean, P., and W. J. Whorton. 1980. New agricultural implement impressing rangeland managers, WR-22-80, News Release. Oakland, CA: U.S. Department of Agriculture, Science and Education Administration.

Dixon, R. M. 1979. Land imprinter, vegetative rehabilitation and equipment workshop, *33rd Annual Report.* Casper, WY: U.S. Department of Agriculture, Forest Service.

———. 1981. Land imprinter activities, vegetative rehabilitation and equipment workshop, *35th Annual Report,* 12–13. Missoula, MT: U.S. Department of Agriculture, Forest Service.

———. 1982. Land imprinting activities, vegetative rehabilitation and equipment workshop, *36th Annual Report,* 20–21. Missoula, MT: U.S. Department of Agriculture, Forest Service.

———. 1983. Land imprinting activities, vegetative rehabilitation and equipment workshop, *37th Annual Report,* 13. Missoula, MT: U.S. Department of Agriculture, Forest Service.

———. 1986. *Infiltration/desertification control research: Principles and practices bibliography.* Tucson, AZ: Imprinting Foundation.

———. 1987. Imprintation: A process for reversing desertification. In *Proceedings: International Erosion Control Conference XVIII,* 290–306. Reno, NV: International Erosion Control Association.

———. 1988. Imprintation: A process for land restoration. Tucson, AZ: Imprinting Foundation.

Dixon, R. M., and J. R. Simanton. 1981. Some biohydrologic impacts of land imprinting. In *Proceedings of American Water Resources Association and Arizona-Nevada Academy of Sciences,* 127. Phoenix, AZ: American Water Resources Association.

Editorial Staff. 1983. Flood damages may exceed 400 million dollars. Tucson, AZ: *Tucson Citizen,* October 8.

Eyre, L. A. 1985. Population pressure on the arid lands: Is it manageable? *Arid Lands Newsletter* 23:11.

Haverkamp, M. R., P. C. Ganskopp, R. F. Miller, F. A. Sneva, K. L. Marietta, and D. Couche. 1987. Establishing grasses by imprinting in the northwestern U.S. In *Proceedings of the symposium on seed and seedbed ecology of rangeland plants,* ed. G. W. Frasier and R. A. Evans, 299–308. Tucson, AZ: U.S. Department of Agriculture/Agricultural Research Service.

United Nations Secretariat on Desertification. 1977. *Desertification: Causes and consequences.* New York: Pergamon Press.

Robert M. Dixon is the Chairman of The Imprinting Foundation, 1231 East Big Rock Road, Tucson, AZ 85718.

Desert Restoration: The Role of Woody Legumes

Ross A. Virginia

Abstract: *Recent ecological studies of plant production, nutrient cycling, and water use by legume woodlands in the California Sonoran Desert and the Chihuahuan Desert of New Mexico have provided useful information for restoration efforts in these and in other arid and semiarid regions of the world. Where possible, nitrogen-fixing woody legumes should be included in desert revegetation programs since they enhance soil fertility. Understanding and working to duplicate or mimic the natural patterns of moisture and nutrient availability found in these desert woodlands may improve the success rate for their revegetation.*

Key words: *soil fertility, nitrogen fixation, symbionts, desert revegetation.*

Introduction

Restoration of disturbed lands is a difficult task. A combination of stresses, including drought, low soil fertility, excessive salinity, and herbivory, often limits plant establishment and growth following disturbance. Relatively few investigations have been conducted aimed at developing restoration (revegetation) management plans for warm deserts (Aldon and Oaks 1982). However, ecosystem-level studies of deserts undisturbed by man provide information useful to the development of revegetation practices for deserts. This paper examines desert systems dominated by woody legumes and shows how information on the function of natural desert ecosystems can be applied to increase the success of restoration efforts.

Woody-Legume-Dominated Systems

Woody legumes are often an important component of arid and semiarid ecosystems throughout the world. Many of these trees and shrubs have the ability to develop very deep root systems (Phillips 1963); symbiotically fix atmospheric N_2 by forming a root nodule association with rhizobia (Allen and Allen 1981); and possess physiological adaptations to aridity (Nilsen et al. 1983) and salinity stresses (Felker et al. 1981). These characteristics, which permit survival and high productivity in arid environments, have increased interest in the cultivation and management of woody legume stands for fuelwood and forage production, and for soil stabilization and soil improvement. (United States National Academy of Sciences [NAS] 1979; 1980).

Woody legumes are important components of the Sonoran and Chihuahuan Deserts of the United States. Woody legumes are also frequent in temporary drainage channels called washes or arroyos. Sonoran Desert washes which dissect mountain slopes (bajadas) are dominated by the small-leaved (microphyllous) woodland which includes a number of woody legumes: smoke tree (*Psorothamnus spinosus*), mesquite (*Prosopis glandulosa*), palo verde (*Cercidium floridum*), cat's claw acacia (*Acacia greggii*), and ironwood (*Olneya tesota*).

A series of ecosystem studies have assessed the contribution of woody legumes to plant production and nutrient cycling processes in warm deserts of the Southwest. These studies have examined the importance of symbiotic N_2-fixation for establishment and growth of tree legumes on nitrogen-deficient desert soils and the affects of woody legumes on the physical and chemical properties of the soil beneath their canopies (Jarrell et al. 1982; Sharifi et al. 1982; Nilsen et al. 1983; Virginia and Jarrell 1983; Jenkins et al. 1987; Jenkins et al. 1988a, b; Virginia 1986; Virginia et al. 1989).

Woody Legumes and Soil Properties

Woody legumes affect soil properties in similar ways. However, the degree to which tree legumes alter soil properties is determined by factors such as their plant production, root distribution, the amount and quality of litter input into the soil, and the water regime of the site. Legume trees frequently are deeply rooted and the main mechanism by which they alter the fertility of surface soil is through root activity. Rooting depths exceeding 50 m have been reported for *Prosopis glandulosa* (Phillips 1963). Nutrients found in low concentration throughout the soil profile may be taken up by the extensive root systems and become concentrated near the soil surface from litter-fall and decomposition, forming the so-called "islands of fertility" in these systems (Barth and Klemmedson 1978; Virginia and Jarrell 1983). These nutrients may eventually become available to shallow-rooted plants growing in association with the woody legumes. Oldeman (1983) has suggested that the design of agroforest systems should incorporate deeply rooted plants (i.e., tree legumes) as a nutrient source for more shallowly rooted herbaceous understory species. Such a strategy might be successful for the restoration of deserts where soil fertility has been reduced by soil disturbance.

Nitrogen-fixing woody legumes (e.g., mesquite, smoke tree, ironwood) can have a very beneficial effect on soil chemical and physical properties. The increased level of soil nitrogen beneath mesquite canopies in a variety of desert systems is largely an accumulation of symbiotically fixed nitrogen (Figure 1). Studies of mesquite woodlands in the Sonoran Desert near Harper's Well, California, found available nitrate concentrations (more than 300 mg/kg) comparable to productive agricultural lands, while soils nearby lacking mesquite cover were considered deficient in nitrogen (Rundel et al. 1982; Virginia and Jarrell 1983).

Nitrogen-fixing root nodules are seldom found on the surface root system of mature mesquite trees (Virginia and Jarrell 1983). However, measurements of the natural abundance of ^{15}N in mesquite tissues (Shearer et al. 1983) and population estimates of mesquite nodulating rhizobia from the deep soil (Jenkins et al. 1987; Virginia et al. 1986) indicate that the deep soil is the site of mesquite root nodulation and symbiotic fixation.

The decomposition of litter which accumulates beneath woody legumes

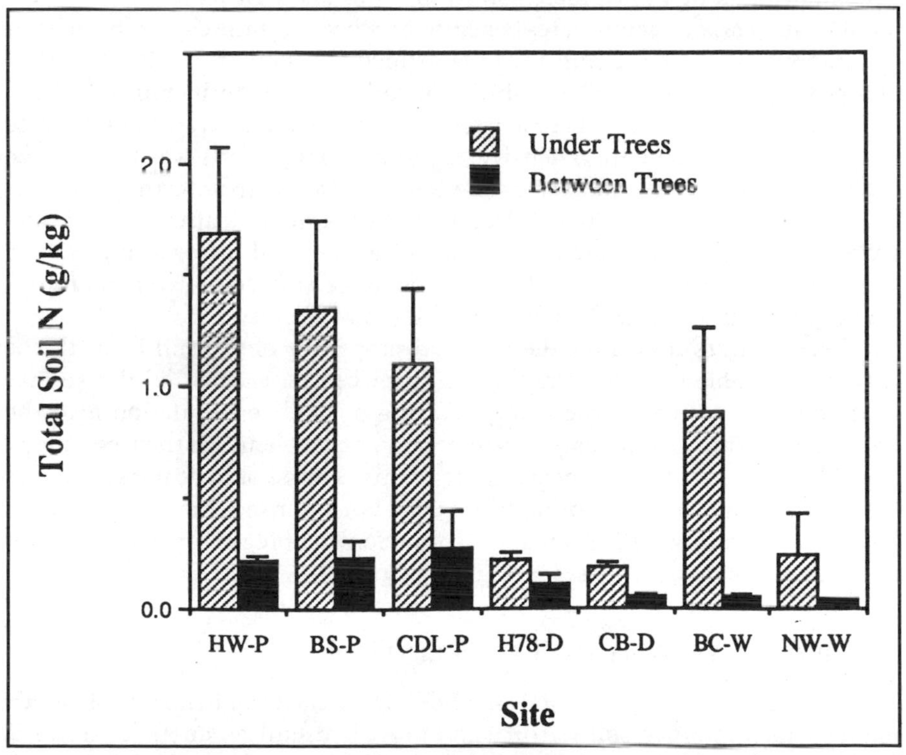

FIGURE 1. *Total soil nitrogen accumulation under mesquite trees and from the unvegetated soil between trees for sites in the Sonoran Desert.; Site locations are: Harper's Well, CA (HW); Borrego Sink, CA (BS); Clark Dry Lake, CA (CDL); Highway 78, CA (H78); Carrizo Badlands, CA (CB); Catavina Baja California (BC); and Nude Wash, CA (NW). Site types are labeled P for playa, D for sand dune, and W for wash. Data are from Virginia and Jarrell (1983). See this paper for methods and site descriptions. Error bars are one standard error of the mean.*

often results in a marked increase in soil nitrogen levels beneath the tree canopy compared to soil found between tree canopies (Figure 1). These woody legumes might act as nitrogen pumps by fixing N_2 at depth and depositing a portion of this nitrogen near the soil surface as litter or through the turnover of surface roots. Woody legumes may improve soil fertility by altering the relative abundance of some nutrients beneath the plant canopy compared with the surrounding soil. For example, many woody legumes are very salt tolerant (Felker et al. 1981). One physiological mechanism imparting salt tolerance to these tree legumes is sodium exclusion by roots (Eshel and Waisel 1965). Since other cations such as calcium and magnesium are not excluded by the plant, leaves have a different ratio of cations than the soil on which they are growing. The decomposition of leaves beneath the plant canopy will alter the cation balance of the surface soil. For example, the sodium absorption ratio (SAR) of the surface 30 cm beneath mesquite canopies was 7.9 compared to 17.3 for soil from between canopies (Virginia and Jarrell 1983). A lower SAR indicates a more favorable soil salinity for plant growth.

Organic carbon accumulates beneath the woody legume canopy from litter decomposition and has a number of beneficial effects on the soil's chemical and physical makeup. The availability of soil nutrients (mineralization) can be enhanced when suitable carbon substrates are available as an energy source for bacteria. The population density of soil microorganisms was found to be highest beneath plant canopies in the Sonoran Desert (Bamforth 1984). Soil bacteria act to increase rates of decomposition, mineralization, and, if free-living N_2-fixing bacteria are present, may increase soil nitrogen levels. The tree canopy also provides shade and moderates surface soil temperatures, making conditions more favorable for microbial activity.

Legume trees may alter the physical properties of the soil beneath the canopy. Wind-blown soil particles and litter can be trapped by the surface root and stem system of the plant resulting in sand accumulation near the base of the plant. The increase in soil organic matter beneath the tree canopy should increase infiltration beneath the plants. This should decrease soil loss by reducing surface flow during infrequent but intense rains. Tree legumes have been used to stabilize soil and to provide wind breaks for establishment of other plants (United States NAS 1979; Zollner 1986).

Establishment of Tree Legumes

A major challenge to the restoration of deserts is the establishment of woody plants which improve soil fertility and provide useful products for humans as well as food and shelter for wildlife. Tree legumes, especially those which increase soil nitrogen levels from symbiotic N_2-fixation, should be part of revegetation plans when possible since they can grow in low-nitrogen soils and they have beneficial effects on various soil properties critical for plant growth.

Conditions for natural reestablishment in deserts are infrequent (Zedler and Ebert 1977) and natural recovery occurs slowly. Conditions for plant establishment may occur only every five to ten years when soil moisture storage is sufficient to allow deep root development, as Went and Westergaard (1949) observed in Death Valley, California.

Successful revegetation of woody legumes depends in part on providing soil capable of supplying growth promoting nutrients and necessary soil organisms, such as root symbionts (N_2-fixing bacteria, mycorrhizal fungi). Soil management practices, such as deep ripping or auguring holes up to 3 m deep to improve moisture infiltration, have improved establishment and survival of plants in arid regions (Goor and Barney 1976). Imprinting the desert soils with a shaped roller has been very effective in some cases. The imprinter forms the soil into small basins that concentrate water and may also improve infiltration (Dixon 1981; this volume). Other tree establishment techniques for arid regions are reviewed by Felker (1986) and Goor and Barney (1976).

Plant establishment may be enhanced by collecting or concentrating natural rainfall onto the revegetation site by constructing catchments (Jain and Singh 1980; Shanan et al. 1970; Sharma 1986). The objective of collecting device and irrigation schemes should be to maximize deep soil water storage to favor deep root development and to reduce evaporation, losses which mainly occur from the upper 1 m of soil.

Disturbance may affect the presence and population density of soil symbionts necessary for the formation of beneficial plant-bacterial or plant-fungal associations. Seedlings for outplanting can be inoculated with native soil collected from under the same species or with commercial inocula (Moorman and Reeves 1979). The most important symbiotic partners are rhizobia, which enable some plants to fix atmospheric nitrogen, and mycorrhizal fungi which improve root characteristics, phosphorus uptake, and plant water relations (Bloss 1986; Roskoski et al. 1986).

Careful planning and a comprehensive understanding of the soil and plant characteristics and the environmental conditions of the revegetation site are the key to success in desert revegetation. Establishment of conditions which mimic natural establishment events (e.g., deep soil water storage) may be necessary to revegetate desert sites. For woody legumes this means that moisture should be available deep in the soil to promote root development and the formation of root symbiotic associations. It may be necessary to inoculate seedlings with root symbionts prior to planting to promote rapid growth in nutrient deficient soils. The beneficial effects of woody legumes on soil fertility and the physical and chemical properties of soil warrant their use for the restoration of disturbed desert systems.

Acknowledgments

This research was supported by NSF grant BSR-8506807.

References

Aldon, E. F., and W. R. Oaks, eds. 1982. *Reclamation of mined lands in the Southwest—A Symposium.* Albuquerque, NM: Soil Conservation Society of America.

Allen, O. N., and E. K. Allen. 1981. *The Leguminosae, a source book of characteristics, uses, and nodulation.* Madison, WI: University of Wisconsin Press.

Bamforth, S. S. 1984. Microbial distributions in Arizona deserts and woodlands. *Soil Biology and Biochemistry* 16:133–37.

Barth, R. C., and J. O. Klemmedson. 1978. Shrub-induced spatial patterns of dry matter, nitrogen, and organic carbon. *Soil Science Society of America Journal* 42:804–09.

Bloss, H. E. 1986. Studies of symbiotic microflora and their role in the ecology of desert plants. *Desert Plants* 7:119–27.

Dixon, R. M. 1981. Rangeland revegetation through land imprinting. *Livestock* 59:6–7.

Eshel, Y., and Y. Waisel. 1965. The salt relations of *Prosopis farcta* (Banks et Sol) Eig. Israel Journal of Botany 14: 150–51.

Felker, P., ed. 1986. *Tree plantings in semiarid regions.* Amsterdam: Elsevier Press.

Felker, P., P. R. Clark, A. E. Laag, and P. F. Pratt. 1981. Salinity tolerance of the tree legumes mesquite (*Prosopis glandulosa* var. *torreyana, P. velutina,* and *P. articulata*), algarrobo (*P. chilensis*), Kiawe (*P. pallida*), and tamarugo (*P. tamarugo*) grown in sand culture on nitrogen free media. *Plant and Soil* 61:311–17.

Goor, A. Y., and C. W. Barney. 1976. *Forest tree planting in arid zones.* New York: Ronald Press.

Jain, B. L., and R. P. Singh. 1980. Runoff as influenced by rainfall characteristics, slope and surface treatment of microcatchments. *Annals of Arid Zone* 19:119–25.

Jarrell, W. M., R. A. Virginia, D. H. Kohl, G. B. Shearer, B. A. Bryan, P. W. Rundel, E. T. Nilsen, and M. R. Sharifi. 1982. Symbiotic nitrogen fixation by mesquite and its management implications. In *Mesquite utilization.* Lubbock, TX: College of Agricultural Sciences, Texas Tech University.

Jenkins, M. B., R. A. Virginia, and W. M. Jarrell. 1987. Rhizobial ecology of the woody legume mesquite (*Prosopis glandulosa*) in the Sonoran Desert. *Applied and Environmental Microbiology* 53:36–40.

———. 1988a. Rhizobial ecology of the woody legume, Psorothamnus spinosus in a Sonoran desert arroyo ecosystem. *Plant and Soil* 105:113–20.

———. 1988b. Depth distribution and seasonal population fluctuations of mesquite-nodulating rhizobia in warm desert ecosystems. *Soil Science Society of America Journal* 52:1644–50.

Moorman, T., and F. B. Reeves. 1979. The role of endomycorrhizae in revegetation practices. *American Journal of Botany* 66:14–18.

Nilsen, E. T., M. R. Sharifi, P. W. Rundel, W. M. Jarrell, and R. A. Virginia. 1983. Diurnal and seasonal water relations of the desert phreatophyte *Prosopis glandulosa* (Honey mesquite) in the Sonoran Desert of California. *Ecology* 64:1381–93.

Oldeman, R. A. A. 1983. The design of ecologically sound agroforests. In *Plant Research and Agroforestry,* ed. P. A. Huxley, 173–206. Nairobi, Kenya: International Council for Res. in Agroforestry.

Phillips, W. S. 1963. Depth of roots in soil. *Ecology* 44:424.

Roskoski, J. P., I. Pepper, and E. Pardo. 1986. Inoculation of leguminous trees with rhizobia and VA mycorrhizal fungi. *Forest Ecology and Management* 16:57–68.

Rundel, P. W., E. T. Nilsen, M. R. Sharifi, R. A. Virginia, W. M. Jarrell, D. H. Kohl, and G. B. Shearer. 1982. Seasonal dynamics of nitrogen cycling for a *Prosopis* woodland in the Sonoran Desert. *Plant and Soil* 67:343–53.

Shanan, L., N. H. Tadmor, M. Evenari, and P. Reiniger. 1970. Microcatchments for improvement of desert range. *Agronomy Journal* 62:445–48.

Sharifi, M. R., E. T. Nilsen, and P. W. Rundel. 1982. Biomass and net primary production of *Prosopis glandulosa* (Fabaceae) in the Sonoran Desert of California. *American Journal of Botany* 69:760–67.

Sharma, K. D. 1986. Runoff behavior of water harvesting microcatchments. *Agricultural Water Management* 11:137–44.

Shearer, G., D. H. Kohl, R. A. Virginia, B. A. Bryan, J. L. Skeeters, E. T. Nilsen, M. R. Sharifi, and P. W. Rundel. 1983. Estimates of N_2-fixation from variations in the natural abundance of ^{15}N in Sonoran Desert ecosystems. *Oecologia* 56:365–73.

United States National Academy of Sciences. 1979. *Legumes: Resources for the future.* Washington, DC: National Academy Press.

United States National Academy of Sciences. 1980. *Firewood crops, shrubs and tree species for energy production.* Washington, DC: National Academy Press.

Virginia, R. A. 1986. Soil development under legume tree canopies. *Forest Ecology and Management* 16:69–79.

———, and W. M. Jarrell. 1983. Soil properties in a *Prosopis* (mesquite) dominated Sonoran Desert ecosystem. *Soil Science Society of America Journal* 47:138–44.

———, M. B. Jenkins, and W. M. Jarrell. 1986. Depth of root symbiont occurrence in soil. *Biology and Fertility of Soils* 2:127–30.

———, W. M. Jarrell, P. W. Rundel, D. H. Kohl, and G. Shearer. 1989. The use of variation in the natural abundance of ^{15}N to assess symbiotic nitrogen fixation by woody plants. In *Stable Isotopes in Ecological Research,* ed. P. W. Rundel, J. R. Ehleringer, and K. A. Nagy, 375–394. Ecological Studies, vol. 68. New York: Springer-Verlag.

Went, F. W., and M. Westergaard. 1949. Ecology of desert plants. *Ecology* 30:26–38.

Zedler, P. H., and T. A. Ebert. 1977. Shrub seedling establishment and survival following an unusual September rain in the Colorado Desert. *Bulletin of the Ecological Society of America* 58:47.

Zollner, D. 1986. Sand dune stabilization in central Somalia. *Forest Ecology and Management* 16:223–32.

Ross A. Virginia is a professor in the Biology Department and Systems Ecology Research Group, San Diego State University, San Diego, CA 92182.

Barrens Restoration in the Cretaceous Hills of Pope and Massac Counties, Illinois

Lawrence R. Stritch

ABSTRACT: *In 1806 the United States Public Land Survey characterized an area of 488 km in southeastern Illinois as barrens. In 1987 the Illinois Department of Conservation began to restore a portion of the degraded barrens at the Cretaceous Hills Nature Preserve. This paper discusses the reasons for the disappearance of this now nearly extinct natural community and the methods being employed to restore it.*

KEY WORDS: *barrens, Cretaceous Hills, restoration, Illinois, Illinois Natural Areas Inventory.*

Introduction

The purpose of this paper is to document restoration work completed at the Illinois Department of Conservation's Cretaceous Hills Nature Preserve in Pope County, Illinois, and to recommend future management activities for barrens communities there. This restoration work is significant in that it represents the first concentrated management effort to assure that this distinct natural community type does not disappear from the state.

Background

"Barrens" was the term used by early settlers, naturalists, and surveyors to describe areas of land that were sparsely timbered (United States Public Land Survey 1806; McInteer 1944; Bourne 1820; Peck 1834; Keith 1980). Bourne (1820), Peck (1834) and Ellsworth (1836) distinguished prairie, barrens and forest. They characterized barrens as lands with scattered, rough, stunted trees interspersed with thickets of hazel and other brushwood and large expanses of grasses and forbs, which is the definition accepted for this paper.

Within the Cretaceous Hills Section of the Coastal Plain Natural Division of Illinois (Schwegman 1973), floral elements exist on barrens that demonstrate this natural community's affinity to the southeastern United States. Narrow-leaved sunflower (*Helianthus angustifolius*), meadow beauty (*Phexia mariana*), silver broomsedge (*Andropogon ternarius*), beardgrass (*Cymnopogon ambiguus*), Elliot's broomsedge (*Andropogon elliotii*), deerberry (*Vacinium stamineum*), and black-eyed susan (*Rudbeckia bicolor*) are southeast extraneous elements (Fernald 1950) that exist in the barrens landscape of Pope and Massac counties, Illinois.

Those barrens, as described by the United States Public Land Survey, covered an area of approximately 488 km (Hutchison, Olson and Vogt 1986). In pre-settlement times, it was a gently rolling landscape of long, loess-mantled ridges, gravel knobs and narrow valleys with wide, shallow, gravel-bottomed streams. The landscape was dominated by a nearly continuous herbaceous layer of grasses and forbs, many of which are commonly associated with prairies and dry forests (Voigt and Mohlenbrock 1964; Hutchison and Johnson 1981). Little bluestem (*Schizachyrium scoparium*), big bluestem (*Andropogon gerardii*) and Indian grass (*Sorghastum nutans*) were dominant with other species of grasses scattered among them, such as wild rye (*Elymus* spp.), Elliot's broomsedge, gamma grass (*Tripsacum dactyloides*), broomsedge (*Andropogon virginicus*) and panic grass (*Panicum* spp.). Forbs commonly encountered in the barrens included mountain mint (*Pycnanthemum* spp.), goldenrod (*Solidago* spp.), bush clover (*Lespedeza* spp.), sunflower (*Helianthus* spp.), tick trefoil (*Desmodium* spp.), black-eyed susan (*Rudbeckia hirta*), meadow beauty, rose pink (*Sabatia angularis*), ironweed (*Vernonia* spp.), milkweeds (*Asclepias* spp.), blazing star (*Liatris* spp.), sedges (*Carex* spp.), and many other colorful wildflowers.

There were thickets of "brush and briers," including such species as catbriars (*Smilax* spp.), grapevines (*Vitis* spp.), poison ivy (*Toxicodendron radicans*), blackberries and dewberries (*Rubus* spp.), willows (*Salix* spp.), hazel (*Corylus americana*), winged sumac (*Rhus copallina*), dogwood (*Cornus florida*), crabapple (*Malus ioensis*) and stunted, twisted, gnarled, fire retarded blackjack oak (*Quercus marilandica*). In many areas these thickets made travel difficult (PLS 1806).

The original barrens landscape was dotted with large open-grown oaks and hickories, including post oak (*Quercus stellata*), blackjack oak, white oak (*Q. alba*), barrens oak (*Q. falcata*), black oak (*Q. velutina*), pignut hickory (*Carya glabra*), scalybark hickory (*C. ovata*) and barrens hickory (*C. texana*). In some areas, trees were concentrated into groves and areas of scattered timber (Engelmann 1866; PLS 1806).

Barrens is a complex natural community that owes its existence to a balance of natural forces preventing its succession to a forest community (Stritch 1987). Keith (1980), McInteer (1944), Hutchison, Olson and Vogt (1986) and Engelmann (1863, 1866) have documented how fire, geology, soils, site aspect, disease, wildlife and aboriginal man interacted to maintain the barrens. With the encroachment upon the barrens by white settlers, the barrens natural community disappeared rapidly. This was primarily attributable to the reduced occurrence of fire and the decimation of the large herds of buffalo, elk, and deer (Bourne 1820; Peck 1834; McInteer 1944; Engelmann 1866; Vestal 1936).

The Illinois Natural Areas Inventory (INAI) was a comprehensive, two-year search for natural land using aerial photographs, low flying aircraft, and ground surveys to examine the entire area of the state. The Inventory identified a total of only 21 ha of barrens in Illinois as retaining its pre-settlement character. In 1986, The Nature Conservancy classified Illinois Barrens as Globally Imperiled.

All of Illinois' rare barrens plants, some of which may be distinct ecotypes, exist as very small populations at a few sites. In the case of marsh blazing star (*Liatris spicata*), ladies tresses (*Spiranthes vernalis*) and blue hearts (*Buchnera americana*), they occur as single individual plants. Fink (1987) has stated that the barrens of Pope and Massac counties may have been an important native habitat of Bachman's sparrow (*Aimophila aestivalis*) and Bewick's wren (*Thryomanes bewickii*). Although no in-depth work has been accomplished on invertebrates native to the barrens natural community in Illinois, in Pennsylvania 11 of 25 listed lepidopterans are found in a barrens natural community (Opler 1985).

In 1986, the Illinois Nature Preserves Commission approved Department of Conservation plans to manage and restore a portion of the degraded barrens at the Cretaceous Hills Nature Preserve, Pope County, Illinois.

The Cretaceous Hills Nature Preserve barrens restoration effort is consistent with the legislative mandate set forth in the Illinois Natural Areas Preservation Act of 1981. The Illinois Natural Areas Preservation Act of 1981, Section 702-2, states:

> Natural lands and waters together with the plants and animals living thereon in natural communities are a part of the heritage of the people. They are of value . . . as habitats for rare and vanishing species . . . and as living museums of the native landscape wherein one may envision and experience primeval conditions.
>
> It is therefore the public policy of the State of Illinois to secure for the people of present and future generations the benefits of an enduring resource of natural areas, including the elements of natural diversity present in the state, by establishing a system of nature preserves.

Prior to the restoration effort, no example of a barrens natural community was being managed in the Cretaceous Hills. The barrens restoration effort of Cretaceous Hills Nature Preserve is the Illinois Department of Conservation's effort to assure that this natural community type, which had all but disappeared from the Coastal Plain Natural Division of Illinois, will be permanently protected and properly maintained.

The Barrens Today

Today, the barrens region of Pope and Massac counties is a patchwork of cultivated fields, orchards, pastures, second growth forests, and abandoned agricultural fields. Bourne (1820), Peck (1834), Ellsworth (1838), McInteer (1944) and Engelmann (1863, 1866) have documented the dramatic changes that occurred in the landscape and have attributed those changes to the arrival of pioneer settlers. It is among the abandoned agricultural fields that one can discover floral elements that were common to the barrens. The old fields that were grazed and not cultivated are the best candidates for restoration because they contain the greatest representation of barrens elements and most closely resemble the barrens as described by government surveyors and early writers.

Prior to implementation of management activities, the restoration site in Cretaceous Hills Nature Preserve was a patchwork of grassy openings and brushy thickets intermingled with small groves of young trees and saplings. By ocular estimate, grassy, brushy areas constituted approximately 60% of the restoration site and were dominated by little bluestem, catbriers, winged sumac and saplings of flowering dogwood, persimmon (*Diospyros virginiana*), eastern red cedar (*Juniperus virginiana*) and other early successional woody species. The more wooded areas were dominated by flowering dogwoods. Relatively large (25-61 cm DBH) shingle oaks, southern red oaks (*Quercus falcata*), river birches (*Betula nigra*) and sycamores (*Platanus occidentalis*) were scattered over the site.

Floristic analysis of the preserve restoration site demonstrated that the greatest concentration and diversity of barrens herbaceous species were within the more open areas, as would be expected. However, individual plants and small patches of barrens vegetation were discovered still existing in areas more heavily shaded by trees.

Examination of the historical sequence of Agricultural Stabilization and Conservation Service (ASCS) aerial photographs revealed that: (1) the site had been abandoned for approximately 50 years; and (2) the site had been pastured but apparently not cultivated prior to 1938. The latter conclusion was reinforced through field observations which revealed no evidence of plow lines similar to those still visible in an "old field woods" located immediately to the southeast.

United States Public Land Survey (PLS) notes were used to help determine pre-settlement character of the site. The north boundary of the restoration site is located at or near the north line of Section 15, T15S-P6F and approximately 600 feet east of the west section line of the same section. The surveyor's notes describe the north line as crossing "brushy barrens—a few scattering post oaks," and the west line as crossing "brushy and briary barrens." Hutchison and Johnson (1981) provided information about the flora and physiognomy of the pre-settlement landscape in the Cretaceous Hills region.

Management Program

The objective of management has been and continues to be the restoration to its pre-settlement character and maintenance of a high quality example of the barrens natural community. The approach has been to achieve this by simulating the disturbance factors which led to the development and maintenance of this natural community.

Following the findings of Anderson and Schwegman (1972), the first phase of restoration included the use of prescribed fire on 12 ha. This was followed by removal on 4 ha of a high percentage of the woody plants with hand tools, power brush cutters, and chainsaws. The removal of this quantity of woody plants is consistent with the goal of achieving a landscape depauperate of trees as described by the PLS. Almost all old field successional trees (Bazzaz 1968), such as sassafras (*Sassafras albidum*), red maple (*Acer rubrum*), silver maple (*A. saccharinum*), tulip tree (*Liriodendron tulipifera*), white ash (*Fraxinus americana*), eastern red cedar and persimmon were removed from the open areas. Several open-grown southern red oaks and one shingle oak were left. In the more wooded successional areas, only those oaks of open growth form were left. All other woody material was cut down and removed from the site. This is consistent with restoring the area with trees possessing a growth form as described by early chroniclers (Bourne 1820; Ellsworth 1838; PLS 1806). The cut stumps were not treated with herbicides and many of the woody species resprouted. The intention is to use prescribed fire to control woody reinvasion and resprouting.

Prescribed burning was accomplished by March 11, 1987 and initial field work was carried out during March and April. Work crews from the Department of Corrections carried out the prescribed management activities under the direction of the District Natural Heritage Biologist. Permanent photo stations and vegetative sampling plots were established to document and monitor changes in the barrens restoration.

Posttreatment Observations

The prescribed burn, in combination with the hand removal of woody vegetation, stimulated recovery of the barrens community at this site. Numerous blazing stars (*Liatris squarrosa*), several butterfly weeds (*Asclepias tuberosa*), green fringed orchids (*Plantanthera lacera*) and one sand milkweed (*Asclepias amplexicaulis*) were observed for the first time on this site following treatment. On the area that was burned but where no hand removal of woody vegetation was carried out, the recovery of barrens vegetation is not as obvious. The prescribed fire alone did not eliminate woody vegetation to the extent accomplished by burning and hand removal of woody vegetation. On this part of the site, shading will continue to hamper the recovery of the herbaceous barrens species, such as wild white indigo (*Baptisia lactea*), blazing star, and little bluestem.

Future Management Needs

Woody encroachment on the Cretaceous Hills Nature Preserve barrens restoration site will need to be controlled on a continuing basis. Saplings and trees, except for those species mentioned in the PLS notes, should be eliminated from the site. Only scattered, open-grown, PLS-listed species of trees should be allowed to remain. To control Japanese honeysuckle (*Lonicera japonica*), an exotic weed species, prescribed burns should be conducted annually for an additional four years. If Japanese honeysuckle persists, treatment with a 1.5% solution of Roundup herbicide after the first killing frost in the fall has been shown to be effective. Serious consideration should be given to the reintroduction of post oak and white oak and to the re-establishment of herbaceous barrens species, such as mountain mint (*Pycnanthemum pilosum*), blazing star (*Liatris spicata*), rough blazing star (*Liatris scabra*), rough goldenrod (*Solidago rigida*), showy goldenrod (*Solidago speciosa*), prairie dock (*Silphium terebinthinaceum*), rosin weed (*S. integrifolium*), French grass (*Psoralea onobrychis*), western sunflower (*Helianthus occidentalis*), green milkweed (*Asclepias viridiflora*), prairie willow (*Salix humilis*), perennial foxtail (*Setaria geniculata*), brown-eyed susan (*Rudbeckia bicolor*), prairie gentian (*Gentiana alba*) and pink milkwort (*Polygala incarnata*) that may be currently absent from the site, presumably because of past grazing and the lack of fire.

References

Anderson, R. C., and J. E. Schwegman. 1972. The response of southern Illinois barren vegetation to prescribed burning. *Illinois State Academy of Science* 64:287–291.

Bazzaz, F. A. 1968. Succession on abandoned fields in the Shawnee Hills, southern Illinois. *Ecology* 49:924–36.

Bourne, A. 1820. On the prairies and barrens of the West. *American Journal of Sciences and Arts* 2:30–34.

Ellsworth, H. L. 1838. *Illinois in 1837–38; a sketch descriptive of the situation boundaries, face of the country, prominent districts, prairies, rivers, minerals, animals, agricultural productions, public lands, plans of internal improvement, manufactures, and c. of the State of Illinois.* Philadelphia, PA: Augustus Mitchell.

Engelmann, H. 1863. Remarks upon the causes producing the different characters of vegetation known as prairies, flats, and barrens in southern Illinois, with special reference to observations made in Perry and Jackson counties. *American Journal of Sciences and Arts* 86:384–96.

———. 1866. Massac County, and that part of Pope County south of Big Bay River. In *Geological Survey of Illinois*, vol. I, ed. A. H. Worthen, 428–55. Springfield, IL: State of Illinois.

Fernald, M. L. 1950. *Gray's manual of botany*, 8th ed. New York: D. Van Nostrand.

Fink, T. 1987. Personal communication.

Hutchison, M. D., and M. Johnson. 1981. *The natural character of Township 15 South, Range 6 East of the third principal meridean in Pope and Massac Counties, Illinois.* Belknap, IL: Natural Land Institute.

Hutchison, M. D., S. Olson, and T. Vogt. 1986. *A survey of the barrens region in Massac and Pope Counties, Illinois.* Belknap, IL: Natural Land Institute.

Keith, J. H. 1980. Presettlement barrens of Harrison and Washington counties, Indiana. In *Proceedings of the Fifth North American Prairie Conference.* Springfield: University of Missouri.

McInteer, B. B. 1942. The barrens of Kentucky. *Transactions of the Kentucky Academy of Science* 10:7–12.

______. 1944. A change from grassland to forest vegetation in the Big Barrens of Kentucky. *American Midland Naturalist* 35:276–282.

Opler, P. A. 1985. Invertebrates. In *Species of special concern in Pennsylvania,* eds. H. H. Genoways and F. J. Brenner. Pittsburgh, PA: Carnegie Museum of Natural History.

Peck. J. M. 1834. *A gazetteer of Illinois.* Jacksonville, IL: R. Goudy.

Schwegman, J. E. 1973. *Comprehensive plan for the Illinois nature preserves system: Part 2 – the natural divisions of Illinois.* Rockford, IL: Illinois Nature Preserves Commission.

Stritch, L. R. 1987. Barrens community classification and restoration in Southern Illinois. In *Proceedings of the Eighth Northern Illinois Prairie Conference – Preserving Our Heritage – Restoring Our Past,* 54–58. Joliet, IL: Will Co. Forest Preserve District.

United States Public Land Survey. 1806. *Field notes and plats for T15S-R6E.* Springfield, IL: Illinois State Archives.

Vestal, A. G. 1936. Barrens vegetation in Illinois. *Transactions of the Illinois State Academy of Science* 29:79–80.

Voigt, J. W., and Mohlenbrock, R. H. 1964. *Plant communities of Southern Illinois.* Carbondale, IL: Southern Illinois University Press.

Lawrence R. Stritch, Ph.D., is a Natural Heritage Biologist, Illinois Department of Conservation, Colconda, IL 62938.

Dune Revegetation at Buhne Point, King Salmon, California

Andrea J. Pickart

ABSTRACT: *The artificially restored 9.3 ha Buhne Point sand spit at Humboldt Bay, California, was revegetated using native plants. An experimental phase was carried out in the spring of 1986 to test the effectiveness of five planting methods and the suitability of nine native dune species for stabilization. In addition, an experimental seed mix was planted and dune ridges were planted with the native dune grass Elymus mollis. The effect of a rooting hormone was tested on five dune species planted as cuttings or divisions. The implementation phase occurred in winter 1985–1986, when 2.2 ha were seeded using the five methods. Follow-up monitoring was conducted in 1987. Density, cover, species diversity, species composition, and size class distribution were monitored for each individual planting method. Density and cover of* Elymus *was also measured.*

KEY WORDS: *dune revegetation, native plants, monitoring, Buhne Point.*

Introduction

THE BUHNE POINT sand spit, located on the east shore of Humboldt Bay, California, has historically been the focal point of wave energy passing through the Humboldt Bay jetties. Following the completion of the jetties in 1899, the spit exhibited considerable instability in response to storm waves, failure and reconstruction of the jetties, channel dredging, and spoil deposition (Tuttle 1985). Erosion of the spit accelerated in 1973, and by 1982 it had disappeared completely, exposing Buhne Drive and the community of King Salmon to storm waves and flooding. This threat to community safety prompted a 1983 Congressional allocation of nine million dollars to develop a shoreline erosion control "demonstration project." The purpose of the project was twofold: to reconstruct the spit so as to prevent future erosion, and to demonstrate state-of-the-art methods that could be applied to other projects.

The new spit was completed by the U.S. Army Corps of Engineers in August 1984, using dredge spoils from the Humboldt Bay entrance channel. In keeping with the demonstration nature of the project, native plant revegetation techniques were developed and used to stabilize the surface without the detrimental ecological impacts associated with nonnative introductions.

Revegetation was conducted by the Humboldt County Public Works Department under contract to the San Francisco District of the Corps of Engineers. The project was divided into an experimental and an implementation phase. Experimental plantings were conducted in spring 1986 and full-scale revegetation was implemented in the winter of 1986–1987. In 1987 the project was turned over to the Humboldt Bay Harbor, Recreation and Conservation District. Remedial work and follow-up monitoring have since been sponsored by that agency.

Experimental Phase

A review of the literature and trips to the major dune revegetation projects on the California coast revealed that very little published or unpublished data on dune restoration existed prior to 1983. Information generated by past projects was largely site specific, due to the differing goals of revegetation attempts and to wide variation in climatic, edaphic and floristic characteristics. For this reason, an experimental phase was developed to test the suitability of native species for dune stabilization, and the relative effectiveness of various planting techniques.

Due to federally imposed scheduling constraints, experimental plantings were carried out during the spring of 1986, necessitating the use of irrigation. Both qualitative and quantitative experiments were conducted. Quantitative experiments tested field germination, survival, and cover of nine dune species. Species were selected which occur in the local, semistabilized community classified as Northern Foredune by the California Native Plant Society (Holland 1986). In addition to quantitative seeding experiments, five species were tested as vegetative propagules experimentally treated with a rooting hormone. Qualitative experiments consisted of planting and photographically monitoring areas of a native plant seed mix and vegetative propagules of the native dune grass *Elymus mollis*.

Quantitative Experimental Plantings: Seeding

Two 19 x 24 m rectangular grids were established to test germination, survival, and cover of nine species under six treatments. Each grid contained nine rows of 1 m^2 plots in blocks of four. Three of those nine rows represented controls seeded at three different sowing densities, while each of the other six rows were subjected to a different planting treatment at the middle sowing density. Within a row, individual species were planted in replicated, randomized plots. Five species were tested in one grid and four species plus a species mix were tested in the other. A more detailed account of experimental methods is contained in Pickart (1986a). Species tested are listed in Table 1.

Seeds were collected from locations around Humboldt Bay and processed if appropriate (Newton 1985a). Seeds of beach morning glory (*Calystegia soldanella*) and beach pea (*Lathyrus littoralis*) were scarified prior to planting. An optimum sowing density was calculated for each species based on known seed counts and available information (Pickart 1986a). This rate was used in the treatment rows and in one control row, but was halved and doubled in the second and third control rows respectively, to test for density effects.

The six treatments were designed to test the effects of two important environmental parameters, nutrient availability and substrate mobility, on germination, survival, and cover of the above-mentioned species. Soil moisture was held constant through the use of irrigation. The treatments used are listed in Table 2.

TABLE 1
SPECIES TESTED FOR GERMINATION, SURVIVAL AND COVER

Sand verbena	*Abronia latifolia*
Beach bur	*Ambrosia chamissonis*
Beach sagewort	*Artemisia pycnocephala*
Beach morning glory	*Calystegia soldanella*
Seaside daisy	*Erigeron glaucus*
Beach buckwheat	*Eriogonum latifolium*
Beach pea	*Lathyrus littoralis*
Dune tansy	*Tanacetum douglasii*

TABLE 2
TREATMENTS TESTED FOR EFFECT ON GERMINATION, SURVIVAL AND COVER

Treatment	Description
Control	Seeds sown at 5 cm depth, optimum density
Compost	Redwood compost with 1% added N, ll.3 kg/m^2
Slow-release fertilizer	Osmocote fertilizer (14:14:14) applied at manufacturer's specification, 4,880 kg/ha
Soluble fertilizer	Ammonium sulfate (21:0:0) applied at 0.8 kg/ha
Jute matting	Coarse fiber netting applied over control conditions
Hydromulch	1680 kg/ha Spra-mulch applied with 488 kg/ha ammonium phosphate (16:20:0)
Depth test	Seeds for each species planted at depths of 2.5, 5.0, 7.5, 15.0 and 23.0 cm

The two grids were monitored throughout the 1986 growing season for germination and survival. Cover was calculated at the end of the growing season by photographing each plot and tracing and planimetering projected images of plants. Results for the three variables are discussed separately below.

Germination

For all species, germination of the controls was highest. Inhibition of germination by fertilizers is recognized (Roberts and Bradshaw 1985) as the result of toxicity and osmotic effects. The reduced germination under fertilizer, compost, and hydromulch treatments can be attributed to this phenomenon. Reduced germination under the jute matting treatment was the result of excessive burial from sand deposition and the physical barrier of the netting. Under control conditions, medium- and large-seeded species (with the exception of *Calystegia*) exhibited relatively high germination, while small-seeded species showed very poor germination. The depth-test treatment provided evidence that the small-seeded species beach evening primrose (*Camissonia cheiranthifolia*), beach sagewort (*Artemisia pycnocephala*) and seaside daisy (*Erigeron glaucus*) have a light requirement for germination. Light requirement has been shown to be an important factor in many small-seeded species (Bidwell 1979), preventing germination at depths where depletion of stored reserves may occur prior to emergence. Low germination may also have been caused by failure to emerge due to insufficient hypocotyl elongation. Despite pre-treatment, *Calystegia* showed poor germination, demonstrating that dormancy was not overcome by the scarification treatment.

Survival

Survival was not affected by treatment for any species, but under control conditions significant differences were found between species. *Artemisia* survival was 47%, while all other species showed survival of 60–100%. The apparent reason for lower survival of *Artemisia* was an inappropriately high planting density, resulting in self-thinning (Pickart 1986a). (Survivorship curves for *Artemisia* followed a generalized pattern described for populations undergoing self-thinning.)

Cover

Cover was calculated at the end of the first two growing seasons. The results showed an expected trend towards increased cover under fertilized treatments. The exceptions to this trend were *Calystegia* and *Lathyrus*. Nodulation by symbiotic bacteria has been demonstrated for *Lathyrus* (Holton 1980), which may explain a lack of response to treatment. Most of the species showed a significantly greater cover under the slow-release fertilizer treatment, and this trend was intensified in the second year (fertilizer was not reapplied).

Quantitative Experimental Planting: Vegetative Propagules

Cuttings or divisions of six species were collected at various locations around Humboldt Bay (Newton 1985b). Each species was divided into a control and an experimental group treated with a rooting hormone. The species tested and the type of propagule are listed in Table 3.

A subsample of propagules of each species were monitored over the 1986 growing season. Height, number of leaves, number of branches and/or rosette diameter were recorded. Net increase over the growing period was calculated and statistically compared. No significant differences were detected between treatment and control groups, indicating that the rooting hormone had no effect on growth. However, overall survival differed between species, suggesting that some are more suitable than others for this type of transplanting. Survival by species is shown in Table 4.

Qualitative Experimental Planting

The qualitative experiments were carried out on a larger scale than quantitative experiments in order to allow an assessment of costs as well as success

TABLE 3
VEGETATIVE PROPAGULES TESTED

Species	Propagule Type
Beach sagewort (*Artemisia pycnocephala*)	Rooted divisions
Beach morning glory (*Calystegia soldanella*)	Rhizomes
Beach strawberry (*Fragaria chiloensis*)	Rooted divisions
Beach pea (*Lathyrus littoralis*)	Rooted divisions/cuttings
Dune tansy (*Tanacetum douglasii*)	Rooted divisions/cuttings

TABLE 4
SURVIVAL OF VEGETATIVE PROPAGULES BY TREATMENT

Species	% Survival, Control	% Survival, Treated
Artemisia pycnocephala		
Divisions	92	95
Cuttings	56	70
Calystegia soldanella	8	18
Fragaria chiloensis	56	56
Lathyrus littoralis		
Divisions	4	2
Cuttings	12	8
Tanacetum douglasii	100	100

of plantings. Two types of planting were investigated: seeding of an experimental mix, and planting of vegetative propagules of dune grass.

Seeding

An experimental mix of seeds was prepared, using the same optimum sowing densities that were used in the quantitative experiments to develop appropriate species proportions in the mix. Methods and costs of seed collection for this portion of the project are documented in the Phase One Planting Report (Newton 1985b). Approximately 0.3 ha was planted, using California Conservation Corps (CCC) labor to apply seeds and rake them to a depth of approximately 5 cm. Seeds were planted at an overall rate of 22.5 kg/ha, with 450 kg/ha of ammonium sulfate applied 8.5 weeks after planting. Irrigation was supplied throughout the growing season. Germination trends closely reflected those exhibited in the quantitative plots.

Dune Grass Planting

Approximately 0.8 ha of graded "dune" ridges were planted in *Elymus mollis*. Three rows of ridges were aligned approximately parallel with the beach face. Dune grass was purchased from the Wave Beachgrass Nursery in Florence, Oregon. Culms were collected in Oregon, trimmed, and shipped to the site, where roots were temporarily buried to prevent desiccation. A CCC crew planted the grass at 2 culms per hill at 61 cm spacing, and 15–30 cm depth. Ammonium sulfate fertilizer was applied at 450 kg/ha . Methods and costs of planting are detailed in Newton (1985b). Growth was vigorous with extensive culm multiplication. At the end of the first growing season, mortality was calculated (Newton 1985b). Overall mortality on the ridges was 39%, with individual dunes ranging from 12–47%. Higher mortality was associated with areas receiving inconsistent irrigation. Long-term cover and density figures for *Elymus* plantings will be discussed in the Long-term Monitoring section.

Implementation Phase

The results of the experimental phase were used, to the extent that they were then available, in developing planting specifications for the implementation phase. Seeding was chosen over transplants due to its high success and relatively low cost, and additional dune grass plantings were carried out. A detailed account of the implementation phase is contained in the Phase Two Planting Report (Bio-Flora 1986).

A total of 2.2 ha were planted in November-December 1986. Once again, labor was supplied by the CCC. In order to fully exploit the experimental value of the project, five seeding methods (hereafter "treatments") were used. All treatments were irrigated throughout the 1986 growing season. The methods were selected based on the results of qualitative and quantitative studies in the experimental phase. The five treatments are listed and described in Table 5. Bio-Flora (1986) contains an analysis of relative costs of treatments.

Seeds for the implementation phase were collected during the summer of 1986. A total of 166 kg of processed seed of 8 species was collected. Documentation of seed collection methods and costs is contained in the Phase Two Seed Collection report (Newton 1985c).

A revised seed mix was prepared based on preliminary results from the experimental phase, with some consideration of seed availability. An important difference between this mix and the experimental mix of Phase One was the addition of *Artemisia pycnocephala.* The mix is shown in Table 6.

In the implementation phase, problems were encountered which greatly influenced the outcome of the project. A discrepancy in the calculated versus actual hectarage for each planting area caused an overabundance of seed mix to be applied. While the specified seed application rate was 28 kg/ha, the

Table 5
Planting Methods (Treatments) Used in the Implementation Phase

Method	Description
Hand-raked	Seeded with broadcast-spreader, raked to 5 cm. Fertilized in July with 28 kg/ha Osmocote 14:14:14.
Hand-raked/Hydromulch	Seeded with broadcast-spreader, raked to 5 cm. Subsequently hydromulched with 1687 kg/ha Spra-mulch and 562 kg/ha Osmocote 13:13:13.
Hydroseed/Hydromulch	Seed applied hydraulically with 562 kg/ha Spra-mulch, second application of 112.5 skg/ha mulch and 562 kg/ha Osmocote 13:13:13.
Harrow	Seed applied with broadcast-spreader, sown with a tractor-drawn harrow. Fertilized in late spring with 28 kg/ha Osmocote 14:14:14.

Table 6
Seed Mix Used in all Treatments of Implementation Phase

Species	kg/ha
Abronia latifolia	7.8
Ambrosia chamissonis	6.9
Artemisia pycnocephala	4.3
Camissonia cheiranthifolia	0.4
Eriogonum latifolium	3.0
Lathyrus littoralis	1.9
Solidago spathulata	3.0
Tanacetum douglasii	1.4

actual rate applied was 46 kg/ha. Slow-release fertilizer, specified at 562 kg/ha, was applied at 900 kg/ha.

This overapplication of seed ultimately controlled the species composition and plant density throughout the implementation area (Pickart 1986b). Differences resulting from planting methods were subordinated to an overall pattern of predominance of early-emerging species. *Artemisia, Eriogonum* and *Camissonia* germinated first and the density of these individuals was intensified by their overrepresentation in the seed mix (Pickart 1986b). In addition, growth and development of the three species represented was poor, despite heavy fertilizer applications. The longer-term outcome of this phenomenon is discussed in the Long-term Monitoring section.

A second problem encountered in the implementation phase was the accidental introduction of weedy grass seed into one of the planting areas. One area was sprayed with hydromulch contaminated with nonnative grass seed. These grasses germinated quickly, and soon overtopped the natives. CCC labor was used to remove grasses after the first winter, but continued management will be necessary to prevent reestablishment (Pickart 1986b).

Long-Term Monitoring

The Corps of Engineers officially concluded the Buhne Point Shoreline Erosion Demonstration Project in 1987. Irrigation had been provided throughout the summer of 1987 to encourage greater development of plants. The Humboldt Bay Harbor, Recreation and Conservation District assumed title to the property and long-term management responsibility. In the fall of 1987 the Harbor District initiated a long-term monitoring program designed to evaluate the degree to which the project meets the full goals of stabilization and community restoration.

Five variables were chosen for monitoring: percent cover, species composition, species diversity, plant density and size class distribution. These variables were selected to permit assessment of community structure as well as stabilization effectiveness. Each of the five variables were monitored separately for each individual planting treatment, including the experimental planting. In addition, monitoring of cover and density of dune grass was carried out separately for each dune. A detailed account of monitoring methods is contained in the 1987 long-term monitoring report (Pickart 1987a). The initial monitoring was carried out three years after the implementation planting. Results will be discussed below by variable.

Density by Treatment

The only treatment which showed consistently significant differences in density was the experimental planting. This treatment showed significantly *lower* densities for *Artemisia, Eriogonum,* and total density. For all other species, densities were significantly *higher* than in other treatments. This trend is due to the initial application rates. As discussed above, the implementation phase was marked by an overapplication of seed, particularly *Artemisia* and *Eriogonum*. These species were applied at much lower rates in the experimental phase (*Artemisia* was not included at all), resulting in significantly lower densities.

Cover by Treatment

Total cover for the harrow treatment was significantly lower (30%) than all other treatments (46–52%). The lower cover in the harrow treatment is probably the result of a combination of factors, including planting treatment, fertilizer rates, and irrigation cover (Pickart 1987a). Cover values for all other treatments are similar to mean cover values reported for Northern Foredune on Humboldt Bay (Duebendorfer 1985; Pickart 1987b). This community is characterized by sparse cover which is believed to be imposed by low nutrient levels.

Size Class Distribution

Size class distributions of individual plants were very heavily weighted towards the lowest size class (0–25 cm^2) for all treatments except the experimental plantings. Frequencies are much more evenly distributed throughout a greater number of size classes in the experimental planting than all other treatments. The high density of seeds applied in the implementation phase resulted in severe competition between individuals, marked by suppressed growth and development.

Species Diversity

Again, the experimental planting was distinguished from all other treatments. The Diversity Index for this treatment (H^1) is .77 (H^1 max = 1), compared to values ranging from .32–.40 for all other treatments. ($H^1 = -\Sigma p_i \log p_i$.) This is once again the result of overapplication of seed in the implementation planting.

Species Composition

Most of the implementation treatments show a strong dominance by *Artemisia* (55–60%) and *Eriogonum* (31–37%). The hydroseed/hydromulch treatment exhibited an even stronger dominance by *Artemisia* (88%). Only the experimental planting is characterized by sizeable contributions by other species. The species proportions parallel the trends revealed through the density and species diversity analyses. Species compositions for the implementation phase are controlled not by the proportion of seed in the original seed mix, but by the pattern of initial establishment which in turn was controlled by application and germination rates.

It is difficult to contrast species composition at the project site with those reported for the Northern Foredune community elsewhere because sampling methods differed. The existing data, however, indicate that the implementation treatments contain uncharacteristically high proportions of *Artemisia*. The experimental treatment, while it is more diverse than the others, is more heavily dominated by *Eriogonum* than the natural community. All treatments lack sufficient quantities of the community dominant, beach goldenrod (*Solidago spathulata*).

Elymus Density and Cover (Vegetative Plantings)

Density of live culms per dune ranged from 1.3–7 culms/m^2. Cover ranged from 8–62%. The highest cover, density and vigor were found to occur on the dune which was located closest to the shoreline. *Elymus* is usually restricted to the primary foredune where growth is stimulated by sand burial. Over time, it is possible that the dune grass on dunes farther from the shoreline may decline in vigor and that other, more competitive species may invade.

Conclusions

It is too early to reach any final conclusions about the success of the Buhne Point revegetation project in meeting all of its stated goals. All areas with the exception of the harrow treatment have achieved suitable total cover values to meet the goals of stabilization and community restoration. At the current time, all planting areas with the exception of the experimental treatments fall short of desired community parameters. Species diversity is low, species composition is weighted towards a few dominant species, and size class distributions are assymetric, with overrepresentation in the smallest size class. The experimental planting shows acceptable species diversity, density, and size class distribution. *Elymus* densities and cover are generally suitable for stabilization, but those areas furthest from the shoreline have exhibited a decline in vigor.

Probably the most important contribution of the Buhne Point Project is the experimental information developed. The project is unique in the amount of preliminary experimentation carried out, the complexity of the planting plan, documentation of techniques, and follow-up monitoring. A great deal of information on species biology and effectiveness of planting techniques was produced. A series of reports covering different aspects of the project was produced by Humboldt County Public Works Department. These reports have been appended to the final report documenting both engineering and vegetation aspects of the project produced by the Corps of Engineers (1987).

References

Bidwell, R. G. S. 1979. *Plant physiology.* 2nd ed. New York: Macmillan Publishing Co.

Bio-Flora Research, Inc. 1986. *Buhne Point Shoreline Erosion Demonstration*

Project, Humboldt Bay, Phase Two Planting: Methods and Costs. Unpublished report. Eureka, CA: Humboldt County Public Works Department.

Duebendorfer, T. E. 1985. *Habitat survey of* Erysiumum menziesii *on the North Spit of Humboldt Bay.* Unpublished report. Eureka, CA: Humboldt County Public Works Department.

Harper, J. L. 1977. *Population biology of plants.* New York: Academic Press.

Holland, R. F. 1986. *Preliminary descriptions of the terrestrial natural communities of California.* Unpublished report. Sacramento, CA: Resources Agency, Department of Fish and Game, Non-Game Heritage Section.

Holton, B. 1980. *Some aspects of the nitrogen cycle in a northern California coastal beach-dune ecosystem.* Ph.D. dissertation, University of California, Davis, CA.

Newton, G. A. 1985a. *Seed collection. Buhne Point Shoreline Erosion Demonstration Project.* Unpublished report. Eureka, CA: Humboldt County Public Works Department.

Newton, G. A. 1985b. *Phase one planting: Methods and cost analysis.* Buhne Point Shoreline Erosion Demonstration Project. Unpublished report. Eureka, CA: Humboldt County Public Works Department.

Newton, G. A. 1985c. *Phase two seed collection: Methods and cost analysis.* Buhne Point Shoreline Erosion Demonstration Project. Unpublished report. Eureka, CA: Humboldt County Public Works Department.

Pickart, A. J. 1986a. *Phase one monitoring report. Buhne Point Shoreline Erosion Demonstration Project.* Unpublished report. Eureka, CA: Humboldt County Public Works Department.

Pickart, A. J. 1986b. *Qualitative evaluation of phase two planting. Buhne Point Shoreline Erosion Demonstration Project.* Unpublished report. Eureka, CA: Humboldt County Public Works Department.

Pickart, A. J. 1987a. *Monitoring results.* Buhne Point Shoreline Erosion Demonstration Project. Unpublished report. Eureka, CA: Humboldt County Public Works Department.

Pickart, A. J. 1987b. *A classification of Northern Foredune and its Relationship to Menzies' wallflower on the North Spit of Humboldt Bay.* Unpublished report. San Francisco, CA: The Nature Conservancy.

Roberts, R. D. and A. D. Bradshaw. 1985. The development of a hydraulic seeding technique for unstable sand slopes, II: Field Evaluation. *Journal of Applied Ecology* 22:979–994.

Tuttle, D. C. 1985. *The history of erosion at King Salmon-Buhne Point, Humboldt Bay, California from 1851 to 1985.* Unpublished report. Eureka, CA: Humboldt County Public Works Department.

U.S. Army Corps of Engineers. 1987. *Buhne Point Shoreline erosion demonstration project,* vols. 1-4 and appendices. San Francisco, CA.

Andrea J. Pickart is the Preserve Manager of The Nature Conservancy's Lanphere-Christensen Dunes Preserve, 6800 Lanphere Rd., Arcata, CA 95521.

Thermal Microenvironments and the Restoration of Rare Butterfly Habitat

Stuart B. Weiss and Dennis D. Murphy

ABSTRACT: *Topography and microclimate are major determinants of habitat suitability for butterflies. The variability of insolation, wind, and fog exposure across habitats affects the distribution and abundance of four endangered butterfly taxa for which habitat restoration is being attempted. Understanding the thermal ecology of these butterfly taxa is critical in developing realistic guidelines for restoration.*

KEY WORDS: *butterfly, endangered species, habitat restoration, microclimate, topography.*

Introduction

Local topography is a fundamental variable in conservation and restoration projects. Topography creates thermal microenvironments at many scales, because exposures to sun, rain, wind, fog, and air temperature—that is *topoclimates*—vary across habitats (Geiger 1966). The variety of temperature and water regimes mediated by topography are responsible for much of the biological diversity supported by landscapes, since many plants and animals have restricted thermal and hydrologic requirements. Indeed, since all ecosystems exhibit topoclimate variation, understanding the interaction of topography and climate at specific sites is a crucial first stage in planning ecologically sound restoration efforts.

The critical roles of topoclimatic considerations in habitat restoration and ecosystem management are richly illustrated by the habitat requirements of several endangered butterfly species which have become the focus of recent conservation activities. Butterflies are the most extensively studied taxonomic group of insects and have proven to be particularly sensitive to environmental changes. Furthermore, they are highly visible indicators of invertebrate diversity. Normal climatic fluctuations have been shown to cause local population extinctions in butterflies, but often the recorded disappearances of butterflies have resulted from urban, agricultural, and industrial expansion that have destroyed native habitats.

Small habitat remnants support vulnerable populations of once more widespread butterfly species in many parts of this country. Several of these species have been listed by the U.S. Fish and Wildlife Service as "threatened" or "endangered." Five butterfly taxa so listed are found in the greater San Francisco Bay Area, a circumstance that reflects both the great biological diversity of the region and the sustained pressure for urban and agricultural development. The conferral of this protective status means that, under sections 7 and 9 of the Endangered Species Act, no action may be taken by the federal government that would "jeopardize" the continued existence of the listed species. Development activities in the habitat of these species must be mitigated through site selection and through restoration of disturbed areas to enhance their values as endangered species habitat.

The first priority of restoration workers focusing on butterflies is to establish appropriate vegetation, including larval host plants and adult nectar sources. Most butterfly species exhibit narrow host ranges, feeding as larvae on one or just a few plant species. The distribution of those host plants may be restricted to certain topoclimates; for example, plants near their northern range limits are often restricted to warm south-facing slopes. Butterfly species subject to conservation efforts usually are further restricted to a small portion of the distribution of their host plants, due to their exacting thermal requirements. This "nesting" of butterfly habitat requirements—host plants grow in a subset of available habitat, and butterflies are dependent on a subset of the host plant distribution—forms the basis for planning butterfly restoration. This paper gives an overview of the conservation and restoration ecology of four endangered and threatened San Francisco Bay Area butterfly species, illustrating the complexity of topoclimatic effects on the success of butterfly restoration projects.

Case Studies

Bay Checkerspot Butterfly: Topoclimates and Population Persistence

The Bay checkerspot butterfly, *Euphydryas editha bayensis,* is restricted to discrete patches of native California grassland on soils derived from serpentine rock. The butterfly was listed as a "threatened species" by the U.S. Fish and Wildlife Service in September, 1987, and is the focus of a conservation agreement at the Kirby Canyon Sanitary Landfill near Morgan Hill. This plan includes a substantial restoration program for areas of the landfill (Murphy 1988).

Topography has a primary influence on persistence of Bay checkerspot butterfly populations in the unpredictable California climate. Topoclimates affect development rates and survival of the immature stages of the Bay checkerspot. When autumn rains stimulate germination of the primary larval host plant, *Plantago erecta,* larvae emerge from the diapause and start feeding. These postdiapause larvae bask in direct sunlight to raise their body temperature into a suitable range for rapid growth. Their growth rates are constrained by the insolation levels of the grassland slopes on which they bask and find food. On the average, larvae and pupae may complete pre-adult development more than one month sooner on warm, south-facing slopes than on very cool north-facing slopes (Weiss et al. 1988).

Population size is mediated by topoclimates. Population fluctuations are largely determined by survival rates of prediapause larvae. Eggs and larvae need 4 to 5 weeks for development, but host plants often senesce before larvae are large enough to enter diapause (Singer 1972). Survivorship thus depends on the phase relationship between adult flight and larval host plant senescence. Females that fly early and lay eggs on cool slopes where host plants senesce later have the highest reproductive success. Offspring of these early females also have some chances of surviving on warm slopes where host plants senesce earliest. Larvae hatchings from eggs laid by late females on warm slopes will not survive; and larvae from the latest eggs laid may not survive even on cool slopes.

The spatial pattern of the prediapause survival across topoclimates changes from year to year (Murphy and Weiss 1988). If the phase relationship is favorable, a Bay checkerspot population will experience "thermal advance"; that is, the next generation of postdiapause larvae will develop on slopes warmer than where their parents developed, and the population will increase in size. If the phase relationship is poor, the population will exhibit "thermal retreat" to cooler slopes and population size will decrease. Populations inhabiting topographically uniform areas without cool slopes that can act as refuges are unable to undergo thermal retreat, and are more vulnerable to extinction than populations in more diverse habitat patches.

Climatic extremes of drought and deluge intensify the effects of thermal retreats and can threaten populations with extinction. The California drought of 1975–77 caused major declines in numbers and numerous extinctions of small Bay checkerspot populations because host plants senesced very early (Singer and Ehrlich 1979; Ehrlich et al. 1980). Extremely wet years with little winter sunshine, such as those caused by the El Ni–o condition in 1982–1983, also can cause large population declines when adult flight is far more delayed than host plant senescence (Dobkin et al. 1987).

The effects of thermal retreat can be partially buffered when postdiapause larvae are able to disperse from cool to warm slopes, which allows them to speed their development to adulthood by up to several weeks (Weiss et al. 1987). Larvae can disperse over 10 m/day, so certain topographic conditions may allow a high proportion of larvae to move to warmer slopes. Narrow east-west running ridgelines (forming adjacent north- and south-facing slopes) and V-shaped gullies in a similar direction provide dramatic thermal variation over short distances. The rounded shapes of California coastal foothills provide thermal changes over longer distances, but still within the dispersal capabilities of late instar larvae.

An understanding of the role of these complex topographic requirements for persistence of Bay checkerspot populations was used in the initial conservation stage and in the planning for eventual restoration of Kirby Canyon. In the conservation phase, consideration of topography in the landfill design at the Kirby Canyon site minimized initial impacts on the Bay checkerspot, because disturbed areas largely consisted of warm south- and west-facing slopes which support larvae only after thermal advance. The most topographically diverse habitat, supporting most of the population, is on the east side of the main ridge, away from the landfill, where 100 ha have been set aside and will be managed specifically for the butterfly by the regulation of grazing and habitat disturbance.

During the restoration phase of the landfill operation, topographic considerations will guide the final contouring of the landfill surface prior to revegetation with larval host plants of the Bay checkerspot and other species native to the serpentine soil-based grassland. Contouring of the landfill surface will create higher quality habitat by establishing areas with cool slope exposures among warmer slopes by building up small hillocks and low ridges. Any such contouring is naturally constrained by drainage and slope stability, but even relatively small areas of cool habitat will enhance habitat quality for the Bay checkerspot in the restored area. Fine-scale landscaping will include rock outcrops for aesthetic reasons and as a specialized plant habitat. Long-term monitoring of butterfly and plant responses to restoration will allow continual adjustment of the process over the more than 50-year active lifespan of the landfill.

An experimental approach to determine proper topsoil treatments and seed applications has preceded revegetation of the serpentine soil. Use of seed collected by mowing and vacuuming on site and applications of stockpiled topsoil have resulted in establishment of the larval host *Plantago*, several nectar source species, and a wide variety of other native plant species in test plots (Koide and Mooney 1987). Revegetation of grassland disturbed by the landfill with native plants, in conjunction with proper contouring, will do much to mitigate impacts from the landfill project.

San Bruno Elfin: Shade and Sun

Many butterfly species are affected by topographically determined habitat factors, from thermal conditions to plant distributions, and, like the Bay checkerspot, require a wide range of topoclimates for long-term population persistence. Other butterflies, in contrast, are adapted to a very narrow range of topoclimates, and small deviations in azimuth and tilt within a site may differentiate between suitable and unsuitable habitat.

The San Bruno elfin (*Incisalia mossii bayensis*) elegantly illustrates this principle. Habitat supporting the San Bruno elfin consists of rock outcrops where thin soils and natural disturbance create favorable conditions for growth of the succulent larval host plant *Sedum spathulifolium*. *Sedum* is generally restricted to cool north-, northeast- and northwest-facing slopes around such outcrops in coastal mountains, and readily colonizes rock-strewn, cut-and-filled slopes.

The butterfly fly from late February through March, avoiding dense late spring and summer fogs which can envelop the California coast. Larvae feed on *Sedum* inflorescences, pupate in June, then enter diapause until the next flight season. Like most butterflies, the San Bruno elfin requires direct sun and relative calm for flight. The early flight season and host plant restriction to cool slopes creates solar exposure constraints in the habitat.

Extensive patches of *Sedum*, as a rule, provide the best habitat for elfin butterflies. Yet, patches of only several hundred plants have supported small elfin populations during seven years of monitoring; indeed the butterfly seems well adapted to the highly patchy nature of its habitat. Two of the most extensive stands of *Sedum*, however, support very few butterflies. One is in an abandoned quarry, and the other is along an old road cut.

The steep faces at these two artificially disturbed sites, which support the densest stands of host plants, are slope exposure virtually unknown in native *Sedum* habitat. At these sites, we have taken solar exposure measurements with a Solar Pathfinder, which determines site-specific shading patterns by superimposing monthly sun paths over a hemispherical reflection of the horizon (to extrapolate solar exposure at different times of the year). The *Sedum* distribution at each site was mapped in detail. Areas receiving no direct sun on March 15 were delineated, and the proportion of host plants available to adult butterflies was calculated from overlays of the plant distribution and shading.

At the quarry site approximately 60% of a total of nearly 125,000 *Sedum* rosettes are deeply shaded during the February-March flight season, hence unavailable as habitat. Sunshine is available after 11 A.M. for most of the remainder of the *Sedum*, but westerly afternoon winds directly strike the habitat leaving only a brief period for adult flight. At the road cut site, 60% of 52,000 *Sedum* rosettes are on the steep road cut face and are deeply shaded in March even though the face is only 1–3 m in width. *Sedum* on natural contours on San Bruno Mountain in San Mateo County, California, are rarely shaded in this manner. Not all contours produced by human disturbance create unsuitable habitat for San Bruno elfin; a northeast-facing road cut near the summit supports one of the largest elfin colonies on the mountain. This site receives direct sun from 8 A.M.–1 P.M. during the flight season.

Restoration work has been proposed to mitigate the destruction of elfin colonies by quarrying activities on San Bruno Mountain and by construction of a highway bypass near Devil's Slide (also in San Mateo County). Observations at previously disturbed areas may prove valuable in development of restoration guidelines for this butterfly. The solar exposure of restored areas will determine whether *Sedum* and nectar sources established there will serve as suitable San Bruno elfin habitat (Figure 1). Due north-facing slopes of greater than 40° in tilt are unsuitable habitat because they receive little direct sun in March. Slopes between 35° and 40° in tilt receive only marginal insolation. Steep northeast-facing slopes receive direct sun until at least noon, while steep northwest-facing slopes receive direct afternoon sun, but prevailing afternoon westerly winds often suppress flight. Cliffs and trees can exacerbate shading. Large patches of *Sedum* reestablished in areas with less than adequate solar exposure will not support San Bruno elfin butterflies; thus will do little to mitigate disturbance to or loss of extant San Bruno elfin colonies.

Callippe Silverspot: Macrotopography and Macroclimate

This picture of the exacting topoclimatic requirements of butterflies may obscure the role of large-scale phenomena in determining local butterfly distributions. The life history of the callippe silverspot (*Speyeria callippe cal-*

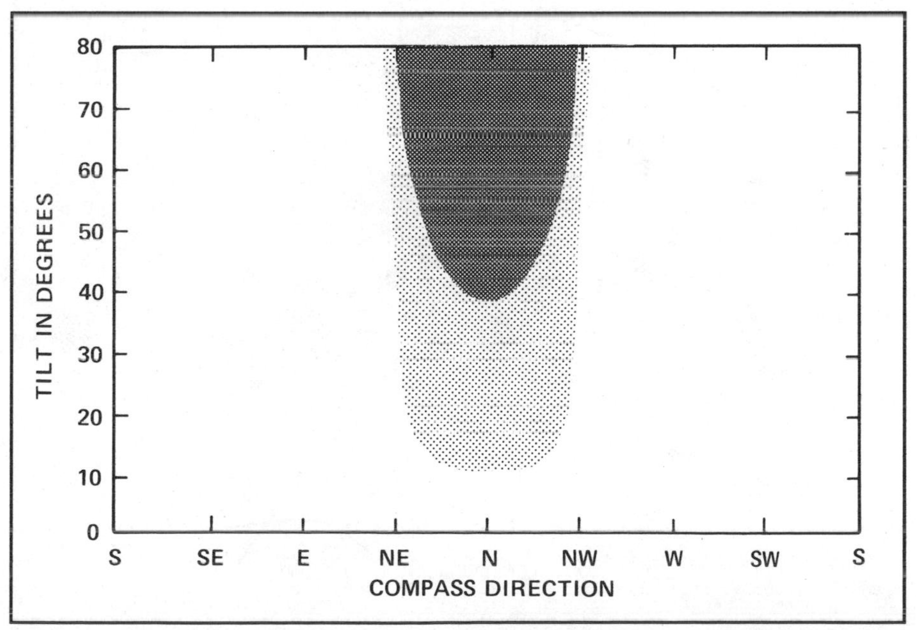

FIGURE 1. *Habitat suitability for the San Bruno elfin butterfly. The entire shaded area is slope exposures where Sedum is found. The darker tone exposures receive no sun in March, and thus cannot support the butterfly. The light gray areas receive adequate direct sun to support the butterfly.*

lippe) for instance, contrasts dramatically with that of the San Bruno elfin. While the small elfin butterfly exists in areas of a few hundred square meters, the silverspot is a large, vagile butterfly with an adult home range covering many hectares of grassland habitat. Whereas elfin habitat is primarily within the fog belt, the silverspot is conspicuously absent from this zone (Figure 2). The elfin avoids fog by flying in February and March, and the silverspot flies late in the Coast Range growing season, from May through July, when fog dominates the macroclimate of San Bruno Mountain.

Prime callippe silverspot habitat, based on the distribution of adults, appears to be in grasslands on largely north- and east-facing slopes that contain dense stands of the larval host plant *Viola pedunculata* and nectar sources including thistle (*Cirsium*), buckeye trees (*Aesculus*), and various native composite species (*Asteraceae*). This restriction to north- and east-facing slopes may reflect thermal requirements of the immature stages, but virtually nothing is known about these stages. Eggs are laid individually near senescent *Viola* and newly hatched larvae eat their eggshells and immediately enter diapause. The larvae break diapause during early spring and feed until they pupate.

During the callippe silverspot's flight season, localized overcast conditions

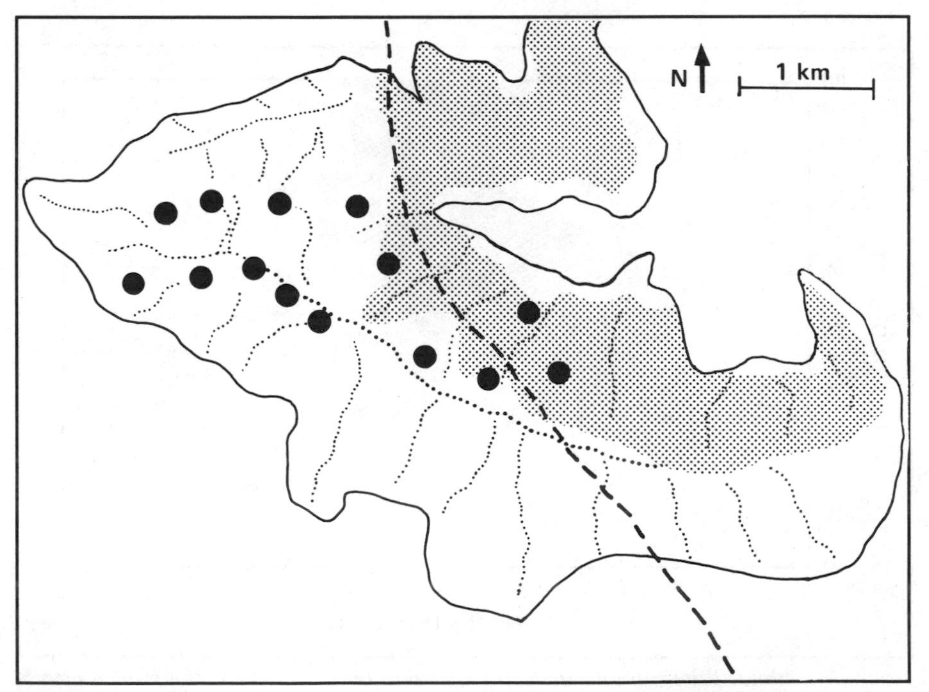

FIGURE 2. *San Bruno Mountain showing the distribution of the callippe silverspot (gray areas) and the San Bruno elfin (black circles). The limit of persistent summer fog is approximated by the dashed line. Fog approaches from the west.*

can prevent flight for days. Fog is most prevalent on the coastward, western portions of San Bruno Mountain, and the silverspot is rarely found west of a distinct fog line, which is determined by topography, even though *Viola* and nectar sources are abundant there. The silverspot appears to be restricted to the sunny, eastern portions of San Bruno Mountain on grassland slopes of appropriate exposure.

Yet, just as these macroclimatic circumstances determine large-scale habitat suitability on San Bruno Mountain, a behavioral trait of the callippe silverspot further complicates its distribution. Male butterflies aggregate on hilltops, creating what are thought to be mating assemblages, in which females seek mates. Hilltops and ridgelines acts as foci for adult butterflies dispersing from surrounding slopes that support plant resources. The presence and orientation of large-scale topographic features, therefore, is an important, if often somewhat subtle, component of habitat suitability which must be considered in restoration efforts.

The habitat requirements of the callippe silverspot greatly constrain the potential effectiveness of restoration efforts directed at this butterfly on San Bruno Mountain. The most promising areas for large-scale revegetation of grasslands containing *Viola* are in the area west of the fog line. Here removal of the invasive alien brush species, gorse (*Ulex europaeus*), is to be followed by revegetation activities. The thermal regime, however, appears to be unfavorable for the callippe silverspot, and new "habitat" created in this area is unlikely to be available to the butterfly except during exceptionally clear periods during the flight season. Restoration of suitable habitat for this species is most likely to succeed along margins of undisturbed habitat, in topographically appropriate situations away from the fog belt.

Mission Blue Butterfly: Early Successional Species

The exacting habitat requirements of the preceding three butterflies suggest that restoration of butterfly habitat is a difficult challenge. Other California butterflies listed as endangered, however, offer more hope to the restoration ecologist. All utilize early successional plant species, which rapidly colonize disturbed habitats. The biology of one of these, the mission blue butterfly (*Plebejus icarioides missionensis*), offers particular promise for successful habitat restoration on San Bruno Mountain.

Mission blue habitat itself is relatively easy to restore because the larval host plants of the butterfly, three species of *Lupinus,* readily colonize all but the most disturbed habitats. Lupine species make up much of the plant biomass on many road cuts and along rock outcrops on San Bruno Mountain. Several nectar sources, particularly coast buckwheat (*Eriogonum latifolium*), occur with lupines in disturbed areas in nearly all topoclimates.

The major topographic restriction on the distribution of the mission blue appears to be exposure of habitat areas to wind. Habitats exposed to the full force of westerly winds during the late-March to early-June flight period of the mission blue support this species. Habitats on opposite cuts along a road, for example, may dramatically differ in suitability where one is exposed to and the other protected from the wind.

Topography also determines the period of mission blue flight, but without complex implications for population dynamics and persistence, as is the case with the Bay checkerspot. Most importantly, the mission blue is not constrained to a narrow phenological window for flight. One larval host, silver bush lupine (*Lupinus albifrons*), is available on both warm south-facing slopes and cool north-facing slopes over a several month period. On warm slopes, butterflies fly in early April, whereas they are often delayed until May on cooler slopes. A second commonly used host plant, summer lupine (*Lupinus formosus*), is restricted to cooler exposures, and mission blue butterflies locally using this plant fly from mid-May into June. The availability of alternate host plants renders much of the San Bruno Mountain landscape suitable as habitat for the mission blue, and buffers the butterfly from the vicissitudes of its climate.

These comparatively broad requirements for habitat suitability for the mission blue suggest that it should respond particularly favorably to restoration efforts, provided that the restored habitats are relatively sheltered from the wind. Development activities on San Bruno Mountain offer numerous opportunities to revegetate cut and filled slopes with Lupinus and nectar source species. Indeed, mission blue butterflies have oviposited on lupines planted from seed and containers in several restored areas on San Bruno Mountain.

Discussion and Conclusions

Not all butterflies are restricted to habitats lacking complex vegetation structure, as are the grassland and brushland species discussed here. Nor are thermal constraints unique to these habitats. Successional vegetational stages within mature woodland, for example, provide wide variation in thermal regimes analagous to the topoclimatic gradient from warm to cool slopes discussed here. Such mosaics of microclimates will shift in distribution over time as woodland succession proceeds. Most woodland butterflies have habitat requirements met in forest clearings or on forest edges. The interaction of topography and forest vegetation offers a particular challenge to the ecologist seeking to restore woodland butterfly habitat, although guidelines for configurations of forest openings which encourage high woodland butterfly diversity have been proposed (Warren 1985).

Further complicating restoration projects for butterflies is the fact that most butterfly populations are small, on the order of hundreds of individuals, and commonly go extinct due to natural events, such as droughts or successional changes in habitats. Most butterflies exist as metapopulations, or "populations of populations." The demographic units making up metapopulations are interdependent upon one another for colonists to insure long-term persistence. Extensive management may be required in restored habitats where butterfly populations exist isolated from sources of colonists—a circumstance especially common to urban restoration projects.

Topoclimate requirements thus represent but one class of diverse abiotic and biotic environmental parameters which have direct impact on the probability of success of restoration projects. In the case of butterflies, provision of host plants and other components of their habitats alone may not result in successful habitat restoration. That habitat suitability has subtle and complex components is a principle likely to apply to many small-bodied ectothermic organisms for which resource requirements are met in relatively small habitat areas. This class of organisms includes most invertebrates which play key roles in nutrient cycling and other ecological processes. Restoration ecologists should not assume that the invertebrate fauna will "take care of itself" once a plant community is established or reintroduced. Presence or absence of certain topographic features may greatly affect the eventual biological diversity of restored landscapes and will prove critical in restoration targeted specifically at organisms with exacting habitat requirements.

Acknowledgments

We thank Anne H. Ehrlich, Paul R. Ehrlich, and Kathy E. Freas for commenting on this manuscript. Waste Management, Inc., has generously provided access to the Kirby Canyon site and funding for much of the recent research on *Euphydryas* butterflies presented here. Additional funds were provided by NSF grant BSR 87-00102. The staff of Thomas Reid Associates graciously allowed access to data on which the San Bruno Mountain Habitat Conservation Plan was based.

References

Dobkin, D. S., I. Oliviera, and P. R. Ehrlich. 1987. Rainfall and the interaction of microclimate with larval resources in the population dynamics of checkerspot butterflies (*Euphydryas editha*) inhabiting serpentine grassland. *Oecologia* 71:161–166.

Ehrlich, P. R., D. D. Murphy, M. C. Singer, C. B. Sherwood, R. R.White, and I. L. Brown. 1980. Extinction, reduction, stability and increase: the responses of checkerspot butterfly (*Euphydryas*) to the California drought. *Oecologia* 46:101–105.

Geiger, R. 1965. *The climate near the ground.* Cambridge, MA: Harvard University Press.

Koide, R. T., and H. A. Mooney. 1987. Revegetation of serpentine substrates: response to phosphate application. *Environmental Management* 11:563–568.

Murphy, D. D., and S. B. Weiss. 1988. A long term monitoring plan for a threatened butterfly. *Conservation Biology* 2:367–374.

Murphy, D. D. 1988. The Kirby Canyon Conservation Agreement: A model for the resolution of conflicts involving threatened invertebrates. *Environmental Conservation,* 1545–1548.

Reid, T. S. et al. 1980–87. San Bruno Mountain Habitat Conservation Plan, and annual updates. San Mateo County, CA.

Singer, M. C. 1972. Complex components of habitat suitability within a butterfly colony. *Science* 176:75–77.

Singer, M. C., and P. R. Ehrlich. 1979. Population dynamics of the checkerspot butterfly (*Euphydryas editha*). *Fortscheritte der Zoologie* 25:53–60.

Warren, M. S. 1985. The influence of shade on butterfly numbers in woodland rides, with special reference to the wood white (*Leptidea synapsis*). *Biological Conservation* 33:147–164.

Weiss, S. B., D. D. Murphy, and R. R. White. 1988. Sun slope, and butterflies: topographic determinants of habitat quality for *Euphydryas editha. Ecology* 69:1486–1496.

Weiss, S. B., R. R. White, D. D. Murphy, and P. R. Ehrlich. 1987. Growth and dispersal of larvae of the checkerspot butterfly *Euphydryas editha. Oikos* 50:161–168.

Stuart B. Weiss is a research assistant in, and Dennis D. Murphy is Director of the Center for Conservation Biology, Department of Biological Sciences, Stanford University, Stanford, CA 94305.

Pest Plants in Woodland Restorations

John Harrington and Evelyn Howell

Abstract: *Pest plants are those species that interfere with restoration goals of community structure, function, and appearance. Interference may involve direct competition with desired plants or may occur through indirect means, including microclimate changes. This paper discusses the origins and characteristics of pest plants and their roles in the establishment and management of woodland restorations. The roles are dependent on the landscape condition at the time of initial restoration. Characteristics of plants that allow pests to flourish are divided into two categories: those that allow for pest plants to be readily available when conditions are appropriate for colonization, and those that provide these species with "staying power" once established. The majority of woody pest plants in the midwestern United States were introduced more than 100 years ago by the nursery and landscaping trade. The paper concludes with case studies of pest removal and pest prevention. Cost per hectare for removing shrubs from deciduous woodlands was $3,115. Implemented policies restricting the use of plant species – Norway maple (*Acer platinoides*), hybrid honeysuckle (*Lonicera x-bella*), and common buckthorn (*Rhamnus cathartica*) – that have documented pest tendencies are discussed.*

Key words: *restoration, woodland, pest, management.*

Introduction

Plant community management involves the determination of strategies and the implementation of techniques in order to influence or direct change in the landscape, and in particular, change in vegetation composition.

Three approaches are used:

1. Manipulation of the environment so that it is conducive to the growth of desired species but inhospitable to others;
2. Direct intervention in the growth of plants (killing some, stimulating the growth of others); and

3. Adoption of policies and use of procedures that influence the availability of species.

This paper discusses how these approaches can be used to meet an objective common to most restorations: the control of pest species through removal or prevention. Using examples drawn from our experience of woodland restorations in the upper midwestern United States, we hope to stimulate discussion and encourage the sharing of information on pest plants and their management.

Pest Plants: Definition and Role in Restorations

Woodland restorations vary in composition and structure depending on their purposes and can be divided into three general types (Howell 1988):

1. Those which attempt to establish complements of native species that mimic as closely as possible the patterns, abundances, proportions, and relationships of pristine communities. The assumption is that in doing so, natural selection and other processes will continue. Restorations at arboreta, or those comprising buffer zones for protected natural areas, are often of this type.
2. Those which establish natural processes, such as nutrient cycling, but which do not necessarily replicate the vegetation structure and composition of natural communities. Individual species are not as important as the roles they represent. An example would be the establishment of essentially self-maintaining vegetation on a mined site.
3. Those which emphasize the human experience of the community. Visual quality and "aesthetic essence" are primary considerations. Plantings around homes or in public parks are often of this type.

Pest species, in this context, are plants that interfere with such restoration goals. The plants may be native or nonnative, and the same species may operate as a pest in one situation but not in another. Interference may involve direct competitive interactions including:

1. Replacing or inhibiting desired plant and/or animal species (altering community composition);
2. Eliminating the desired proportions of species, often with the pest plants establishing themselves in relatively high numbers within their life form class (altering community structure);
3. Changing the desired natural processes and functions; and
4. Slowing down the restoration process or the achievement of restoration goals.

These effects are reinforced by microclimate changes brought about by the presence of pest species. In woodlands, such changes can include increased shade, reduced moisture availability, and changes in soil fertility. For example, in the Midwest, Eurasian species often have a longer growing season than American species. Woodland spring ephemerals require high sun levels to bloom and produce food. Such levels typically occur under the leafless April and May canopies of native trees and shrubs. These herbs are weakened when their habitat is invaded by Eurasian species which produce solid canopies as early as April in parts of the Midwest. Table 1 lists three Eurasian and four native species, indicating the months in which they have foliage. As noted in the table, the Eurasian species typically begin foliage growth several weeks earlier than our native species and retain their foliage for several weeks longer in the fall.

Examples of Pest Species in Midwestern Woodland Restorations

Pest interactions with other species are dependent on the landscape condition at the time of restoration. Woodland restorations are of two types:

1. Those starting without a canopy.
2. Those with some form of canopy in place. An example typical of the Midwest would be a grazed oak woods. The canopy may be largely intact but the midstory and herb layer are generally quite degraded.

Pests operate differently in these situations. For example, native woodland shrubs, such as hazelnut (*Corylus americana*) and grey dogwood (*Cornus racemosa*), may operate as pests in the first, seldom in the second.

Characteristics that support pest tendencies – behaviors that interfere with restoration and management goals – can be divided into two categories, those that allow for pest plants to be "readily available" when conditions are appropriate for colonization and those that provide these species with "staying power" once established.

Characteristics of pest plants that support their availability for colonization and establishment include:

Table 1
Foliage Phenology

Species	Native	Apr	May	Jun	Jul	Aug	Sep	Oct	Nov
Trees									
Acer saccharinum	x								
Acer platinoides									
Midstory									
Corylus americana	x								
Lonicera x-bella									
Rhamnus cathartica									
Herbs									
Isopyrum biternatum	x								
Erythronium albidum	x								

1. Efficient reproduction methods through high seed production.
2. Long viability and high germination rates for seed (Table 2). Tartarian honeysuckle (*Lonicera tatarica*) has a potential germination rate of 85%. Within 33 days after a 30-day dormancy period at 5oC 50% of the seed has germinated. Amur privet (*Ligustrum amurense*) has a 77% viability for seed and common buckthorn (*Rhamnus cathartica*) between 54% and 60%. Buckthorn seed under appropriate conditions can begin germinating within 18 days (U.S. Forest Service 1974).
3. Annual intervals of seed production. The majority of pest species found in the Midwest produce seed crops each year.
4. Rapid maturation. Both honeysuckle and buckthorn begin producing seed at an early age, often within three years.
5. Long distance seed dispersal by birds, ground animals, water, and wind as portrayed in Table 3. Buckthorn seeds, for example, are distributed by birds who remove the fruit pulp and scarify the seed as it passes through their digestive systems.
6. Tolerance of a wide range of environmental conditions. This characteristic is also the first trait that contributes to the staying power of a pest plant once established in a restoration.

Many of these species also have the ability to adapt environments to their needs. Japanese honeysuckle (*Lonicera japonica*), for example, tends to establish itself in forest gaps and then proceeds to climb and outgrow young shrubs and saplings, thereby maintaining open to semishaded light conditions in which it will actively grow (Evans 1984).

In addition to broad environmental tolerance, traits that encourage staying power include:

1. The ability to sprout and spread vegetatively. This trait can make management particularly difficult. Many control methods, including cutting, mowing and burning, which damage a plant, can act as stimulants for profuse suckering.
2. A relatively low level of natural controls in the environment.
3. Hybrid vigor; some pests have been "created" by natural interbreeding between species brought to the United States from separate parts of

TABLE 2
SEED NUMBERS AND VIABILITY

Species	Seed	Viability	% Day	Stratification (days)
Acer ginnala	17,000	52	50/10	41/90-150
Ligustrum vulgare	18,600	77	–	36/60-90
Lonicerna tatarica	142,000	85	58/33	41/30
Rhamnus cathartica	19,100	54	54/18-40	70/60
Rhamnus frangula	27,000	60	55/18	40/60
Viburnum opulus	13,600	60	–	40/30-60

TABLE 3
DISPERSAL MECHANISMS OF PEST PLANTS

Species	Wind	Songbird	Wildlife	Vegetative
Exotic				
Acer platinoides	———			
Acer ginnala	———			
Euonymus alata		———		
Lonicera japonica			———	———
Lonicera x-bella		———		
Rhamnus cathartica		———	———	
Viburnum opulus		———	———	
Populus alba	———			
Native				
Acer negundo	———	———	———	
Robinia pseudoacacia				———
Zanthoxylum americanum		———		———
Cornus stolonifera		———		———

the world. One of the major pest plants in Wisconsin is the hybrid *Lonicera x-bella,* which developed by the natural hybridization of *Lonicera tatarica* and *Lonicera morrowii.*

ORIGINS OF PEST PLANTS

Stopping the spread of many of our existing pest plants and preventing the introduction of new ones is aided by knowing the origins of such plants. Since European settlement, large numbers of plant species from foreign shores have been introduced to America. Many of these species have naturalized, spreading into remnant forests, prairies, and wetlands. In 1980 it was estimated that 20% of Illinois' naturalized flora had exotic origins (Ebinger et al. 1984). The majority of woody pest plants in the midwestern United States were introduced more than 100 years ago by the nursery and landscaping trade. Table 4 lists a number of midwestern pest species and their origins. Some of these species were originally intended for rural hedgerows and windbreaks, from which their propagules were easily dispersed. More recently, ornamental landscaping is expanding to suburbs and has brought these species into more frequent contact with natural areas and restoration settings.

Pests may be native as well as exotic (Table 5). Both are problems largely through the agency of humans – exotics because we brought them here, many natives because we have changed the environment by stopping fire and grazing, and by manipulating water regimes so that habitats inviting to pest species have been created.

TABLE 4
ORIGINS OF PEST PLANTS

Species	Date	Country	Origin/Use
Acer platinoides	1800s	Europe	Nursery/shade
Euonymus alata	1860	Asia	Nursery/landscaping
Ligustrum vulgare	pre-1700	Europe	Nursery/hedge/landscaping
Lonicera japonica	1806	Asia	Nursery/conservation
Lonicera tatarica	pre-1700	Eurasia	Nursery/hedge
Lonicera x-bella	–	–	Hypbridization
Lythrum salicaria	1700s	Europe	Shipping/gardening
Populus alba	1784	Eurasia	Shade/windbreaks
Rhamnus cathartica	pre-1700	Eurasia	Hedgerows
Vinca minor	1789	Europe	Nursery/ornamental

TABLE 5
MAJOR PLANT SPECIES IDENTIFIED AS POTENTIAL MANAGEMENT PROBLEMS IN MIDWESTERN NATURAL AREAS (WEST 1984)

Exotic	
Chinese bittersweet	*Celastrus orbiculatus*
Autumn olive	*Elaeagnus umbellata*
Winged euonymus	*Euonymus alata*
Wintercreeper	*E. fortunei*
Common privet	*Ligustrum vulgare*
Japanese honeysuckle	*Lonicera japonica*
Amur honeysuckle	*L. maackii*
Tartarian honeysuckle	*L. tatarica*
White poplar	*Populus alba*
Common buckthorn	*Rhamnus cathartica*
Guelder rose	*Viburnum opulus*
Common periwinkle	*Vinca minor*
Hybrid honeysuckle	*Lonicera x-bella*
Norway maple	*Acer platinoides*
Wayfaringtree	*Viburnum lantana*
Bluegrass	*Poa* spp.
Sweet clover	*Melilotus* spp.
Reed canary grass	*Phragmites australis*
Native	
Box elder	*Acer negundo*
Aspen	*Populus tremuloides*
Sumac	*Rhus* spp.
Prickly ash	*Zanthoxylum americanum*
Black locust	*Robinia pseudoacacia*

Case Studies

As discussed in the Introduction, the control of pest species involves both removal and prevention. The following case studies illustrate these approaches as they have been used in the Midwest.

Removal

Although removal can be accomplished by changing the environment, the most commonly used approach is active intervention in the growth of plants through cutting, burning, herbicide use, etc. One example of such a project for which cost figures are available involved the removal of honeysuckle (*Lonicera x-bella*) from a small deciduous woodland on the campus of the University of Wisconsin–Madison (Henderson and Howell 1981). Approximately 70% of the woods had dense areas of honeysuckle (70–100% cover; 2–4 m high), 20% had moderate stands (20–70% cover; 1–3 m high) and 10% had sparse cover (0–20% cover; 1.5 m high). Shrub removal was accomplished by cutting and applying a 1:5 solution of Roundup (glyphosate) herbicide to the cut stumps. The brush piles were then chipped by a three-person crew.

The per-ha costs were as follows:

Hand Labor (cutting, stacking, herbicide treatment)	170 hr @ $6/hr	$1,020
Chipper and crew	30 hr	900
Roundup	3 gal @ $65/gal	195
		$2,115

The cost is significant and varies with density. Time estimates of required hand labor for the three shrub density categories involve about 30 hours per ha for sparse cover and 210 hours for dense cover.

Prevention

Probably the best means to manage pest species is to prevent their introduction in the first place. One approach is followed by the City of Madison Conservation Parks, which practices active control of seed sources, rather than attempting complete removal of all exotic or natural disturbance species. For example, female box elder (*Acer negundo*) trees are destroyed to prevent the spread of seeds while fewer resources are expended on control of male box elders.

Through its planning commission, the City of Middleton, Wisconsin, has taken an even more active role in limiting the spread of pest plants. In 1985 Middleton adopted parking lot standards which include provisions for restricting the use of plant species which tend to invade their conservancy areas (Middleton City Planning Commission 1985). The Middleton City Planning Commission reviews all landscape plans for parking areas. Approval of these plans depends on a number of criteria, one of which requires that Norway maple (*Acer platinoides*), buckthorn (*Rhamnus cathartica*), and honeysuckle (*Lonicera x-bella*), all invasive plants of dry-mesic woodlands, not be planted in any parking area development that requires commission approval. The seeds of Norway maple, for example, are wind blown and replace sugar maple (*Acer saccharum*) seedlings in mesic woods. Recent observations of ground and sapling layers in sugar maple woods of southern Wisconsin have found Norway maple establishing itself as both a ground- and middle-story component of these woods.

The provision restricting certain species from landscape use was developed by a group of landscape architects and botanists. The potential significance of such a standard is realized if one recognizes that Norway maple is one of the top selling urban landscape trees in the middle and eastern United States. To date the city has received few complaints about this policy. Most landscape architects asked to substitute species for those on the list are willing to comply.

Summary

Eradication of pest plants that impact restorations is not realistic but development of reliable control techniques is. This will require greater communication between restorationists who have studied and documented pests and an understanding of the autecology of these species. Moreover, it will require an understanding by the public, landscapers, and even restorationists of the existing and potential repercussions of such species.

In summary, pest plants are those species that significantly alter the biologic composition or aesthetic quality of a natural community. They can be naturalized or native species but they interfere with desired functions, uses, purposes, and goals of a site. Pest plants have the potential for outcompeting and eradicating plants of natural communities and then establishing themselves within the community, often in a dominant role. Their populations grow entirely or predominantly in situations disturbed or impacted by humans.

References

Ebinger, J., J. Newman and R. Nyboer. 1984. Naturalized winged Euonymus in Illinois. *Natural Areas Journal* 4:20–29.

Evans, J. 1984. Japanese honeysuckle (*Lonicera japonica*): A literature review of management practices. *Natural Areas Journal* 4:4–10.

Henderson, R. and E. Howell. 1981. Time and cost figures for honeysuckle control in a disturbed southern Wisconsin forest. *Restoration and Management Notes* 1:18–19.

Howell, E. A. 1988. The role of restoration in conservation biology. *Endangered Species Update* 5:1–4.

Middleton City Planning Commission. 1985. *City parking areas specifications and standards.* City of Middleton.

U.S. Forest Service. 1974. *Seeds of woody plants in the United States.* Washington, DC.

West, K. A. 1984. Major pest species listed, control measures summarized at natural areas workshop. *Restoration and Management Notes* 2:34.

Professor John Harrington and Professor Evelyn Howell, Department of Landscape Architecture, University of Wisconsin-Madison, Room 25 Agriculture Hall, 1450 Linden Drive, Madison, WI 53706.

Restoration of Dune Habitat at Spanish Bay

Marylee Guinon and David Allen

Abstract: *As part of the Coastal Commission development permit, the Pebble Beach Company agreed to restore approximately 160 ha of sand dunes within a golf course and provide a native species protection program at the Spanish Bay Resort development on the Monterey Peninsula in California.*

The dune restoration program began in 1984 with construction of an experimental dune designed to test various restoration and erosion control techniques. Over 383,000 m³ of material was imported to create dunes where they had been previously removed by sand mining operations. Since 1985 several implementation programs have been initiated, including exotic species eradication, plant salvage, local seed collections, plant propagation, sand dune construction, hydromulching, erosion control, sensitive species management, and long-term management. Dune revegetation was accomplished by hydromulching of 27 species native to the coastal strand community. The paper discusses specific restoration methods, and, in the context of a multiple-use project, the significance of scheduling, restoration guidelines, specifications, performance standards, maintenance, and monitoring. The authors suggest that restorationists adopt the successful strategies of the construction industry in which projects are designed, engineered, tested, and constructed by codes, regulations, and performance standards.

Although intended to be a dune "restoration" project, where dune habitat is maintained over the long-term, Spanish Bay evolved into a dune "revegetation" project, where the landscaped dunes are adjacent to the greens, tees, and fairways or in some cases serve as the golf course's rough. Due to the multiple-use nature of the project, significant biological and physical pressures are placed on the dunes, hence making veritable habitat establishment questionable.

Key words: *mitigation, multiple-use, revegetation, dune(s).*

Introduction

As part of a California Coastal Commission development permit, the Pebble Beach Company of Monterey agreed to restore approximately 160 ha of sand dunes and provide a native species protection program at Spanish Bay on the Monterey Peninsula in California. The 580-ha resort development includes a hotel, a condominium and an 18-hole championship golf-links course, where the sand dunes interlace with the greens, tees, and fairways of the golf course. Dunes on the site had been removed by a previous sand mining operation from the 1930s through the 1970s.

Much of the site therefore required dune restoration, as well as enhancement of riparian and forested areas. In addition to removing the network of dunes, mining lowered surface elevations and removed sand to granite bedrock in many locations, creating impervious surfaces which supported little or no native vegetation. Over 383,000 m^3 of material was imported from an abandoned mine to rebuild the dunes removed by the former quarry operations.

The project consultants prepared a resource management plan for the Pebble Beach Company and were then hired to implement the initial phases of dune, forest, and riparian restoration. The plan presents the policies, planning concepts, implementation, and long-term management criteria according to which the natural habitats of the site were created, restored, and enhanced. Because of the project's size and complexity, multiple-use components, and the simultaneous construction of a golf course and dunes, the restoration plan required continual modification. Restoration schedule delays were a direct result of construction schedule delays.

The dune restoration program began in 1984 with construction of an experimental dune designed to test various native habitat restoration and sand stabilization techniques. Since 1985 several implementation programs have been initiated including: exotic species eradication, whole plant salvage, local seed collections, plant propagation, sand dune construction, hydromulching, outplanting, erosion control and sensitive species management. The final and perhaps most critical phases of implementation are monitoring and maintenance of the dunes. Because the golf course maintenance staff elected to conduct these programs, no monitoring data is presented herein.

This paper discusses the exotic plant eradication, native plant protection, seed collection, hydromulching and dune plant propagation program. The authors address objectives, materials, methods, and preliminary results for these programs well as a critique of the Spanish Bay permit process and restoration program.

Exotic Plant Eradication

Several exotic plant species were removed from the Spanish Bay site including ice plant (*Carpobrotus edulis*) and its hybrid with the native sea fig (*Carpobrotus chilensis*), pampas grass (*Cortaderia atacamensis*), kikuyu grass (*Pennisetum clandestinum*), acacia (*Acacia longifolia* and *A. verticillata*), French broom (*Cytisus monspessulanus*), and a number of herbaceous species, including plaintain (*Plantago cornucopifolia*). Eradication, by a combination of chemical and mechanical means, was primarily targeted at ice plant and pampas grass, since they were the most abundant and aggressive exotic species on the site and are expected to pose significant on-going maintenance problems. Most pampas grass, acacia, ice plant, and French broom were mechanically removed during the clearing operation. The remaining exotics were either eradicated by using Roundup, a surfactant and dye, or by hand weeding.

A 2% solution of Roundup was applied either manually with backpack sprayers or broadcasted with hoses or booms mounted on a truck. Spraying was conducted during the spring of 1985 when wind velocities were less than 8 km/hr and when no rain was anticipated. Approximately 210 ha were mechanically sprayed with booms or hoses. These areas supported little or no native vegetation; after three weeks, missed areas were resprayed. The bright dye aided in ensuring coverage of target plants. Approximately 37 ha were hand weeded and hand sprayed by field crews. Native plants in these areas were protected from chemical drift by careful application and plastic shields. An ice plant mass was easily pulled away from adjacent natives and sprayed. The removal rate of the exotic species was high. Mature pampas grass required several applications of Roundup and was best removed by construction equipment.

Coastal Strand Community Seed Collection

In preparing its resource management plan, the Pebble Beach Company reviewed the relevant literature on the ecological and physical conditions of sand dunes for information on general habitat characteristics, microenvironments, cultural requirements, rooting configurations, and successional status of the species used in the restoration program. The consultants also made field observations at all state beaches along Monterey Bay. The combined knowledge obtained from the literature and the field enabled the biologists to design an effective seed-collection program and seeding and planting plans that reflect biological knowledge of species and their successional trends.

Species selected for the restoration program are members of the coastal strand community and were thought to have once inhabited the Spanish Bay site. Certain species were included because they are dune-colonizing or early successional species, or because they are considered valuable sources of inoculant seed which will be maintained in the dune complex and will eventually become dominant. All of the species collected were native with the exception of sea rocket (*Cakile maritima*), an annual species which was included because it is an early successional species capable of rapidly establishing on the dunes; it is a nonaggressive and noninvasive species that will not compete with the natives, and it stabilizes sand.

The biologists conducted seed collections in 1984, 1985, and 1986. Seed from several of the species, though available commercially, was collected from dunes adjacent to and on the Monterey Peninsula in order to maintain

TABLE 1
COASTAL STRAND COMMUNITY SPECIES DISTRIBUTIONAL RANGES

Species	Distribution
Abronia latifolia (yellow sand verbena)	Santa Barbara Co. to British Columbia (B.C.)
Abronia umbellata (pink sand verbena)	Los Angeles Co. to Sonoma Co.
Achillea borealis (yarrow)	S. California to Washington
Armeria maritima var. *californica* (sea pink)	San Luis Obispo to B.C.
Artemisia pycnocephala (beach sagewort)	Monterey Co. to Humboldt Co.
Baccharis pilularis (coyote bush)	Monterey Co. to Sonoma Co.
Cakile maritima (sea rocket)	Monterey Co. to Mendocino Co.
Camissonia cheiranthifolia (beach evening primrose)	Santa Barbara Co. to Coos Co., OR
Carex spp. (dune sedge)	San Luis Obispo Co. to WA; S. Rosa Island
Castilleja latifolia (seaside painted cup)	Monterey Co. to Santa Cruz Co.
Convolvulus soldanella (beach morning glory)	San Diego Co. to Washington
Corethrogyne californica (beach aster)	Monterey Pen. to San Francisco Pen.
Dudleya farinosa (liveforever)	Los Angeles Co. to Oregon
Erigeron glaucus (seaside daisy)	San Luis Obispo to Oregon
Eriogonum latifolium (coast buckwheat)	San Luis Obispo to Oregon
Eriogonum parvifolium (coast buckwheat)	San Diego Co. to Monterey Co.
Eriophyllum staechadifolium (lizardtail)	Monterey Co. to Santa Cruz Co.
Eschscholzia californica var. *maritima* (poppy)	Santa Barbara Co. to Mendocino Co.
Franseria chamissonis var. *bipinnatisecta* (beach bur)	S. California to B.C.
Grindelia stricta var. *venulosa* (gum plant)	Monterey Co. to Coos Co., OR
Haplopappus ericoides (mock heather)	Los Angeles to Marin Co.
Lathyrus littoralis (beach pea)	Monterey Co. to B.C.
Lotus scoparius (deerweed)	S. California to Humboldt Co.
Lupinus arboreus (yellow bush lupine)	Ventura Co. to Del Norte Co.

From: Munz, P. A., and Keck, D. D. 1973. *A California Flora*, combined edition. Berkeley: University of California Press.

the genetic integrity of the region's flora. With the assistance of the California Department of Parks and Recreation, which issued a collection permit, field crews collected local seed collections on state beaches, private property, and on the Spanish Bay site.

Since genetic architecture information is unavailable for the coastal strand community species, the Spanish Bay seed collection sites were simply restricted to the local Monterey Bay beaches. It is likely that genetic populations for several species were mixed, since within-species morphological variation is apparent. Polymorphism is frequently used as an indicator of genetic variation when precise data is unavailable. In the absence of genetic architecture analysis through extensive laboratory analysis, such as electrophoresis, it is critical to err on the conservative side when defining collection areas. Genetic architecture data indicates historical biogeographical patterns only and does not predict species adaptability.

The 1984 collection was designed to provide an adequate sample of species to be applied onto the experimental dune. Seed from 17 species was collected in 1984, yielding approximately 45 kg of clean seed. Fourteen of the 17 species were either hand seeded, hydroseeded, or hydromulched onto the experimental dune. Biologists collected data on clean seed weight, percent pure seed, percent germination, percent live seed, seeds per weight, and number of live seed.

An intensive seed collection program conducted in 1985 yielded over 182 kg of cleaned seed. Continual monitoring of the dune collection sites was required to determine the optimal seed harvest time for each species. In order to minimize impacts to donating plants, quantities of seed collected from each species was determined by the density of the donating plants. Collection site monitoring included location of desired species and determination of optimal seed harvest dates. Collection efforts were intensified for the dominant species in order to avoid depletion of the species with limited seed supplies.

Due to unforeseen resort construction delays, the first hydromulching phase did not commence until June 1986. Fresh seed from the 1985 collection was used and continual germination tests were made to determine the effects of storage due to the delays. Because seed from the dune species do not store well, a supplemental seed collection effort was made in 1986 to ensure fresh seed for 1986 and 1987 hydromulching.

The native plant seed collected during the summer months of 1986 yielded approximately 90 kg of cleaned seed from nine species. This seed was stored only 2–7 months and subsequently hydromulched in October 1986 and January 1987. Also hydromulched onto the dunes in 1986 and 1987 were chaff and seed from the 1985 collection. Although this seed was not fresh, seed from many of the desired species was available only from the earlier collection.

Hydromulching

The two main objectives of the hydromulching operation were to establish a native plant cover and to stabilize the sand dunes. Because the dunes surround the greens, tees, and fairways, the dynamic nature of the dunes was arrested and the sand was not allowed to move. Native plant rehabilitation was one of the primary permit conditions set forth by the Coastal Commission to allow construction of the Spanish Bay Resort. Since sand stabilization was critical, the normally mobile sand dunes were stabilized so that the golf course is protected from sand movement.

Because the native plant species are comparatively slow to germinate and produce extensive root systems, a nurse crop was added to the hydromulching mixture to stabilize the dunes temporarily until the native plants become established. Three annual species, zorro fescue (*Festuca megalura*), blando brome (*Bromus mollis*), and crimson clover (*Trifolium incarnatum*) comprised the nurse crop. These plants do not normally occur in sand dunes and were not expected to persist long in this environment. The hydromulching mixture consisted of water, mulch, fertilizer, tackifyer, native plant seed, and nurse crop seed. The mixture was constantly agitated, which is critical to ensure even dispersal of the (minority) native seed component. The mixture was applied to most of the work site using hoses, due to poor accessibility.

The native coastal strand species utilized were divided into three species groups—foredune, middune, and hinddune—to emulate natural dune plant associations observed in Monterey Peninsula coastal strand communities. A planting plan of Spanish Bay was used to define the fore-, mid-, and hinddune sites. Along with distance from the ocean, site exposure was critical in determining location for each group. Usually sand movement is the most important criterion in determining the location of these groups, the foredunes being dynamic and the hinddunes stable. However, the dunes at Spanish Bay are to remain stable, so this factor is of no great consequence, hence exposure to wind and salt was of primary importance in creating the planting plan.

In January of 1984 an experimental sand dune was constructed at Spanish Bay to determine optimum stabilization methods, irrigation and fertilizer rates, planting regimes, and maintenance requirements. Approximately 45 kg of native plant seed collected from 14 species in 1984 was hydromulched onto experimental strips on the dune. The experimental strips included combinations of excelsior blankets, jute netting, straw tacked beneath plastic netting, woodchips, hydroseeding, hydromulching, and handseeding. Based upon test dune results, the biologists chose hydromulching as the optimal revegetation and erosion control method and subsequently developed several variations of the hydromulching mixture. Mulch rates were set at 368 kg/ha or 276 kg/ha. The mixture of the nurse crop remained constant but the density of the nurse crop stand was either 538 plants/m^2 or 269 plants/m^2. The fertilizer formulation also remained constant and was applied at 36 kg/ha and 268 kg/ha. A tackifyer was applied at 15 kg/ha. From these variables, several possible hydromulching treatment combinations were developed.

An impact-head irrigation system was installed throughout the newly created dunes. This system was employed to help stabilize the barren dunes from wind erosion and to provide water to the young dune plants. The irrigation system also permitted us to initiate hydromulching at any time of year, which was essential given the demanding golf course construction schedules. The quantities of water applied to the dunes varied with weather conditions, but initial cycles were set for 15 minute periods three times daily. Watering was reduced and eventually eliminated as the native plants became sufficiently hardy.

Native seed mixes were applied in accordance to dune aspect and situation (fore-, mid-, and hinddune). Evaluation was based on:

- stabilizing effect;
- growth of nurse crop;
- growth of native plants with respect to growth in another similar situation but with differing treatment;
- growth of natives in a control situation; and
- transects to determine densities and species composition.

In addition, the biologists significantly increased the number of plants to be propagated and outplanted to compensate for the delays in hydromulching and consequent lower germination rates.

During the first phase of hydromulching, it became apparent that the mulch rate of 276 km/ha was too light for use on the dune slopes but provided adequate coverage on level ground, so the mulch rate of 368 kg/ha was used almost exclusively. A critical decision made during the first hydromulching phase was to use the lower nurse crop seeding rate, which would result in 269 established plants per m^2. This decision was based on the quality of the imported fill material used to recreate the dunes and to compensate for the unexpected prolific establishment of the nurse corp. The fill material contained a considerable percentage of silts and clay and these edaphic factors meant that the nurse crop would thrive and possibly persist for a number of years. The nurse crop was intended to exist for only one or two seasons. In addition, the nurse crop species showed remarkable germination (90–100%) and a mortality rate that was substantially lower than expected (50%). An almost uniform, thick, and verdant nurse crop which competed with the natives resulted. Many native plants perished and those that survived became etiolated and weak. In areas where the nurse crop was thin, the natives flourished. The thick nurse crop cover was very successful at providing stabilization, but this effect could have been achieved at much lower nurse crop seeding densities. The nurse crop was thinned and natives were reseeded. Native plant nursery stock grown from our own propagation yard was also planted into these areas to augment the seeding.

In the months following hydromulching, the high seedling density of the native plants throughout the dunes proved very satisfactory. Beach sagewort, yarrow, the buckwheats, and lizardtail were especially successful in producing a great number of seedlings. In sample quadrats, an average of 89 native plant seedlings per m^2 were found. The number of native seedlings ranged from 22–228 plants/m^2. This number was greatly influenced by the density of the nurse crop. The quadrats with the thickest nurse crop stand showed the least number of native seedlings. The growth of the nurse crop and amount of cover produced by it varied greatly within a single sow rate. The determining factor appeared to be the quantity of water received by any given area, and this was heavily influenced by drainage patterns. The hydromulching process produced a good and even distribution of seed and fertilizer.

The following two hydromulching phases in 1986 and 1987 benefited from the observations of the first phase. The nurse crop seeding rate was dramatically reduced and, in some areas with low erosion potential, was done away with completely. The mulch rate remained at 909 kg/ha and the fertilizer was primarily applied at the lower rate of 49 kg/ha. This again was due to the physical properties and high organic components of the fill material.

Native plant seed was collected during the summer months of 1986 and cleaned, yielding approximately 90 kg of seed. This seed was stored for two to seven months and subsequently hydromulched in October 1986 and January 1987. The sow rate for the natives varied from 8.9–11.1 kg/ha for each grouping. The seed hydromulched in 1986 and 1987 was augmented with seed remaining from the 1985 collection; therefore, sow rates were slightly higher than for the first hydromulching phase. The seed collected in 1985 had greatly reduced germination potential by 1986. Germination test results ranged from less than 1% to 50%. Mock heather was affected the most from storage and showed the lowest germination. The sow rates used in the latter two hydromulching phases produced a good display of native plant seedlings. In any other situation where sand stability is not such an overriding concern, the nurse crop seeding should be eliminated.

Native Plant Propagation

A small plant propagation yard including a greenhouse and mist house was constructed to supply container stock of dune plants to augment the hydromulching efforts, to grow trees and shrubs which will replace those removed during construction and to propagate rare plants and other species for which seed supply were limited. Additional off-site propagation facilities included 100 m^2 raised beds and a 1,130 m^2 sand bed used as a dune sedge propagation bed.

Over 150,000 plants have been raised, primarily dune species. An additional 100,000 plants were grown in 1988. The majority of the plants were raised in supercells which lend themselves ideally to the long root systems of the dune plants.

The dune species were grown to supplement hydromulching by increasing native plant density in sparse areas. The objective of the revegetation program was to establish the dunes with an overall average density of one plant every 1.5 m^2. Outplanting took place as soon as hydromulching success could be evaluated, i.e., after approximately five months. In some instances irrigation failures produced washouts and these locations were planted and reseeded immediately. For most species it was best if they were outplanted from supercells no later than six to seven months after germination, because they would outgrow the container.

The only two species requiring scarification were beach morning glory (*Convolvulus soldanella*) and coffeeberry. Species such as sea rocket (*Cakile maritima*) and beach evening primrose (*Camissonia cheiranthifolia*) were successfully reestablished from the hydromulching, and they reseeded themselves so readily on-site that it was not necessary to grow and outplant these species. Mock heather (*Haplopappus ericoides*) and the liveforever (*Dudleya farinosa*) are somewhat slow growing. The sand verbenas (*Abronia* spp.), poppy (*Eschscholzia californica* var. *maritima*), and beach pea (*Lathyrus littoralis*) require significant space for their roots. The buckwheats (*Eriogonum* spp.) propagated exceptionally well from seed. Beach sagewort (*Artemisia pycnocephala*) has an extremely high germination percentage so only a light seeding is necessary to produce a large number of plants. Seaside painted cup (*Castilleja latifolia*) is a partial parasite and was seeded both by itself and with other species in flats; both flats showed excellent growth. *Carex* spp. showed good growth from seed, but seed supplies were limited, and *Carex* was propagated vegetatively. Enough *Carex* was propagated to cover 50 ha of dunes, since it will withstand trampling better than the other natives.

After becoming sufficiently hardy, the plants were moved from the greenhouse to a shade house. On the coast with foggy summers and mild temperatures, shade cloths were detrimental to many of the dune species. However, shade cloths would undoubtedly be invaluable at inland propagation sites.

Preliminary container outplanting results were excellent, with little or no mortality to report. The planting program was completed in the summer of 1987. Of course, continual monitoring, exotic eradication, and general maintenance will be critical to success. The Pebble Beach Company golf course maintenance staff is responsible for monitoring and maintenance.

Dune Maintenance Programs

Long-term maintenance management of the Spanish Bay dunes is essential to ensure their continued stability, integrity, and natural beauty. The project biologists outlined long-term management procedures to golf course maintenance personnel in a maintenance manual. Maintenance should include the ongoing exotics eradication program, repairing dune blowouts, revegetating, monitoring dunes, and pedestrian traffic control.

Discussion

Restoration involves both construction and biological recovery. Private developers and construction contractors operate in a world of clearly defined engineering specifications and regulations. Restoration ecology, however, is not yet an exact science, and biologists do not yet enjoy the luxury of having restoration codes and standards. Restorationists must adopt the successful strategies of the construction industry, of projects designed, engineered, tested and constructed by codes, regulations, and performance standards. Habitat construction drawings, specifications, guidelines, and performance standards should be clearly defined for implementation, maintenance, and monitoring. Just as a hotel is built "to code," habitats will be restored "to code."

Permit conditions are the language developers understand. Although the Spanish Bay permit conditions were well researched and detailed, hindsight allows us to suggest how future permitting could benefit from our experiences at Spanish Bay. The following lists specific ideas:

1. If a habitat restoration project co-occurs with a construction project, as it did at Spanish Bay, construction schedules for the habitat restoration and impacts of schedule delays must be clearly defined. Planning and measures should be required to compensate for schedule delays.
2. Where restored habitats are exposed to multiple use, they should be physically protected until the habitat is established.
3. Habitat restoration includes both initial construction and continual maintenance until the habitat is firmly established and functioning. Habitat, unlike hotels, cannot be built and immediately occupied upon completion. Ongoing maintenance and monitoring are essential.
4. Verifiable permit conditions should be written in a language which private developers understand; that is, in terms of measurable performance standards. Where feasible, specific implementation materials and methods should be described, provided that they do not hinder the flexibility required by the inexact science of restoration ecology. Where specifications for restoration are not definitive, emphasis should be placed on evaluation and measurement of the end product, i.e., on performance standards. Maintenance, including rebuilding as necessary, and continual monitoring of the habitat, must be related to the performance standards.
5. Conservation of local gene pools should be stressed in addition to species conservation in the course of a habitat restoration.

Conclusion

Spanish Bay is a unique mitigation project in that the dune habitat adjoins and frequently overlaps the 18-hole golf course. This multiple-use project was further complicated by the construction of dunes concurrent with development. Although intended to be a dune "restoration" project, where the dune ecosystem is maintained over the long term, Spanish Bay evolved into a dune "revegetation" project, in which the landscaped dunes are adjacent to the greens, tees, and fairways, and in some cases the dunes serve as the golf course's rough. Dune plants have successfully adapted to harsh environmental conditions; however, these natives do not tolerate trampling. Plant casualties will result from visitors utilizing the site's public access, hotel guests strolling the site, golfers chasing errant golf balls, etc. Due to the multiple-use nature of the project, significant biological and physical pressures are placed on the dunes, hence making veritable habitat establishment questionable.

Marylee Guinon is a principal of Sycamore Associates, 1612 Rose St., Berkeley, CA 94703. David Allen is a botanist with Sycamore Associates, 1612 Rose St., Berkeley, CA 94703.

Natural Systems Agriculture: Developing Applications for the Prairie Areas of Minnesota and Surrounding States

George M. Boody

Abstract: *Modern agriculture has severely altered the ecosystems of the prairie pothole region in the upper midwestern United States. Ecosystem restoration has been attempted with programs such as cropland retirement programs and prairie pothole restoration efforts. However, they have met with limited success, and the restored acres are vulnerable to renewed row cropping as a result of policy changes. A natural systems agriculture approach is described that links cropland retirement efforts, direct restoration, and sustainable agriculture. The paper describes the design of analogues to natural ecosystems that are also economically productive to the landowner. Examples of related research efforts in Africa, Kansas, and the western United States are discussed. Six basic principles of a natural systems agriculture are outlined. A strategy to focus on marginal lands that were previously drained potholes is discussed. Natural systems agriculture research projects and supportive policy changes are suggested.*

Key words: *prairie pothole, prairie wetlands, cropland retirement, sustainable agriculture, conservation reserve, perennial polyculture, restoration agriculture.*

George M. Boody is president of Agri-Systems Analysis, 2420 - 23rd Ave. So., Minneapolis, MN 55404.

Restoration on the Ocean Shore

Joel W. Hedgpeth

Abstract: *The shores of the sea are common to all mankind, and no one is forbidden to approach the seashore* (Nemo igitur ad litus maris accedere prohibeture Emperor Justinian) *as long as he respects public buildings and monuments subject to the law of nations. If he has a house near the sea, and it is taken away by storms, the shore is returned to the public domain. Since the days of Justinian, however, the world population has been increasing at such a rate that by the early 1990s half the population of the United States will be living within a short journey's distance from some shore on the ocean or the Great Lakes. The pressure to build private dwellings on the shore is increasing, yet few realize that the forces of the sea can obliterate the land on which they dwell, and the trend of the rising sea level will eventually force evacuation of many sites, especially in southern California. Broad sandy beaches offer some protection against wave action, but diversion or holding back this sand by inland dams increases the danger to structures. While the shore may not be restored, it may be protected by removal of dams and coastal structures that inhibit sand movement that replenishes the sand beaches.*

Key words: *coastal zone, coastal protection, coastal erosion, beach sand, shoreline restoration.*

Joel W. Hedgpeth is Emeritus Professor, School of Oceanography, Oregon State University. He resides at 5660 Montecito Avenue, Santa Rosa, CA 95404.

Temperate Forests and Watersheds

Restoration Reforestation

Howard Horowitz

Abstract: *Reforestation, as generally practiced on both public and private lands, is geared towards the establishment of uniform, fast-growing coniferous plantations rather than towards the restoration of healthy forest ecosystems. Commercial reforestation is too often based on the excessively dense planting of seedlings from a single species, and on the overuse of herbicides to control "unwanted" shrubs and hardwoods. The real goal of forest restoration—the reestablishment of economically productive and ecologically sustainable forests—can be achieved more successfully using entirely different approaches to reforestation. This paper suggests some of the key elements of such a program: careful site evaluations; mixed-species planting at wider spacings than currently prevalent; respect for biotic diversity and local site variation; the utilization of integrated pest management concepts and techniques; the cultivation of a professional labor force; and "stewardship contracting." These concepts are proving to be successful where they are implemented, and point toward the emergence of the art and science of "restoration reforestation."*

Key words: *reforestation, vegetation management, tree planting, stewardship, herbicides.*

Introduction

"Reforestation" is a nice-sounding word; it is also an integral part of modern forest management. However, this "reforestation" has little resemblance to an ecologically based method of forest restoration. A decade of work in the reforestation business has led this author to conclude that the goals and methods of commercial reforestation programs are narrowly conceived and short-sighted. This paper describes these problems, and suggests approaches to reforestation that are more compatible with the long-term health and productivity of forest lands.

How can concepts of restoration ecology be introduced into the management of commodity production lands? These areas encompass far more acreage than potential parks or wilderness preserves, but they have too often been overlooked, or perhaps even been given up as lost, by well-meaning environmentalists. Those who advocate increasing timber production by intensifying management of "high-yielding lands," as a way to preserve potential wilderness areas, fail to understand the dynamic appetite of modern resource consumption; harvest increases are absorbed into an ever-rising allowable cut, without reducing the pressure to transform roadless areas into "multiple-use" tree farms. Improvements in forestry technique and waste utilization are offset by continuing movement into steeper and more marginal terrain, where soil erosion is more severe and reforestation is more difficult. My focus, therefore, is on the "working forests"—a fog-enshrouded crazy-quilt of clearcut, old growth patches, and "logging shows" where rural workers earn their scarce cash and their plentiful firewood.

A Thumbnail History of Reforestation

Over the last half century, commercial reforestation has become a seemingly efficient "science," geared toward the establishment of uniform, fast-growing conifer stands on freshly cutover areas. The history of American lumbering was rapacious from the outset; despite massive logging, from the Northeast across the Great Lakes region to the Pacific, there was little effort at reforestation until well into the twentieth century. The "seed trees" left behind in cutover areas were usually scrawny and misshapen, and provided a poor genetic base for the natural regeneration of new forests. Tree planting was first undertaken on a large scale in the 1930s, to provide jobs in the Depression and to restore eroded lands. Seedling mortality was often high, because the nursery technologies for the mass production of vigorous seedlings were still in their infancy, storage was poor, and trees were often planted in locations far beyond their appropriate geographic range. Aerial seeding emerged as a widely used "labor-saving" alternative to hand planting in the 1950s, but this technology had significant shortcomings. It was wasteful of seed and depended on rodent poison to prevent nearly total seed loss after aerial dispersal. In addition, aerial seeding resulted in uneven seedling distribution. Capricious breezes, for example, would result in some areas becoming grossly overstocked, while adjacent areas would get no seedlings at all. Federal restrictions on the broadcast use of poisoned seed in 1970 eliminated aerial seeding as an option. However, the attempt to eradicate (or severely reduce) important elements of the forest ecosystem and the concern for appropriate spatial distribution of the seedlings continue to be dominant problems in contemporary reforestation. Some of the greatest differences between "commercial reforestation" and "restoration reforestation" involve their approaches to dealing with these two issues.

Great advances in nursery techniques have resulted in higher seedling survival rates than in former years, but the philosophical underpinnings of reforestation have barely emerged out of the "dark ages." Foresters are now much better at preventing "backlog" areas, where repeated replantings fail to establish a new plantation. However, the prevailing goal—a monocultural stand of crop conifers—and the prevailing methods of achieving that goal—the overplanting of seedlings and the overuse of herbicides to control shrub and hardwood species—constitute a travesty of sound forest management. The real goal of forest restoration—namely the reestablishment of economically productive and ecologically sustainable forests—can be achieved more successfully using entirely different approaches.

"Reforestation" has become necessary as a consequence of even-aged timber production practices. Alternative methods that manage and maintain "all-aged" forest stands have been successfully practiced for centuries in many regions, and these stands reproduce themselves without requiring any "reforestation" at all. However, we will not soon see the cessation of clearcutting in commercial forest regions, and burned-over areas will still need reforestation if rapid coniferous reestablishment is desired. Even where clearcutting is the predominant method of timber harvest, it is possible to implement creative forest regeneration strategies using affordable and appropriate methods.

Restoration Reforestation

Key elements in an ecologically sound reforestation program include:

1. The use of more thorough field evaluation methods than those routinely employed, to acquire detailed locally generated information about vegetation interactions and site conditions;
2. Careful planting of healthy seedlings, selecting appropriate species for each microsite, and generally spacing them at wider intervals than is usually done in commercial plantations today;
3. Management of regenerating stands to maximize the development of "edge" relationships between tree and shrub species, and to respect the value of natural openings in the stand;
4. Use of seedling protection devices, such as cardboard mulches, shadeblocks, and animal browse barriers, when cost-effective;
5. A fundamental revision of the prevailing paradigm of "vegetation management," this involves rejecting the overly simplistic reliance on the eradication and suppression of "competing" brush and hardwood species, and replacing the paradigm with the far more positive "creation of optimum stand composition";
6. The cultivation of a professional, responsible labor force to assume greater responsibility for the assessment and performance of reforestation contracts.

Having listed these key elements of a sound reforestation program, I will briefly examine how each of them can transform and improve the prevailing commercial reforestation practices.

More Thorough Site Evaluation Methods

Forests are remarkably complex, with great differences in composition and character, even within the same region. The imposition of standardized management formulas onto diverse sites is undoubtedly easier than the development of site-specific strategies, but it is not sound forestry. Virtually all reforestation programs involve the use of stocking surveys to determine plantation survival, and subsequent stand exams to determine need for thinning, brush control, and other activities. However, these stocking surveys and stand exams are generally inadequate, in that they do not gather the kinds of information needed to evaluate the ecological interactions on each site. Poorly designed stand exams perpetuate bad management; for example, surveys that simply record the presence of brush species on a plot, without measuring the growth interactions with adjacent conifers, may falsely indicate that brush control is needed.

A related problem is the misapplication of research results obtained from study sites chosen specifically for certain uniform characteristics, to diverse operational sites which may have little resemblance to the research sites. As a consequence, the operational management results may be far less satisfactory than these obtained on the research sites. Ironically, forest managers often never discover the shortfall, because of the inadequate pre- and post-treatment site evaluations. In fact, the same wasteful or damaging activities will get perpetuated into the future and passed on as dogma to a younger generation of foresters.

Seedling Selection, Planting, and Spacing

There are four distinct components here, so I will briefly discuss each of them.

1. Careful Planting

This involves the proper handling, microsite selection, and planting of each seedling. It is best accomplished by a skilled labor force that cares enough to slow down and do a good job, but it is frequently accomplished by workers who have been "browbeaten" by rigid inspection procedures. Unfortunately, the "stuff-and-stomp" style of tree planting, which once dominated the industry, is still all too common, and the socioeconomic forces that gave rise to it (and even to the widespread "burial" of trees) have not really disappeared.

2. Healthy Seedlings

Vast improvements in nursery practices, seedling storage, and the timing of reforestation projects have resulted in great increases in seedling viability since the 1960s, although all veteran tree planters still can tell "horror stories" of planting sickly or dead stock.

3. Appropriate Species

The great majority of plantations are monocultures, such as 100% Douglas-fir (*Pseudotsuga menziesii*) in the Cascade foothills or 100% Ponderosa pine

(*Pinus ponderosa*) in regions with less moisture. Some districts intersperse two or three species of conifers on planting sites, or prescribe a strip of cedars in riparian zones, but this is less common that one would expect, given the present state of forestry knowledge. Douglas-fir has such inherent intra-species diversity that its "monoculture" today is not really comparable to the vast stands of hybrid corn in agriculture, but the genetic selection of a few "super-trees" for future seed stock is moving forestry in that dubious direction. Forest restoration must include respect for *all* species, even the so-called "trash" trees that may have no marketable value.

As the recent discovery of the potential medicinal value of yew (*Taxus brevifolia*) bark reminds us, today's "weed" may indeed become tomorrow's "resource," provided that it is still around!

4. Wider Spacing

Most commercial reforestation involves the planting of trees in grid patterns ranging from 1.5 m x 1.5 m to 3 m x 3 m, with the objective of having 1,000–1,500 trees per hectare by the time the stand is five years old. Stand densities in this range will require a precommercial thinning within 10–15 years; otherwise, the stand will suffer from overcrowding and reduced growth. It makes sense to plant trees at wider spacing than those usually specified. These are the reasons: (1) It allows the selection of the best individual planting spots, whereas the tighter grids often require cramming trees into far less satisfactory microsites. (2) It conserves valuable nursery stock. (3) In many regions there is already a backlog of undone pre-commercial thinning. Thinning is expensive, and it is foolish to create an excessive dependency on it, considering the stagnant stands that develop if the work does not get accomplished.

Unfortunately, commercial reforestation has no tolerance for even the smallest "holes" in the spacing grid. This "uniform green blanket" may result in a slight increase in wood fiber production (at least for one or two rotations), but these unbroken immature coniferous canopies do *not* resemble healthy forest ecosystems. Natural young forests have openings here and there, hardwood patches, and riparian zone vegetation, but these important habitats have no place in the tunnel vision of commercial reforestation.

Seedling Protection Devices

The switch from tight spacing to wide spacing is predicated on greater attention being given to each seedling, to increase the likelihood of its survival. Various devices are used to protect planted seedlings from particular kinds of stress. Shadeblocks, for example, are often used on south-facing slopes, and plastic tubing is often used (at great cost) to protect seedlings from rodent damage in high-risk regions. These and other protection measures are appropriate if they are found to enhance seedling survival; however, systematic evaluations of their effectiveness on operational reforestation programs are very rare. I tested several different devices on one harsh, grassy Douglas-fir plantation, and found that the shadeblocks did not increase survival significantly above the "control" survival level. However, cardboard mulches, which are hardly ever used in operational programs, were associated with dramatically higher survival rates. This indicates the need for much more investigation of the effectiveness of the various protection methods.

Vegetation Management

A fundamental revision of the prevailing paradigm of "vegetation management" is needed. The selective control of individual trees or shrubs may enhance seedling survival and growth, but broadcast suppression should be avoided. In some places, noncrop trees, shrubs, or grasses may be selectively planted, seeded, or retained on site, for their beneficial impacts on soil fertility, wildlife, and erosion, and for other reasons.

Although large-scale aerial herbicide application has been the most frequently used practice for controlling noncrop shrub and hardwood species on forest lands, other approaches to vegetation management are likely to result in more sustainable forest productivity in the long term. The author's extensive field measurements of conifer and shrub growth on operational treatment sites in national forest lands in Oregon (Horowitz 1980, 1982) strongly suggest that the substantial economic benefits assumed from the aerial herbicide programs will not be realized at harvest time. Nonetheless, these defective assumptions have managed to persist because of a complex of confusions, including inadequate site evaluations, deeply ingrained negative attitudes towards "brush" which get reinforced by misapplied research results, and false economies of scale that promote "big" projects suitable for aerial technologies even when the actual problem areas are quite localized. The economies of scale are false when clusters of marginal sites are treated because they are adjacent to each other; the technology is governing the size of the treatment area.

Unbalanced negative attitudes towards target brush species are evident in the documentation required for federal brush control projects. Almost invariably, the species interactions are summarized as "competition for light, moisture, and nutrients." Although the elements of this phrase are not necessarily empty, it has been rendered much less meaningful through constant repetition; the simple act of reciting that phrase has too often become a substitute for adequate field analysis. In fact, the body of research on conifer-shrub relationships does not lend itself to such simplified conclusions. In places where the target species are nitrogen-fixers, such as the *Alnus* and *Ceanothus* species, the release opportunity should be weighed against the substantial long term beneficial impact on site fertility that would accrue from a period of brush occupancy. Target brush species may produce other important benefits, including harsh site amelioration, slope stabilization after clearcutting, high quality forage for browsing mammals that might otherwise damage young conifers, resistance to root-rot pathogens in the soil, and others. Of course, suppression of conifers by overtopping or moisture-demanding brush has caused some plantations to fail; however, release should be limited to areas where suppression is occurring or threatening to occur. Too often, the forest dynamics have been reduced to a simple policy: locate areas where brush is present, and spray with the appropriate herbicide. As long as the mere *presence* of brush species is the basis for control programs, rather than the occurrence of conifer *suppression* by brush species, then hectarage greatly in excess of that actually able to derive significant benefit will be included for treatment. The deceptively low per-ha cost in large-scale aerial programs actually disguises the high total costs, and includes much land that can provide little return on the investment. Once the area that actually stands to benefit from vegetation manipulation has been determined, a variety of methods can be used to accomplish the job. These methods are based on a different philosophy than current programs—instead of trying to eradicate or suppress "undesirable competitors," they use selective treatment, or protection of individual trees and shrubs to establish a more optimal stand structure and composition.

The specific method utilized to create structure and composition is dictated by the species involved, the age of the stand, the labor force, and the specific objectives of the land manager. All of these methods are practical and cost-effective in some situations, though they can be mis-prescribed at great cost, and no method is a panacea. These appropriate technologies include:

1. *Brush pulling,* to manipulate vegetation composition on young sites occupied by conifers intermixed with fire-stimulated germinated shrubs (such as *Ceanothus* and *Arctostaphylos*). Specifically designed pulling tools are used to uproot individual shrubs that threaten to overtop crop conifers, while other shrubs (located somewhat further from the crop trees) are retained to grow and provide benefits to the site.
2. *Brush cutting with chain saws,* to selectively cut shrubs and hardwoods too large to be pulled, or that have sprouted from old stumps. This "brushing" operation may be performed in conjunction with a precommercial thinning; a well-timed entry with reasonable and cost-effective technical specifications can be accomplished in many situations. In some instances, the cut stumps could be treated with herbicide to prevent resprouting; in many situations, this is not necessary, or could be accomplished by hand-cutting in future re-entry.

Other methods are also being developed within the general rubric of appropriate technology in forest vegetation management. They are all distinguished from broadcast aerial herbicide application by these characteristics:

1. *Emphasis on manipulating individual plants to achieve better stand conditions,* rather than uniformly treating the entire stand in a uniform manner.
2. *A labor-intensive approach which creates jobs for woods workers.* Rural forested areas often have high unemployment, and in the Pacific Northwest, brush management can provide some good work oppor-

tunities. Labor-intensive brush management is not necessarily more expensive than conventional methods; for example, *Ceanothus*-pulling in the Willamette National Forest was successfully accomplished at far lower per-ha prices than aerial herbicides application would have cost.

3. *Reduced usage of pesticides to accomplish vegetation management objectives.*

These technologies, when adopted in widespread fashion, will go a long way toward the practical realization of integrated pest management concepts in forestry.

Use of a Professional Labor Force and Stewardship Contracts

The quality of reforestation depends on the people who are responsible for establishing the job specifications, and on the people who are responsible for actually performing the work. Traditionally these have been separate groups; the former are foresters employed by the government or industry, while the latter are usually transient laborers hired by contractors who are awarded specific job contracts. The awarding of jobs to the lowest bidder has tended to encourage sloppy performance, and tree planting has acquired a notorious reputation, although there have always been some honorable contractors. The labor force is usually migratory, and too often consists of alien workers who are exploited by their contractors. Contract specifications are based on the assumption of the need to control an unprofessional, untrustworthy work-force. These conditions are not likely to result in sound forest restoration!

The emergence of the Hoedads and other forest worker cooperatives in the 1970s was the first step in restructuring the reforestation work force (Hartzell 1986). Although several dozen forest work cooperatives were formed in the western United States, they have not been able to build an organized labor force because of the large population of transient tree planters in the industry. The physically hard work, unstable income, irregular weather, and migratory lifestyle make reforestation a difficult job for people with families. Despite these limitations, the advent of the cooperatives brought about significant changes in the reforestation industry. These include: (1) relatively stable, increasingly professional workers, which has resulted in higher standards, slower planting, and higher bid prices; (2) the entry of women into the reforestation workforce—outside of the cooperatives, tree planting is overwhelmingly a male occupation; (3) increased activism regarding worker exposure to toxic chemicals on the worksite, and continued efforts to develop new forest worker opportunities, such as watershed rehabilitation and manual brush control.

If the 1970s were highlighted by the rise of cooperatives, then the 1980s have been highlighted by the emergence of "stewardship contracting." They are linked, because stewardship is predicated on the assumption of a responsible work force; in fact, the cooperatives pressed the federal government to consider the stewardship model, and most of these contracts have been awarded to them. While contractors with conventional contracts have little or no responsibility for seedling survival, in stewardship the contractor bears financial responsibility for providing an established healthy stand at the end of the 3–5 year contract period. In this system, the bidders develop and schedule the forestry activities needed to achieve the target goals, and the government selects the best bid (which is not necessarily the lowest priced bid). Although the per-ha cost is higher, the advantages of stewardship are enormous: the establishment of common goals between the agency and the contractors, real incentives for high quality performance, and the professionalization of the work force. I worked on the first stewardship contract (for the Eugene District of the Bureau of Land Management in 1981) and recall that the crew took extra time to cull marginal tree stock and planted with exceptional care.

There are problems and risks associated with the stewardship model as used by federal forestry agencies during the 1980s. For example, the contractors may suffer financially because of severe drought, defective nursery stock, or other factors beyond their control. A more profound problem is that the final goal is still a uniform grid of densely spaced crop conifers, with no holes in the spacing. The methods are more professional, but the outcome is still the "fiber-factory" vision of reforestation.

Conclusion

We have seen some exciting progress towards building a more sustainable foundation for reforestation. Emerging practices of integrated pest management are beginning to transform vegetation management from the simplistic "eradication" and "suppression" to the more ecologically sound paradigm of "optimum stand composition." The complex roles of soil organisms, small forest animals, and other noncommercial components of the forest are beginning to get more attention (see for example Perry 1988). The stewardship concept, as it encourages professional responsibility in the reforestation workforce, is another positive development. These ideas, where implemented by a few progressive forest districts, have proven their value "on the slopes" (Turpin 1988). Perhaps they are early steps in the emergence of "restoration reforestation."

References

Hartzell, H. 1986. *Birth of a cooperative.* Eugene, OR: Hulogos'i.

Horowitz, H. 1980. *An evaluation of conifer growth, stocking, and associated vegetation on North Umpqua B.L.M. release sites.* Eugene, OR: Groundwork.

———. 1982. Conifer-shrub growth interactions on proposed brush control sites in the Western Oregon Cascades, Ph.D. dissertation. Eugene, OR: University of Oregon.

———. 1985. Manual pulling of brush. NCAP News 3.

———. 1986. *Close to the ground.* Eugene, OR: Hulogos'i.

Perry, D. 1988. An overview of sustainable forestry. *Journal of Pesticide Reform,* Fall.

Turpin, T. 1988. Successful silvicultural operations without herbicides in a multiple-use environment. *Journal of Pesticide Reform,* Fall.

Howard Horowitz is assistant professor, Theoretical and Applied Science, Ramapo College, Mahwah, NJ 07430.

The Plant-Soil Bootstrap: Microorganisms and Reclamation of Degraded Ecosystems

David A. Perry and Michael P. Amaranthus

Abstract: *Plants divert large amounts of energy belowground, where it creates favorable soil structure and supports soil organism—such as mychorrhizal fungi—that benefit plant growth. Disrupting the positive links between plants and soils may have contributed to the degradation of numerous ecosystems, including high elevation forests, both moist and dry tropical forests, and grasslands. Reestablishing beneficial soil organisms has facilitated reclamation of some sites, and is likely to aid reclamation of others as well.*

Key words: *reclamation, mycorrhizal fungi, rhizospheres.*

Introduction

In widely varying ecosystems throughout the world, removal of indigenous vegetation through clearcutting, shifting cultivation, or overgrazing has led to deterioration in productive capacity, often accompanied by alterations of plant communities that appear permanent. Degradation of moist tropical forests (MTF) with intensified shifting cultivation is well documented (Arnason et al. 1982; Armitsu 1983). Site degradation is not restricted to MTF, however, but also occurs in the semiarid Miombo woodlands of southern Africa (Maghembe 1987), in the montane coniferous forests of northern India (Sharma 1983), in montane forests of North America (Amaranthus and Perry 1987), and in numerous areas where overgrazed rangelands no longer support perennial grasses. Despite widely varying community types and environmental conditions, a common condition links these ecosystems and contributes to their decline when the dominant plants are too intensively utilized. That condition is the close interdependence between the indigenous plant community and both the biological and physical characteristics of soil (Perry et al. 1989).

Depending on the environment in which they are growing, plants may divert up to 80% or more of the net carbon fixed in photosynthesis to belowground processes. Some of this carbon goes into root growth; however, a relatively high proportion may be used to feed mycorrhizal fungi and other organisms. This is not energy that is lost to the plant. On the contrary, organisms living in the rhizosphere (the zone of influence of the root) improve plant growth through effects on nutrient cycling, pathogens, soil aeration, and soil nutrient- and water-holding characteristics.

Of the various soil organisms that benefit plants, the most is known about mycorrhizal fungi. Roughly 90% of plant species are thought to form mycorrhizae (the combination of fungal and root tissue is called the mycorrhiza: the fungal partner is termed a mycorrhizal fungus). There are basically two broad groups of mycorrhizal fungi: those forming "ectomycorrhizae" (EM)—so termed because there is an obvious external modification of root morphology and color, and those forming what are termed "vesicular-arbuscular mycorrhizae" (VAM)—the name describing structures produced within plant cells by the fungus. Unlike EM, no external modification of the root accompanies VAM. The two types of mycorrhizae also differ in the nature of their infection of the root. In VAM, fungal hyphae always penetrate root cells, while in EM the hyphae wrap tightly around cell surfaces, but seldom penetrate the cell interior. By far the majority of mycorrhizal plant species form VAM. EM plants are almost exclusively woody perennials, including, in particular, members of the families *Pinaceae, Fagaceae, Ericaceae,* and *Dipterocarpaceae.* A few plant species—e.g., alders (*Alnus* spp.) and eucalypts (*Eucalyptus* spp.)—form both EM and VAM; however, this is exceptional. Weeds, especially annuals, often do not form mycorrhizae of either type, or do so only occasionally. (This is probably because annuals complete their life cycle during relatively brief, favorable periods and do not have to survive drought, etc.)

Mycorrhizae improve seedling survival and growth by enhancing uptake of nutrients (particularly phosphorus) and water, lengthening root life, and protecting against pathogens (Harley and Smith 1983). Trees forming EM, including most of the important commercial species in the temperate and boreal zones and 70% of the species planted in the tropics (Evans 1982), virtually always have mycorrhizae. The same is probably true of VAM-forming dominants in the canopy of tropical forests (Janos 1980). EM are generally thought to be crucial, if not for survival, at least for acceptable growth. Pines and oaks planted on sites that have not previously supported EM die or do not grow well unless inoculated (Mikola 1970, 1973; Marx 1980). For trees forming VA mycorrhizae, including the majority of tropical and many temperate deciduous families, the symbiosis often is critical. Janos (1980) hypothesizes that late successional trees of moist tropical forests (MTFs) are usually obligately mycorrhizal, while trees of earlier successional stages and those on relatively fertile soils are facultative (i.e., they may or may not form mycorrhizae, depending on factors such as soil fertility). EM trees often predominate on highly infertile tropical soils (Janos 1980).

Many soil organisms other than mycorrhizal fungi influence plant growth. Tissue sloughing and organic exudates from roots, mycorrhizae, and mycorrhizal hyphae support a complex community of bacteria, and protozoa and invertebrates that graze the bacteria. These organisms influence plant growth in various ways: by fixing nitrogen; by competing for nutrients but enhancing nutrient cycling rates; by decomposing minerals, by welding mineral elements together into organically sustained soil aggregates; and by releasing a complex of hormones, allelochemicals, and chelators that probably affect plants both positively and negatively (chelators are organic compounds that increase the solubility, therefore availability, of certain nutrients – particularly iron). Much remains to be learned about the role of various rhizosphere organisms in plant growth, but it seems likely that their net effect on the plant is positive.

What happens to the mycorrhizal fungi and rhizosphere community when energy inputs from plants are reduced (e.g., by overgrazing, defoliation) or eliminated (by clearcutting)? In very general terms, the same thing that happens to any living system that is cut off from its energy source: it begins to die. Of course, there are many organisms in degraded soils. But so are there many organisms in a corpse; they are the agents of decay. When plants are removed, the mycorrhizal fungi – and probably rhizosphere organisms dependent on those plants – begin to die, and decay organisms come to predominate (Perry et al. 1989). This may be accompanied by changes in soil structure that reduce the soil's ability to store water and nutrients. Comparisons of soils under fallow with those under permanent pasture have shown that the former have far fewer large (0.25 mm to 10 mm) aggregates (Low 1955). Clearcuts that have remained unforested for several years also lost large aggregates (Borchers and Perry in press), have reduced mycorrhizal formation on tree seedlings, and have lower levels of iron chelators than adjacent forest soils (Perry et al. 1982; Parke et al. 1984; Perry and Rose 1983). Although many species of EM fungi produce aboveground fruiting bodies (the familiar mushrooms), others – including some of the more common and important mycorrhizae of conifers – produce belowground fruiting bodies (truffles). Almost all VAM fungal species fruit belowground. Once lost from a site through death or in eroding soil, spores of these belowground fruiters re-enter very slowly.

Little is known about the timing of changes in the makeup of the soil community following removal of trees or as a consequence of overgrazing. Nor do we know enough about how these changes influence the ability of plants to reestablish and grow. However, it seems clear that, in at least some cases, reforestation failures are linked to changes in soil biota following clearcutting. Seedling survival has been dramatically increased in both western North America and northeastern India by reinoculating clearcuts with very small amounts of soil from healthy forests (Amaranthus and Perry 1987; Sharma 1983).

Loss of mycorrhizal fungi and rhizosphere organisms will probably have the greatest negative impact on those droughty or otherwise stressful sites where seedling survival depends on rapid exploitation of soil (Perry et al. 1987). On these sites deterioration of soil physical structure is likely to have the greatest negative impact on plant recovery. When plants are removed and not quickly reestablished, biological and physical changes in the soil lead to decreased seedling survival, which leads, in turn, to further deterioration in the soil. Often such sites become occupied by annual weeds, many of which do not require mycorrhizal fungi. Loss of important rhizosphere organisms due to overgrazing probably contributed to the widespread conversion of perennial grasslands to annuals. In the Miombo woodlands of southern Africa, sites cleared of trees for as little as two years no longer support tree growth (Maghembe 1987), suggesting that, in at least some situations, alteration in essential soil characteristics may occur very rapidly following removal of the dominant vegetation.

Can mycorrhizal fungi and the proper rhizosphere organisms be reintroduced as an aid to reclaiming degraded ecosystems? In at least some cases, the answer appears to be yes. In the following section we briefly discuss our experience on one unreforested clearcut in the Siskiyou Mountains of southwestern Oregon.

Utilizing Mycorrhizal Fungi and Rhizosphere Organisms in Reclamation: An Example

A broad band of granitic bedrock caps higher elevations of the Klamath Mountains of southern Oregon and northern California. Despite sandy soils and short, droughty growing seasons, the granites support productive forests, some of which were clearcut in the 1960s. Despite numerous attempts, these clearcuts have not been successfully reforested. One of these sites (Cedar Camp)—intensively studied over the past few years (Perry and Rose 1983; Amaranthus and Perry 1987)—illustrates the role of rhizosphere organisms in maintaining system integrity, and how reintroducing these organisms can aid reforestation.

Cedar Camp is a 10- to 15-ha clearcut dating from 1968. It is on a 30% slope with a southerly aspect at 1,720 m elevation. The adjacent forest, on the same slope and aspect, is dominated by 80-year-old white fir (*Abies concolor*), and is classed as "Site I," or the highest productivity level for the species and elevation (Amaranthus and Perry 1987). The clearcut has been planted four times, all failures. Current vegetation consists of about 30% cover of annual grass (*Bromus tectorum*), scattered patches of bracken fern (*Pteridium aquilinum*), and an occasional manzanita bush (*Arctostaphylos viridissima*). Manzanita are more frequent at the clearcut forest boundary, where they apparently were protected from slash burning and herbicides used to prepare the site for planting. The only encroachment of natural conifer seedlings from the forest into the clearcut is in association with manzanita.

Despite having less than 4% clay, soils under the forest at Cedar Camp are highly aggregated. Clearcut soils, in contrast, are reminiscent of beach sands. Loss of soil structure is not due to lower total organic matter, which does not differ significantly between forest and clearcut soils, but apparently to the removal of living tree roots and associated EM hyphae (electron micrographs of soils in forest and clearcut are shown in Perry et al. 1987).

Soil microbial communities differ dramatically between forest and clearcut at Cedar Camp. The ratio of bacterial-to-fungal colonies is nearly ten times greater in the latter, and mycorrhizal formation on planted seedlings is reduced (Perry and Rose 1983). Actinomycetes (a class of filamentous bacteria of which many species are known to produce chemicals that inhibit the growth of plants and other microbes) are more abundant in clearcut than in forest soil, and a higher proportion of colonies express allelopathy in bioassay (Perry and Rose 1983; Friedman et al. 1988). Concentration of hydroxymate siderophores – microbially produced iron chelators that are important in plant nutrition and resistance to pathogens – are also reduced in clearcut soils (Perry et al. 1984).

Various factors may have contributed to reforestation failures at Cedar Camp, but two in particular seem likely to underlie the inability of seedlings to secure a "foothold" on this stressful site. First, loss of aggregation reduced the capacity of soils to store and deliver resources – particularly water. Second, reductions or outright loss of essential soil organisms diminished the capacity of tree seedlings to exploit sufficient soil volume to compensate for the lowered resource levels per unit of soil volume.

Had the first planting been successful, or had sprouting manzanita not been sprayed with herbicide, soil structure and microbial communities might have been stabilized at Cedar Camp. Without the proper plants, however, the system entered a positive feedback loop in which deterioration within the soil resulted in further planting failure which in turn led to further soil deterioration. A productive forest was converted to a desert in less than 20 years. Rehabilitation of such a site requires that proper soil be reestablished. At Cedar Camp, adding less than half a cup of soil from the root zone of a healthy conifer plantation to each planting hole doubled growth and increased survival of conifer seedlings by 50% in the first year following outplanting (Amaranthus and Perry 1987). By the third year, only those seedlings receiving soil from the plantation were still living.

Where and How to Reintroduce Mycorrhizal Fungi and Rhizosphere Organisms

Under what conditions might restoring the mycorrhizal fungi and rhizosphere organisms aid reclamation, and how does one go about it? Unfortunately we do not yet have clear answers to either question. With regard to the first, the most likely candidates are sites where (a) soil resources – water or nutrients – are limiting, or (b) growing seasons are short, which means that plants must exploit soil resources quickly in order to successfully establish (Perry et al. 1989).

Because of the important role of mycorrhizal fungi and other rhizosphere organisms in creating favorable soil structure, sites with either very sandy or very clayey soils may be especially susceptible to improvement.

With respect to nutrients, mycorrhizal fungi are especially important in gathering elements that tend to be insoluble in water, hence relatively immobile in soils. This is a particular problem with phosphorus and iron, especially in acid soils such as occur in much of the moist tropics and in many (but not all) coniferous forests. Phosphorus is a primary limiting nutrient in the moist tropics, and reduced levels of soil phosphorus have been implicated in site degradation resulting from intensified shifting cultivation (Arnason et al. 1982). It seems probable that losses of mycorrhizal fungi and perhaps other soil organisms have contributed to degradation of moist tropical forests (Janos 1980). Sedge (*Carex* spp.), the dominant plant in many pastures that have been converted from forest in Central and South America, is nonmycorrhizal.

Given a site where reintroducing the proper soil organisms might help restoration, what is the best way to do it? Seedlings can be inoculated using: (a) whole soil from established plant communities (the approach we used at Cedar Camp); (b) pieces of root containing mycorrhizal hyphae; or (c) pure cultures of desirable organisms (e.g., spores or hyphae of mycorrhizal fungi).

VA mycorrhizal fungi have not yet been successfully grown in pure culture; however, spores can be produced by growing mycorrhizal plants in pots. Many EM fungi can be grown in pure culture, hence inoculation with either spores or hyphal fragments is possible.

Each of the above approaches – whole soil versus some form of inocula – has advantages and disadvantages. Both techniques can be used either in the field at the time of planting or in the nursery. The paper by T. V. St. John in this volume discusses nursery inoculation in some detail.

Whole soil contains an entire suite of soil organisms rather than just one, and this is true to a certain extent with root fragments. There are both advantages and disadvantages to this. The collection of organisms contained in whole soils will probably include many that benefit the plant and that would not be included in pure culture, and perhaps not in root fragments. However, whole soil may also contain pathogens or other organisms detrimental to plants.

The source of soil for inoculation is a critical determinant of its net benefit to the plant. Soil must, of course, come from the root zone of a healthy plant that supports beneficial organisms. This does not necessarily mean the same species of plant as is being used in reclamation; many mycorrhizal fungi – particularly VA's – are compatible with a wide variety of plant species. However, to be safe – especially when working with EM plants – soils should probably be collected from the vicinity of the same plant species. Even this, however, doesn't guarantee success. Age of the plants from which soil is collected may be a factor. Mature trees often do not have the same types of mycorrhizae as younger trees of the same species, and this could influence the success of whole-soil inoculation. Given all the uncertainties, the best approach is to experiment with various sources of soil to see which, if any, works on a particular site.

The technique for whole-soil inoculation is simple, but certain precautions are necessary. Little soil is needed. At Cedar Camp we added about one-half cup per seedling at the time of planting—simply tossing the soil into the planting hole with no particular attention to placement. The time between gathering the soil and adding it to planting holes should be minimized, and the soil should not be allowed to overheat or dry out. Clearly the logistics of an operation of this type could be complicated. Careful planning is essential.

Conclusion

Much remains to be learned about the potential benefits of reintroducing soil organisms that benefit plants into degraded soils. Some, perhaps many, reclamation projects will be aided by this approach. However, though the evidence for this optimism is strong, it is also based on relatively few studies. The best advice for those working in reclamation is to try. One does not have to be a scientist, or even have a high school diploma, in order to experiment. All it takes is common sense—perhaps backed up by a little intuition.

Finally, some largely philosophical musings. Even where it does benefit reclamation, reintroducing the beneficial organisms is not some kind of magic bullet that will allow the rest of the ecosystem to be ignored. The idea that we can find, or create, some single organism or technique that will solve all of our problems is one of the great fallacies of the twentieth century. If humans are to find some proper balance with Nature and persist as a species, we must go beyond biotechnology—the search for the perfectly engineered organism—to ecotechnology, the management and stewardship of whole ecosystems. Not that understanding and perhaps even improving on individual species has nothing to offer. But the approach that has characterized much of modern agriculture and forestry—to consider the crop in isolation from its host community and ecosystem—is folly. As we stated elsewhere (Perry et al. 1987): "It is increasingly clear that the rhizosphere, the soil community as a whole, and the entire ecosystem . . . form a coherent, dynamic unit, and that stability and resilience must ultimately be understood in terms of patterns arising from interrelations within this unit. The real challenges for the future lie in the largely untilled ground of holism."

References

Amaranthus, M. P. and D. A. Perry. 1987. Effect of soil transfer on ectomycorrhiza formation and the survival and growth of conifer seedlings on old nonreforested clearcuts. *Canadian Journal Forest Research* 17:944–50.

Arimitsu, K. 1983. Impact of shifting cultivation on the soil of the tropical rain forest in the Benakat District, South Sumatra, Indonesia. In *IUFRO symposium on forest site and continuous productivity*, ed. R. Ballard and S. P. Gessel, 218–22. U.S. Department of Agriculture, Forest Service, General Technical Report PNW-163.

Arnason, T., J. D. H. Lambert, J. Gale, J. Cal, and H. Vernon. 1982. Decline of soil fertility due to intensification of land use by shifting agriculturists in Belize, Central America. *Agro-ecosystems* 8:27–37.

Borchers, J., and D. A. Perry. In press. Loss of soil aggregates in old, unreforested clearcuts in southwest Oregon. In *Maintaining long-term productivity of Pacific Northwest forests*, eds. D. A. Perry et al. Portland, OR: Timber Press.

Evans, J. 1982. *Plantation forestry in the tropics*. Oxford, United Kingdom: Clarendon Press.

Friedman, J., A. S. Hutchins, C. Y. Li, and D. A. Perry. 1988. Phytotoxic actinomycetes and their possible role in regeneration failure of Douglas-fir in the Siskiyou Mountains of southern Oregon. *Abstracts*. 14th International Congress of Biochemistry. July 15, 1988. Prague, Czechoslovakia.

Harley, J. L. and S. E. Smith. 1983. *Mycorrhizal symbiosis*. London and New York: Academic Press.

Janos, D. P. 1980. Mycorrhizae influence tropical succession. *Biotropica* 12:56–64.

Low, A. J. 1955. Improvements in the structural state of soil under leys. *Soil Science* 6:179–199.

Maghembe, J. 1987. Personal communication. (International Center for Agroforestry. Nairobi, Kenya).

Marx, D. H. 1980. Ectomycorrhiza fungus inoculations: a tool for improving forestation practices. In *Tropical mycorrhiza research*, ed. P. Mikola, 13–71. Oxford, United Kingdom: Oxford University Press.

Mikola, P. 1970. Mycorrhizal inoculation in afforestation. *International Review Forest Research* 3:123–196.

———. 1973. Application of mycorrhizal symbioses in forestry practices. In *Ectomycorrhizae—their ecology and physiology*, ed. Marks, G. C., and T. T. Kozlowski, 383–411. London, New York: Academic Press.

———. 1984. Inoculum potential of ectomycorrhizal fungi in forest soil from southwest Oregon and northern California. *Forest Science* 30:300–304.

Parke, J. L., R. G. Linderman, and J. M. Trappe. 1984. Inoculum potential of ectomycorrhizal fungi in forest soil from southwest Oregon and northern California. *Forest Science* 30:300–304.

Perry, D. A., M. P. Amaranthus, J. G. Borchers, S. L. Borchers, and R. E. Brainerd. 1989. Bootstrapping in ecosystems. *BioScience* 39:230–237.

Perry, D. A., M. M. Meyer, D. Egeland, S. L. Rose, and D. Pilz. 1982. Seedling growth and mycorrhizal formation in clearcut and adjacent undisturbed soils in Montana: a greenhouse bioassay. *Forest Ecology Management* 4:261–273.

Perry, D. A., R. Molina, and M. P. Amaranthus. 1987. Mycorrhizae, mycorrhizospheres and reforestation: current knowledge and research needs. *Canadian Journal Forest Research* 17:929–940.

Perry, D. A., and S. L. Rose. 1983. Soil biology and forest productivity: opportunities and constraints. In *IUFRO symposium on forest site and continuous productivity,* ed. R. Ballard and S. P. Gessel, 229–239. U.S. Department of Agriculture, Forest Service, General Technical Report PNW-163.

Perry, D. A., S. L. Rose, D. Pilz, and M. M. Schoenberger. 1984. Reduction of natural ferric iron chelators in disturbed forest soils. *Soil Science Society American Journal* 48:379–382.

Sharma, G. D. 1983. Influence of jhumming on the structure and function of microorganisms in a forested ecosystem. *Hill Geography* 2:1–11.

David A. Perry is a Professor in the Department of Forest Science, Oregon State University, Corvallis, OR 97331. Michael P. Amaranthus is a Soil Scientist, Siskiyou National Forest, Grants Pass, OR 97526.

Mycorrhizal Inoculation of Container Stock for Restoration of Self-sufficient Vegetation

Theodore V. St. John

ABSTRACT: *Mycorrhizal symbiosis promotes nutrient uptake, resulting in faster growth and better transplant survival. Propagules of the symbiotic fungi may be absent from soil of restoration sites and must be reintroduced along with the host plants that require them. The best strategy usually is inoculation of container stock in the nursery. In most cases the most cost-effective option is to purchase inoculum from a commercial source. Inoculation is best carried out at a propagation or transplant stage. Nursery procedures must be modified to favor development of the symbiosis.*

KEY WORDS: *container plants, ectomycorrhiza, inoculation, mycorrhiza, nursery, vesicular-arbuscular mycorrhiza.*

The Two Major Kinds of Mycorrhiza

Mycorrhiza is a collective term for several distinct kinds of plant-fungus associations. One group is the sheathing or ectomycorrhizae (EM), which may be seen with a hand lens on the rootlets of most conifers, oaks, willows, and certain tropical trees. Mycorrhizae of the vesicular-arbuscular type (VAM) are more difficult to observe but more widespread; they are the most abundant microorganisims in most soils (Hayman 1978). VAM are the mycorrhizae of crop plants and most annual and woody natives. The fungal partners of VAM are markedly unselective with regard to host plant; for example, the same species of fungus may be associated with liverworts, ferns, palms, and garden vegetables. Mycorrhizae that are distinct from the two main kinds are found in the *Ericaceae, Orchidaceae,* and several small tropical families. A few plant species can have both EM and VAM.

Mycorrhizal Fungi Connect the Root to the Soil

The mycorrhizal symbiosis is, in most experimental conditions, beneficial to the host plant. This is often demonstrated by comparing growth of inoculated and uninoculated plants in sterilized soil. At the end of an experiment, mycorrhizal plants may exceed nonmycorrhizal plants in size by a factor of thirty or more. Examples can be found in papers by Kleinschmidt and Gerdemann (1972), Kormanik et al. (1982) and countless others. The magnitude of the mycorrhizal growth response depends heavily on such factors as the species of plant, the species of fungus, and soil fertility. The way in which the growth response is brought about is clearly nutritional in most cases. Mycorrhizae can greatly improve uptake of phosphorus and certain micronutrients, especially zinc and copper.

The mycelium, the mass of thread-like hyphae that make up the body of the fungus, penetrates into the root, passing between the cortical cells (EM) or entering the cortical cells (VAM and other "endomycorrhizae," such as the *Ericaceous* mycorrhizae). The external mycelium extends into the soil, where it aids in acquisition of phosphorus.

The remarkable improvement that mycorrhizae bring about in phosphorus nutrition is a result of the spatial distribution of the hyphae. Since phosphate moves very slowly in soil, it cannot diffuse to the root as rapidly as it is taken up. A zone of depletion forms, isolating the root surface from most of the soil volume. Mycorrhizal hyphae cross the depletion zone, take up phosphorus that is beyond the reach of the root, and transport it to the cortex (Tinker 1978).

The external hyphae consist of coarse, thick-walled trunk hyphae, and fine, short-lived side branches. These lateral branches proliferate locally near nutrient-rich microsites, such as decomposing insect remains or subterranean deposits of decomposing organic matter (Nicolson 1960; St. John et al. 1983). This localized allocation of absorbing organs, or "selective exploitation" of rich microsites, provides a second mechanism by which mycorrhizae aid in nutrient uptake. The localized concentration of hyphae in effect shortens the average pathway along which phosphorus ions must diffuse, greatly improving the rate of uptake over that which would be possible if the hyphae were distributed randomly through the soil (St. John et al. 1983).

The symbionts are not thought to play a significant role in the uptake of nitrogen from soil (Bowen and Smith 1981). However, mycorrhizae play an important indirect role in nitrogen nutrition. Nitrogen fixation is a very phosphorus-dependent process and both legumes (Hayman 1983) and actinorrhizal (Rose and Youngberg 1981) nitrogen fixers have been shown to perform markedly better when mycorrhizal.

Among the important effects of mycorrhizae is improved ability of the host plant to withstand drought. While the fungi of some EM can transport significant amounts of water, claims that VAM can do so are based on misinterpretation of experimental results (see discussion by Fitter 1985). Instead, VAM aid uptake of phosphorus, contributing to phosphorus-mediated drought resistance. Another important effect of VAM is the less severe transplant injury suffered by mycorrhizal plants (Menge et al. 1980). Mycorrhizal plants can also perform better in saline conditions (Pond et al. 1984), and may be more resistant to certain pathogens (Schenck 1981).

In addition to their importance in the mineral nutrition of plants, mycorrhizae are among the most important agents of soil aggregation (Tisdall and Oades 1979). The role of VAM in stabilization of dunes has received a significant amount of attention (Koske and Polson 1984). Their role in soil aggregation is perhaps one of the most important reasons to introduce a rich inoculum of VAM in ecosystem restoration work.

Correlation of Habitat with Mycotrophy

The role of mycorrhizae in phosphorus uptake is most important in host plants that are poorly supplied with root hairs. Such plants are mycotrophic: they respond well to inoculation (Baylis 1975), and tend to be mycorrhizal in the wild state (St. John 1980). Even with a relatively high concentration of phosphorus, mycotrophic plants would not be able to obtain sufficient phosphorus without the aid of their symbionts. Phosphorus is more difficult to extract from dry soil than from moist soil because the rate of diffusion slows with decreasing soil moisture (Fitter 1985). This difference adds to the importance of mycorrhizae in the western United States.

There is good reason to believe that early successional floras include proportionately fewer mycotrophic species than late successional floras. Late successional species often live in an environment of intense competition for soil resources. The soil is colonized not only by a large population of roots, but by microbes of every description, all of which require mineral nutrients for growth and reproduction. In mature vegetation, there may also be a network of mycorrhizal hyphae that in effect connects all plants into a single, community-wide, nutrient-absorbing system (Francis, Finlay and Read 1986). The environment of early successional species, on the other hand, differs in several fundamental ways that make the mycorrhizal condition less advantageous. The soil is often temporarily enriched in easily available forms of mineral nutrients after disturbance (St. John 1988), and there is less requirement for the high efficiency of mycorrhizae in the uptake of nutrients. There may not be a large population of fungal propagules in disturbed sites.

The natural mycorrhizal condition of some of the native plants that we use in restoration is unknown. Surveys of mycorrhizal condition in natives of the western United States include Bethlenfalvay et al. (1984), Kummerow and Borth (1986), and Trappe (1981). Since most desirable plant species for restoration work are late successional species, mycorrhizal inoculation would be a prudent precaution when the natural mycorrhizal status is not known with certainty.

Lack of Native VAM Mycorrhizal Fungi at Restoration Sites

The fungi of VAM are present in almost all undisturbed soils, but may be lost following mechanical disturbance and removal of the vegetation. Miller et al. (1985) reported that loss of propagules was faster in moist than in dry storage, apparently because spores germinated in the absence of suitable hosts (no host root is required for spore germination), and because active microbial populations exerted inhibitory effects.

VAM fungi do not quickly invade restoration sites because the spores are large and not easily carried by the wind. VAM move slowly into an area with the growth of host plant roots from undisturbed areas, and by gradual multiplication of surviving propagules. Spores can be moved by insects, earthworms, foot traffic, and splashing of soil. The time required for reinvasion may be only a few months, as in the eastern United States (Medve 1984), or very slow, as in the arid west (Miller 1979; Reeves et al. 1979). Plants newly introduced on the site will have to obtain nutrients without their symbionts unless they are already mycorrhizal when planted.

Difficulties of Inoculation at Field Sites

With currently available methodology, inoculation at the field site is not likely to be cost effective. Maximum benefit is realized when container plants have been pre-inoculated with fungi selected for their effectiveness.

A potential way of inoculating with EM in the field is the duff inoculation method that has been used in nursery beds. Forest duff, collected from mature stands of the target plant species, can be broadcast and plowed in at the rate of 2–4 kg/m^2 of nursery bed. A mixture of EM is introduced with the duff. Because the inoculum can be obtained for no more than the cost of labor and transportation, it may be one of the more feasible methods for inoculating a restoration site. However, no choice of desirable fungi is possible, and pathogens may be introduced along with the mycorrhizal fungi (Riffle and Maronek 1982). In addition, collection of duff from native forest sites can potentially cause unacceptable environmental consequences at the collection site. In other methods, pure cultures or commercial inoculum have been used, also at very high rates.

Hayman, Morris, and Page (1981) tested four ways of inoculating with VAM in a field experiment. Simple broadcasting and raking in the material produced a level of root colonization no better than that obtained by using autoclaved inoculum. Two other methods, placing inoculum below the seed in furrows and fluid drilling (insertion into the soil in a liquid medium), gave reasonably good colonization. Multiseeded pellets were used with moderate success in the same experiments, and variations on seed and mycorrhizal pellets bonded with clay have been successfully used elsewhere (Hall 1984).

Table 1 shows the amounts of inoculum that have been used in various VAM field experiments. All required large amounts of short-lived, often expensive material. The method used by the author and a student in Brazil (LaTorraca and St. John, unpublished) incorporated a locally collected, wild-type inoculum that imposed only a modest labor cost. In some treatments, inoculation resulted in a doubling of cowpea (*Vigna unguiculata*) yield over the corresponding fertilization rate. An understanding of the ecology and distribution of soil fungi may make the use of wild inoculum feasible in certain specialized circumstances.

It may be of interest to consider a theoretical minimum amount of inoculum for large-scale application. If pure spores could be obtained, as may be possible with the VAM *Glomus versiforme* (*G. epigaeum*; Daniels and Menge 1981), 500 spores for each plant on a hectare containing 40,000 plants would weigh only 3.7 grams. Future reduction in weight and bulk of inoculum will rest heavily on elimination of nonfungal materials, such as roots and soil.

In certain circumstances, before introducing the desired container stock, it may be possible to plant a mycorrhizal host as a nurse crop—an annual species that cannot survive winter conditions at the site (and thus is unlikely to become a pest later). The nurse crop might be used to increase any residual VAM inoculum in the disturbed soil. Alternatively, such a species might be made mycorrhizal in containers and planted on the site to bring inoculum to desired plants that are introduced as seed.

The most practical means of mycorrhizal inoculation is often through

TABLE 1
AMOUNTS OF INOCULUM USED IN SELECTED VAM FIELD EXPERIMENTS

Form of inoculum	Inoculum per ha (metric tons)	Authority
Top soil	75	Powell et al. 1980
Pellets	55	Hall 1980
Top soil	30	Black and Tinker 1977
Pellets	25	Abbott et al. 1983
Pellets	20	Powell 1981
Greenhouse	9.8	Clarke and Mosse 1981
Greenhouse	2.97	Kuo and Huang 1983
Greenhouse	2.5	Owusu-Benoah and Mosse 1979
Pellets	1.5–4.2	Hall 1984
Chopped roots	0.67	LaTorraca and St. John unpublished

the use of mycorrhizal container plants. This means that inoculation is done in the controlled conditions of the nursery, early in the life of the plant, and at a cost of much less labor and much less inoculum than field inoculation.

Some Fungi are Better Than Others

The native soil, which includes pathogens as well as mycorrhizal fungi, may contain only ineffective fungi. Fungi that produce little growth response in host plants are widespread in nature (Fitter 1985).

The fungi of EM vary greatly in suitability for sites and for host plant species (Riffle and Maronek 1982). *Telephora terrestris* is an EM fungus noted for its ability to turn up uninvited in nurseries, where it establishes and grows quickly on the host's root system. Mitchell et al. (1984) found *T. terrestris* to be an effective fungal symbiont with oaks, but the containerized plants received weekly fertilization and may not have provided an effective test of *T. terrestris* for restoration work. Marx et al. (1984) suggested that *T. terrestris* is well adapted to the fertile soil of nurseries but poorly adapted to harsh conditions. One of the most widely used fungi in the hotter parts of the country is *Pisolithus tinctorius*, a fungus often found naturally in harsh sites. However, success with *P. tinctorius* has been rather limited in other areas of the country (Riffle and Maronek 1982). Trappe (1977) proposed initial screening of EM isolates *in vitro*, followed by nursery and field trials.

VAM fungi also vary widely in effectiveness (Mosse 1975). They can be quite specific for the kinds of soil; a fungus that is beneficial in one soil may be useless in another (Mosse 1975; Hayman and Tavares 1985).

Management of Mycorrhizae in the Nursery

Nursery potting mixes are often soilless and contain no natural inoculum, even though mobile, rapidly growing fungi such as *Telephora terrestris* can invade spontaneously. It is desirable to inoculate nursery stock with a mixture of species or with fungi selected for particular hosts or habitats (Perry et al. 1987). Riffle and Maronek (1982) suggested three possible times to inoculate with EM in a forest nursery: before sowing seed, while sowing, and after emergence. They recommend the first two possibilities because they require the least inoculum. Inoculation during a propagation stage is also quite practical. At that time cuttings are already being handled, and relatively little inoculum is required. Inoculum can have a side benefit of stimulating root production.

Nursery planting mixes often contain no native VAM inoculum. If soil is used in the mix, any native fungi have been diluted and weakened in storage and handling. While animal vectors and splashing of soil will initiate some sporadic VAM colonization, the best strategy is to introduce symbionts of known effectiveness. VAM inoculation may be carried out at the germination or rooting stage, at an early transplant, in the final container, or at the time of outplanting. The earlier stages are preferable from most perspectives, for reasons similar to those given for EM. Inoculum can also be introduced at the propagation stage, a method that holds much promise for both timely inoculation and economy of inoculum.

The inoculum is placed a short distance below or beside the roots, with the objective of inducing roots to grow through the inoculum. It is generally preferable to concentrate the inoculum in a single layer rather than to mix it evenly through the soil.

In order to achieve successful colonization of container stock before it goes into the field, environmental conditions must be adjusted to favor the symbiosis. The conditions of light, temperature, and moisture, which promote photosynthesis and root growth in host plants, are generally the conditions that promote root colonization by VAM. High levels of phosphorus and nitrogen, while stimulatory to host plant growth, are inhibitory to mycorrhizae and close control has to be maintained on fertility during the establishment phase. One form of fertilization in the nursery that has worked well is use of Osmocote 18-6-12 (Sierra Chemical Company, Milpitas, CA). The common practice of supplying a large dose of fertilizer at transplant time is a serious mistake from the perspective of microbial symbiosis. This topic was considered in detail by St. John (1988).

Pesticides may be used in various ways in a nursery, and most have detrimental effects on mycorrhizae. In certain circumstances, however, they may become selective tools for manipulating or enhancing mycorrhizal development (Trappe et al. 1984).

While it may be feasible to produce inoculum in-house with the construction of certain specialized facilities, it is unnecessary because VAM inoculum of high quality and reasonable price may now be purchased from commercial sources.

Summary

Mycorrhizal fungi are best introduced at the nursery stage, where relatively small amounts of inoculum may be used to prepare mycorrhizal container plants. In a nursery environment, fungi can be selected for effectiveness and suitability for the planting site. Mycorrhizal container stock will survive transplanting and grow more rapidly than nonmycorrhizal material. These plants will carry inoculum to plants started on site from seed, and will contribute to the important role of mycorrhizal hyphae in soil structure.

References

Abbott, L. K., A. D. Robson, and I. R. Hall. 1983. Introduction of vesicular-arbuscular mycorrhizal fungi into agricultural soils. *Australian Journal of Agricultural Research* 34:741–49.

Baylis, G. T. S. 1975. The magnolioid mycorrhiza and mycotrophy in root systems derived from it. In *Endomycorrhizas,* ed. F. E. Sanders, B. Mosse, and P. B.Tinker, 374–89. London: Academic Press.

Bethlenfalvay, G. J., S. Dakessian, and R. S. Pacovsky. 1984. Mycorrhizae in a southern California desert. *Canadian Journal of Botany* 62:519–24.

Black, R. L. B., and P. B. Tinker. 1977. Interaction between effects of vesicular-arbuscular mycorrhiza and fertiliser phosphorus on yields of potatoes in the field. *Nature* 267:510–11.

Bowen, G. D., and S. E. Smith. 1981. The effects of mycorrhizas on nitrogen uptake by plants. In *Terrestrial nitrogen cycles,* ed. F. E. Clark and T. Rosswall, 237–47. Stockholm: Ecology Bulletin.

Clarke, C., and Mosse, B. 1981. Plant growth responses to vesicular-arbuscular mycorrhiza. *New Phytologist* 87:695–705.

Daniels, D. S., and J. A. Menge. 1981. Evaluation of the commercial potential of the vesicular arbuscular mycorrhizal fungus *Glomus epigaeus. The New Phytologist* 87:345–354.

Fitter, A. H. 1985. Functioning of vesicular-arbuscular mycorrhizas under field conditions. *New Phytologist* 99:257–65.

Francis, R., R. D. Finlay, and D. J. Read. 1986. Vesicular-arbuscular mycorrhiza in natural vegetation systems. *New Phytologist* 102:103–111.

Hall, I. R. 1980. Growth of *Lotus pedunculatus* Cav. in an eroded soil containing soil pellets infested with endomycorrhizal fungi. *New Zealand Journal of Agricultural Research* 23:103–105.

———. 1984. Field trials assessing the effect of inoculating agricultural soils with endomycorrhizal fungi. *Journal of Agricultural Science* (Cambridge) 102:725–31.

Hayman, D. S. 1978. Endomycorrhizae. In *Interactions between nonpathogenic soil microorganisms and plants,* ed. Y.R. Dommergues and S. V. Krupa, 401–22. Amsterdam: Elsevier Scientific Publ. Co.

———. 1983. The physiology of vesicular-arbuscular endomycorrhizal symbiosis. *Canadian Journal of Botany* 61:944–63.

Hayman, D. S., E. J. Morris, and R. J. Page. 1981. Methods for inoculating field crops with mycorrhizal fungi. *Annals of Applied Biology* 99:247–53.

Hayman, D. S., and M. Tavares. 1985. Plant growth responses to vesicular-arbuscular mycorrhiza. *New Phytologist* 100:367–77.

Kleinschmidt, G. D., and J. W. Gerdemann. 1972. Stunting of citrus seedlings in fumigated nursery soil related to the absence of endomycorrhizae. *Phytopathology* 62:1447–53.

Kormanik, P. P., R. C. Schultz, and W. C. Bryan. 1982. The influence of vesicular-arbuscular mycorrhizae on the growth and development of eight hardwood tree species. *Forest Science* 28:531–39.

Koske, R. E., and W. R. Polson. 1984. Are VA mycorrhizae required for sand dune stabilization? *Bioscience* 34:420–24.

Kumerow, J., and W. Borth. 1986. Mycorrhizal associations in chaparral. *Fremontia* 14:11–13.

Kuo, C. G., and R. S. Huang. 1983. Effect of vesicular-arbuscular mycorrhizae on the growth and yield of rice stubble-cultured soybeans. *Plant and Soil* 64:325–330.

LaTorraca, S. M., and T. V. St. John. 1978. Unpublished manuscript.

Marx, D. H., C. E. Cordell, D. S. Kenney, J. G. Mexal, J. D. Artman, J. W. Riffle, and R. J. Molina. 1984. Commercial vegetative inoculum of *Pisolithus tinctorius* and inoculation techniques for development of ectomycorrhizae on bare-root tree seedlings. *Forest Science* 30:1–101.

Medve, R. J. 1984. The mycorrhizae of pioneer species in disturbed ecosystems in western Pennsylvania. *American Journal of Botany* 71:787–794.

Menge, J. A., J. LaRue, C. K. Labanauskas, and E. L. V. Johnson. 1980. The effect of two mycorrhizal fungi upon growth and nutrition of avocado seedlings grown in six fertilizer treatments. *Journal of the American Society of Horticultural Science* 105:400–404.

Miller, R. M. 1979. Some occurrences of vesicular-arbuscular mycorrhizae in natural and disturbed ecosystems of the Red Desert. *Canadian Journal of Botany* 57:619–23.

Miller, R. M., B. A. Carnes, and T. B. Moorman. 1985. Factors influencing survival of vesicular-arbuscular mycorrhizal propagules during topsoil storage. *Journal of Applied Ecology* 22:259–66.

Mitchell, R. J., G. S. Cox, R. K. Dixon, H. E. Garrett, and I. L. Sander. 1984. Inoculation of three *Quercus* species with eleven isolates of ectomycorrhizal fungi. *Forest Science* 30:563–572.

Mosse, B. 1975. Specificity in VA mycorrhizas. In *Endomycorrhizas*, ed. F. E. Sanders, B. Mosse, and P. B. Tinker, 469–484. London: Academic Press.

Nicolson, T. H. 1960. Mycorrhiza in the Gramineae II. Development in different habitats, particularly sand dunes. *Transactions of the British Mycological Society* 43:132–145.

Owusu-Bennoah, E., and B. Mosse. 1979. Plant growth responses to vesicular-arbuscular mycorrhiza. *New Phytologist* 83:671–79.

Perry, D. A., R. Molina, and M. P. Amaranthus. 1987. Mycorrhizae, mycorrhizospheres, and reforestation. *Canadian Journal of Forestry* Research 17:929–40.

Pond, E. C., J. A. Menge, and W. M. Jarrell. 1984. Improved growth of tomato in salinized soil by vesicular-arbuscular mycorrhizal fungi collected from saline soils. *Mycologia* 76:74–84.

Powell, C. Ll. 1981. Inoculation of barley with efficient mycorrhizal fungi stimulates seed yield. *Plant and Soil* 68:3–9.

Powell, C. Ll., M. Groters, and D. Metcalfe. 1980. Mycorrhizal inoculation of a barley crop in the field. *New Zealand Journal of Agricultural Research* 23:107–109.

Reeves, F. B., D. Wagner, T. Moorman, and J. Kiel. 1979. The role of endomycorrhizae in revegetation practices in the semiarid west. I. A comparison of incidence of mycorrhizae in severely disturbed vs. natural environments. *American Journal of Botany* 66:6–13.

Riffle, J. W., and D. M. Maronek. 1982. Ectomycorrhizal inoculation procedures for greenhouse and nursery studies. In *Methods and principles of mycorrhizal research,* ed. N. C. Schenck, 147–155. St. Paul, MN: The American Phytopathological Society.

Rose, S. L., and C. T. Youngberg. 1981. Tripartite associations in snowbrush (*Ceanothus velutinus*). *Canadian Journal of Botany* 59:34–39.

Schenck, N. C. 1981. Can mycorrhizae control root disease? *Plant Disease Reporter* 65:231–234.

St. John, T. V. 1980. Root size, root hairs, and mycorrhizal infection. *New Phytologist* 84:483–487.

———. 1988. Soil disturbance and the mineral nutrition of native plants. In *Proceedings of the Second Native Plant Revegetation Symposium* (San Diego), ed. J. Reiger and B. K. Williams, 35–40.

St. John, T. V., D. C. Coleman, and C. P. P. Reid. 1983. The association of vesicular-arbuscular mycorrhizal hyphae with soil organic particles. *Ecology* 64:957–59.

Tinker, P. B. H. 1978. Effects of vesicular-arbuscular mycorrhizas on plant nutrition and plant growth. *Physiologie Végétale* 16:743–51.

Tisdall, J. M., and J. M. Oades. 1979. Stabilization of soil aggregates by the root systems of ryegrass. *Australian Journal of Soil Research* 17:429–441.

Trappe, J. M. 1977. Selection of fungi for ectomycorrhizal inoculation in nurseries. *Annual Review of Phytopathology* 15:203–222.

———. 1981. Mycorrhizae and productivity of arid and semiarid rangelands. In *Advances in food producing systems for arid and semiarid lands,* ed. J. T. Manassah and E. J. Brishley, 581–99. New York: Academic Press.

Trappe, J. M., R. Molina, and M. Castellano. 1984. Reactions of mycorrhizal fungi and mycorrhiza formation to pesticides. *Annual Review of Phytopathology* 22:331–59.

Theodore V. St. John is Scientific Director, Land Restoration Associates, San Diego, and Laboratory of Biomedical and Environmental Sciences, University of California, Los Angeles, CA 90024.

Restoration of Degraded Tropical Forests

Lawrence S. Hamilton

Abstract: *An attempt is made to clarify the issues, point out semantic problems, and more narrowly identify targets in programs of restoration involving tropical forest lands. An oversimplified picture of a somewhat homogeneous tropical forest biome (equated with rain forests) is not helpful. The widespread use of the term "deforestation" clouds the problem and must be replaced by descriptions that more precisely delineate the kind and extent of degradation. Restoration goals must also be clarified before initiating projects. The problem of restoration following damage from introduction of alien biota is emphasized.*

Key words: *Restoration, tropical forests, deforestation, degradation, endangered species.*

Introduction

The earth badly needs healing and replenishment. Nowhere is there more need than in the tropics on forested and formerly forested lands. Native tropical plant and animal communities and many tropical soils have a propensity for degradation that exceeds that of their counterparts in the temperate zone. The nature of this special fragility has been documented by many (Gomez-Pompa et al. 1972; Dasmann et al. 1973; Hamilton 1976, 1984), and I shall not elaborate here.

It is important that in tackling this challenge of restoration in the tropics, we not be carried astray by simplification of the problems. The problems are complex and the healing solutions are not simple, especially in populous developing countries. I shall not attempt here to present the techniques of restoration. Some of these have recently been set forth by a Working Group of the International Union for the Conservation of Nature and Natural Resources (IUCN) Commission on Ecology of which I was a member (Lovejoy et al. 1985). Rather, I wish to clarify the task, and deal with the dangers of semantic fuzziness, so that the effort we mount is more likely to succeed.

One needs to begin by a clearer delineation of the key words "restoration" and "degradation," and even "tropical forests."

Tropical Forests

Much environmental literature and rhetoric give the impression that the tropical forests are areas of dense, wet, species-rich, evergreen, multi-storied trees with a rich association of other plants and animals. Indeed, these descriptors can be generally applied to the world's tropical *rain* forests. Appropriately, most attention in popular and scientific writing has been given to this biome. The rain forests are the world's "richest and most exuberant expressions of life on land" (Hamilton 1976). The alteration or removal of this kind of tropical forest is causing a phenomenon which has been described as a "species mega-extinction spasm" (Myers 1986). Dramatic reduction, and in many places, cessation of logging and conversion of these rain forests to cropping and grazing is required. And restoration is needed for those millions of acres that are in depauperate forest due to logging, for grasslands (e.g., *Imperata* grass) degraded due to clearing and periodic burning (perhaps some 16 million ha in Indonesia alone), and for soils impoverished by marginal ranching or cropping.

There are many kinds of tropical rain forests ranging from dwarf cloud ("elfin") forests at high elevation to tall lowland riverine forests. Whitmore (1975), one of the world authorities, recognizes 14 different types. Each of these will require special knowledge and treatment in restoration.

Moreover, there are many classes of tropical forests other than rain forests—ranging from closed-canopy tropical moist deciduous forests (e.g., the famous monsoon forests) to open woodland (e.g., the African *miombo*); and from montane tropical forests with a coniferous tree component to dry thorn forests and semiarid tree savannas. Salt-tolerant coastal mangroves are another important kind of tropical forest. Each of these major classes has its own special problems of degradation and restoration. These dry, moist, or coastal types of tropical forests do not have as many press agents as do the rain forests, but merit urgent attention. In fact they have probably been impacted by human activity more than rain forests. Examples of pristine, moist, deciduous forests in the tropics are rare because of their long history of use and conversion.

Unfortunately many popular writers mix concerns and statistics about rain forests and other tropical forests. It is not helpful, for instance, for two writers in the magazine *Asia 2000* to state that subsequent to logging in rain forests, deserts can develop with great speed (Sharp and Sharp 1982). Even the World Wildlife Fund/International Union for Conservation of Nature and Natural Resources (WWF/IUCN) Tropical Forests Campaign suggested these forests are turned into useless deserts by logging (WWF/IUCN 1982). This is nonsense, for there cannot be deserts in rain forest areas receiving upwards of 230 cm of annual rainfall. Such scare statements are not helpful in correctly identifying problems and coming up with solutions. Actually in tropical rain forest when trees are cut, one can hardly jump out of the way fast enough because of the rapid regrowth of vegetation under these hot humid conditions (see later section on "degradation"). By contrast, the arid and semiarid forests of the desert margins have been increasingly "desertified" by failure of the rains in the 1970s and early 1980s, accompanied by increased over-exploitation by grazing and tree removal (Grainger 1983). Tree planting and soil stabilization may be an appropriate restoration technique in this latter situation. In the case of tropical rain forest logging, tree planting is not usually a sensible activity. Although enrichment planting may be helpful in some cases, it is doubtful if the full complement of species can be restored artificially, and the rapid natural regrowth outcompetes most planted trees, except on the nutrient-poor ancient soils. Mangroves have their own unique variety of problems (Hamilton and Snedaker 1984). Let us be clear about precisely which forest we are targetting.

Degradation

Clarity is also needed when describing the nature of "degradation" for which "restoration" is the cure. The term "deforestation" is widely used and is a term which, in my opinion, should be forthwith eliminated from the language (Hamilton 1987). It obscures more than it reveals, unless it is followed by a series of phrases or whole sentences that denote specifically what activities have, in fact, occurred. It has been variously used to include such widely and wildly separate activities with respect to tropical forests as:

- fuelwood cutting for household use using no mechanical equipment;
- fodder lopping;
- commercial logging with heavy machinery;
- shifting cultivation, both sustainable and nonsustainable (i.e., short or no fallow versus long fallow);
- shifting grazing;
- flooding by reservoirs;
- clearing for conversion to continuous cropping or grazing;
- clearing for conversion to food, beverage, or extractive crops (i.e., plantations of rubber, coffee, tea, banana, cocoa, quinine, etc.);
- clearcutting and conversion to forest tree plantations;
- burning, or felling and burning; and
- even harvesting of wildlife or other nonwood forest products (Bowonder 1982).

It is important to define the term "degradation" if we are to plan sound restoration. Some of the previously listed activities alone or in concert can indeed almost destroy the inherent productivity of the land of a tropical forest. Forest logging, followed by burning, followed by abusive agriculture or grazing in tropical semiarid tree savannas or even in woodlands, can result in barren lands, sometimes with moving wind-blown soil, where natural revegetation does not occur or occurs only over centuries. Even in tropical moist or rain forests, a regime of forest removal followed by repeated burning and cropping or grazing can almost destroy the basic productivity of the land where the soils are ancient and of low nutrient status. Here the nutrients are largely in the vegetation and, when it is removed, the remaining nutrients and organic materials are rapidly leached and oxidized under burning and exploitive agronomy. But in rain forests, mere removal of forest products or especially non-wood forest products (fruits, vines, nuts—actually very economically important activities in some areas—[see Prescott-Allen 1982]) does not degrade the basic productivity of the land if it is done with care. The forest remains forest, though the extraction trails, roads and landings (in the case of commercial logging) may require many months or years to revegetate with woody plants. However, the species composition will be altered. In many cases a rare species may be greatly reduced in numbers (rendered endangered) or even be lost. This, too, is degradation, and, if endemic species are lost from an area, it is planetary impoverishment.

Other degraded areas may be created by nature, e.g., a new lava flow or ash fall from tropical volcanos, or a hurricane or tsunami in coastal mangroves. More common, however, is the degradation caused by humans in tropical forest or former tropical forest lands. Extreme examples of degradation include mine waste disposal sites (e.g., tin mine waste in Thailand) and salinized sites resulting from inappropriate management. Instances of tropical forests damaged by air pollution are not well documented, but are well known in temperate forests, and undoubtedly the same factors are operative in the tropics, though on a reduced scale.

Those who use the term "deforestation" are lumping many of these actions (or all of them) together, each with different impacts, each with different socioeconomic causes, and each requiring different kinds of restorative actions. Moreover, there is conservation farming and abusive farming, careful logging and destructive logging. The aesthetic, productive, hydrologically benign, erosionally safe, terraced paddy fields of the deep volcanic soils on Java are "deforested" lands just as are the large areas of former rain forest now in repeatedly burned *Imperata* grass in Sumatra, or the devastated areas of heavy logging followed by burning on serpentine soils in Palawan (Philippines). We need to describe accurately the process that has degraded the forest land. Statistics given that we annually "deforest" an area of tropical rain forest or tropical moist forest equal in size to the country of Cuba are not helpful. Alterations must be distinguished from clearing and conversion; and sustainable, needed conversions separated from short-lived, unsustainable catastrophes.

The introduction of alien species into native ecosystems causes an important kind of degradation prone to escape attention in discussions about tropical forests. Island ecosystems are particularly at risk to disruption from introduced plants and animals, and remote oceanic islands are the most fragile of all in this respect. The degradation of native tropical ecostyems in Hawaii by alien introductions has been as severe as anywhere in the world. Here forest plants and animals developed on once-barren, new, volcanic slopes, isolated by about 4,000 km of ocean from the nearest source of colonization. The flora and fauna that evolved did so in the absence of mammalian predators and herbivores and therefore did not possess the usual defense mechanisms. Consequently, the introduction of the herbivores—goat, pig, cattle, sheep, mouflon (a wild sheep), and deer—ravaged the native vegetation and simultaneously destroyed habitat for many native birds and insects. The introduced predators—mongoose, rat, feral cat, and avian malaria (along with the vector mosquito)—added to the toll on forest birds. Around 4,600 alien plant species have been introduced (Smith 1987), among which are such disasters as banana poka (*Passiflora mollissima*), blackberry (*Rubus Penetrans*), lantana (*Lantana camara*), strawberry guava (*Psidium cattleianum*), and Koster's curse (*Clidemia hirta*). These five alone have degraded or decimated thousands of hectares of native forest lands. Of the 894 species of plants proposed nationwide for the endangered list, 556 are Hawaiian (Herbst 1986). There are currently 30 bird species endangered or threatened (Jacobi and Scott 1985). In the past 200 years since European contact, 28 taxa of birds (Berger 1981) and 273 of plants (Fosberg and Herbst 1975) have become extinct. Hawaii's next threat may be the brown tree snake (native of Papua New Guinea) which has already almost eliminated the native bird population in three decades on Guam. Three of these snakes have been found at the Honolulu airport and nearby military property. Obviously the vector is the island-hopping aircraft.

Grainger (1988) has recently estimated that there are 758 million ha of degraded forest lands in the tropics with a potential for forest restoration. This does not include forest lands that have lost species or suffered adverse shifts in species composition; nor does it include forests disturbed by alien species introductions. It may well be, therefore, a quite conservative estimate of the restoration challenge.

Restoration

Before restoration action begins, there needs to be a clarification of restoration goals. Restoration has as its ultimate goal, perhaps, the achieving of a status something very close to the ecosystem's original conditions. But this will depend somewhat on the appraiser, his or her time perspective, and whether the framework of appraisal is financial or ecological. In many cases, the restoration goal may be to shift toward a more desirable "managed" rather than a natural state. To the farmer of a formerly forested hillslope on which he now grows rainfed maize, this might mean bringing back the productivity of his field to what it was immediately after land clearing and during early years of cropping. Soil erosion and nutrient removals in harvest may have so degraded his field that he may even consider abandoning it and moving on to clear another piece of forest (shifting agriculture). Restoration might involve addition of fertilizer (if it were available and cheap—not usually the case in the tropical developing world) and investments in soil conservation practices to reduce erosion. Additions of nitrogen-fixing trees in rows in a contour, alley-cropping agroforestry system might be a valid restorative step toward sustainability in some situations.

However, when dealing with the topic of *restoring tropical forests,* the concern is appropriately a return to forest cover successively closer to the original. This is surely an imperative action in view of the very rapid rates of forest degradation and outright forest removal that currently characterize resource development in the tropics.

There are a few general types of forest restoration action that span an array of problems in the tropics. The *first* necessary step is to halt the processes or activities responsible for the degradation. For example, restricting fire in degraded *Imperata* grassland gradually allows woody plants, and eventually forest, to reoccupy the site. These are human-maintained grasslands through the use of fire. Where feral animals, such as goats or pigs, are preventing forest regrowth, programs of hunting and then fencing them out can permit a great deal of recovery, as in Hawaii Volcanos National Park. In Hawaii in a recent court victory, a native Hawaiian bird, the palila (*Loxiodes bailleui*), was given standing to sue, and it then won against the State of Hawaii. The suit demanded that the state remove feral sheep and goats from critical palila habitat on Mauna Kea. In late 1986, the palila went to court again to have mouflon (*Ovis musimon*) removed from its home turf on this mountain, and again won against the State and a powerful hunting lobby. While the situation in the Hawaiian Islands is extremely grave, most other tropical islands face serious problems from deliberate or accidental human introductions of alien plants, animals, diseases, and viruses.

A *second* action that may go hand in hand with the first is simply to allow a period of natural healing, rather than to intervene with artificial revegetation. This is particularly appropriate in the humid tropics on sites that are not severely degraded. It has been instructive and somewhat amazing to note the forest recovery in the Gogol Valley of Papua New Guinea, where clearcutting of lowland rainforest was at one time rated an environmental disaster (Seddon 1984). I visited one of the clearcut areas eighteen months after cutting and could hardly force my way through the woody regrowth. A subsequent study by Saulei (1984) showed that six months after logging, 90% of the area was covered with vegetation, and that the 10% bare area was on compacted tractor trails and landings. While initially the vegetation was dominated by secondary species, at the end of 10 years only 7% was small secondary tree species, 23% was large secondary tree species and 70% was primary species. The primary species are returning, though it is almost assured that biological diversity will have been lost, perhaps forever. Nonetheless, an amazing amount of healing has occurred naturally. The age-old nostrum of restoration that suggests planting a tree for every one cut is sheer nonsense in the tropical rain forest (yet it was a policy prescription in the Philippines until about five years ago). It is not so much the cutting of the rain forest that seriously degrades it, but the way in which products are removed, i.e., poor location, design, and maintenance of logging access and extraction routes, and the use of heavy, large equipment on them. Species impoverishment, however, is probably from the cutting. Creative restoration actions that we know little about as yet may be needed to get back forest closely resembling the original. Indeed it is doubtful that we can achieve this total ecological restoration, and, hence, the need for total protection of large areas of this biome, before degradation.

In the drier parts of the tropics, regeneration of forest does not occur readily, and getting out of nature's way may not be enough. Fire protection to reduce fire incidence is often needed along with local community education and involvement, because fire and grazing may be used by them to hold back succession. Much will be learned in Guanacaste National Park as Janzen attempts to restore a deciduous dry tropical forest to that part of Costa Rica (Janzen 1986).

Especially if sites have been kept clear of forest for some time by human activity, a *third* type of activity may be required. This involves artificially revegetating (reforesting) the site. There is a large literature on this topic (see, for example, Food and Agricultural Organization 1985; Weber and Stoney 1986; Lundgren 1980). Non-tree vegetation may need to be established first before trees will become established. The tree species used should be native to the area. Restoring species of either plants or animals by reintroduction requires good science and even artistry to do effectively. Sometimes, where there has been productivity loss, pioneer, fast-growing, and low-nutrient-demanding or nitrogen-fixing species must be used first as a preparatory community. The Nepal/Australian Forestry Project is currently doing this effectively with chir pine (*Pinus roxburghii*) as a nurse crop for the more desirable native broad-leaved trees in the Middle Hills of Nepal.

More extreme kinds of degradation, such as advanced erosion, total nutrient depletion, or soils rendered saline or toxic, require a *fourth* class of drastic and expensive measures. Specialized fields of reclamation have developed to meet these needs, though most of the work has been done in the temperate zone. A useful book is *Ecology and Reclamation of Devastated Land* (Hutnik and Davis 1976). The initial goal in the tropics should probably be to restore to such sites the environmental services of vegetated land (reduction of erosion and sediment, improvement of wildlife habitat wildlife habitat and amenity), and to produce some economic products for human welfare. Production of useful goods and services on these sites can help to relieve the pressure on unaltered forest.

Finally, a *fifth* type of action needing much more attention is the removal of alien species that have become established and are aggressively degrading native ecosystems. This can be very expensive if the alien is well established, as the people in coastal Florida know from experience with Australian pine (*Casuarina*) and tea tree (*Melaleuca*). The watchword is to act promptly when a problem is first recognized. In Hawaii, a recent invader from the Azores and Canary Islands, the small tree *Myrica faya,* has probably been allowed to get too great a foothold before action was taken. Simultaneously, action must be taken to reduce or shut off the supply of new aliens, unless they are very carefully screened.

Conclusion

When planning for tropical forest restoration, there must be a clear assessment of what kind of forest one is dealing with and of what kind of degradation has occurred or is occurring. (These categories and their combinations can be quite numerous and complex.) Then we must define the restoration state as some previous datum. This could be economic or ecological or some combination. Sometimes it could be a definite date, as it was in National Park Service policy: "The condition that prevailed when the area was first visited by white man" (U.S. Department of Interior Advisory Board on Wildlife Management 1963). For Hawaii, Lamoureaux (1985) speculated that an appropriate datum might be post-Polynesian settlement but pre-Captain Cook (January 1778), if we only knew what that state was. The question has cultural and philosophical as well as biological dimensions.

We must go about the more easily accomplished and more immediately practical actions of soil stabilization, rebuilding of fertility, and reforestation. But we must also consider the formidable challenge of rebuilding a forest ecosystem that closely resembles the one which tends to occupy a particular environment in the absence of human *(but not natural)* disturbance. This would involve reestablishing something akin to the natural biological diversity that developed under evolutionary processes at the gene, species, and community levels. The task will be complex and costly, but so was putting a man on the moon. We ought to try, in a few places at least, to get some experience in being positive intervenors instead of destroyers in nature. It is high time we got started in the tropics.

References

Berger, A. J. 1981. *Hawaiian birdlife.* Honolulu, HI: University of Hawaii Press.

Bowonder, B. 1982. Deforestation in India. *International Journal of Environmental Studies* 18(3 and 4):223–236.

Dasmann, R. F., J. P. Milton, and P. H. Freeman. 1973. *Ecological principles for economic development.* New York: John Wiley and Sons.

Evans, J. 1982. *Plantation forestry in the tropics.* Oxford: Clarendon Press.

Food and Agricultural Organization. 1985. Sand dune stabilization shelterbelts and afforestation in dry zones. *FAO Conservation Guides* 10. Rome.

Fosberg, F. R. and D. Herbst. 1975. Rare and endangered species of Hawaiian vascular plants. *Allertonia* 1:1–72.

Gomez-Pompa, A. C. Vasquez-Yanes, and S. Guevara. 1972. The tropical rain forest: A non-renewable resource. *Science* 177:762–765.

Grainger, A. 1983. *Desertification.* London: Earthscan.

———. 1988. Estimating areas of degraded tropical lands requiring replenishment of forest cover. *International Tree Crops Journal* 5(1/2):31–61.

Hamilton, L. S. 1976. *Tropical rain forest use and preservation: A study of problems and practices in Venezuela.* International Series No. 4. San Francisco, CA: The Sierra Club.

———. 1984. A perspective on forestry in Asia and the Pacific. *Wallaceana* 36:3–8.

———. 1987. What are the impacts of Himalayan deforestation on the Ganges-Brahmaputra lowlands and the delta? Assumptions and facts. *Mountain Research and Development* 7(3):256–263.

Hamilton, L.S. and S.C. Snedaker, eds. 1984. Handbook for mangrove area management. EWC/IUCN/UNESCO. Honolulu, HI: East-West Center.

Herbst, D. 1986. Plants may be only a picture. *Honolulu Star-Bulletin,* December 11, 1986.

Hutnik, R. J. and G. Davis, eds. 1976. *Ecology and reclamation of devastated land.* New York: Gordon and Breach.

Jacobi, J. D., and J. M. Scott. 1985. An assessment of the current status of native upland habitats and associated endangered species of the island of Hawaii. In *Hawaii's terrestrial ecosystems: Preservation and management,* ed. C. P. Stone and J. M. Scott, 3–22. Honolulu, HI: Cooperative National Park Resources Study Unit, University of Hawaii.

Janzen, D. 1986. *Guanacaste National Park: Tropical ecological and cultural restoration.* San Jose, Costa Rica: Editorial Universidad Estatal a Distancia.

Lamoureaux, C. H. 1985. Restoration of native ecosystems. In *Hawaii's terrestrial ecosystems: Preservation and management,* eds. C. P. Stone and J. M. Scott, 442–431. Honolulu, HI: Cooperative National Park Resources Studies Unit, University of Hawaii.

Lovejoy, T. E., and Working Group. 1985. *Rehabilitation of degraded tropical forest lands.* IUCU Commission on Ecology Occasional Paper no. 5.

Lundgren, B. 1980. *Plantation forestry in tropical countries: physical and biological potentials and risks.* Swedish University of Agricultural Sciences, Rural Development Studies No. 8.

Myers, N. 1986. Tropical deforestation and a mega extinction spasm. In *Conservation biology: the science of scarcity and diversity,* ed. M. E. Soulé, 394–409. Sunderland, MA: Senauer Association.

Prescott-Allen, R., and C. Prescott-Allen. 1982. *What's wildlife worth?* London: Earthscan/International Institute for Environment and Development.

Saulei, S. M. 1984. Natural regeneration following clear-fell logging operations in the Gogal Valley, Papua New Guinea. *Ambio* 13(5 and 6):351–354.

Seddon, G. 1984. Logging in the Gogol Valley, Papua New Guinea. *Ambio* 13(5 and 6):345–350.

Sharp, D. and T. 1982. The desertification of Asia. *Asia 2000* 1(4):40–42.

Smith, C. 1987. Lecture presented to training program on protected areas and biological diversity (unpublished). Honolulu, HI: East-West Center.

U.S. Department of the Interior Advisory Board on Wildlife Management. 1963. *Wildlife management in the national parks.* Washington, DC: U.S. Department of Interior.

Weber, F. R., and C. Stoney. 1986. *Reforestation in arid lands.* Arlington, VA: Volunteers in Technical Assistance.

Whitmore, T. C. 1975. *Tropical rain forests of the Far East.* Oxford: Clarendon Press.

World Wildlife Fund/IUCN. 1982. *Tropical forest campaign.* WWF booklet. Gland, Switzerland: World Wildlife Fund.

Lawrence S. Hamilton is Emeritus Professor (Cornell University) and a Research Associate in the Environment and Policy Institute, East-West Center, Honolulu, HI 96848.

Restoration in the Feliz Creek Watershed, California

Dahinda Meda

ABSTRACT: *Watershed restoration efforts began in 1975 in the upper Feliz Creek watershed near Hopland, California, in response to a storm which threatened to wash out a road. A buttress was built and trees were planted. Historical information was gathered from government records and by talking to long-time residents and former residents to better understand the ecological deterioration in the watershed. In 1981 my plan for fish habitat and forestland improvement was approved and funded for a 154 ha, one year project in the lower reaches of the upper watershed.*

Stream improvements included:

—Treatment of 17 sites where the creek had been blocked by debris jams, had jumped its channel and was actively undercutting sidebanks;

—Stabilizing the toes of eight landslides with boulders and planting their slopes with willow and incense cedar;

—Stabilizing 122 m of actively undercutting streambanks with rock-filled gabions, boulders and plantings of willow;

—Planting an additional 2,800 trees along the stream corridor.

Forestland improvements included:

—Planting 4,300 trees on 8 ha (including site prep and follow-up);

—Manual conifer release (removal of competing hardwood canopy) on 10 ha.

Additional erosion control work was done. While most of these efforts were apparently successful (slides stabilized; erosion stopped), the plantings were not. In 1983 a breakthrough in seedling survival was made using a curtain fence. Rather than using posts for support, the fence hangs from a high-strength cable stretched between trees. While the complete restoration of the entire watershed may never occur, the process of deterioration has been slowed and even reversed in critical zones.

KEY WORDS: *stream restoration, forest restoration, watershed restoration, erosion control.*

Dahinda Meda is director, Terrarium Institute, 5900 Feliz Creek, Ukiah, CA 95482.

Islands of Life: Restoring and Maintaining Presettlement Plant and Animal Communities in the Ohio Metropolitan Park District

David B. Nolin

ABSTRACT: *The park district of Dayton-Montgomery County in Ohio is actively managing its eight metropolitan park reserves to provide habitat for the region's native plant and animal life. A variety of techniques is being utilized to establish or maintain hardwood forests, prairies, edges, ponds, vernal pools, and conifer stands.*

KEY WORDS: *stewardship, parks, land use planning and management, forests, wetlands, prairie, ponds, vernal pools.*

David B. Nolin is the land stewardship specialist for the Park District of Dayton and Montgomery County, 1375 East Siebenthaler Avenue, Dayton OH 45414.

Urban Woodland Management

Susan Antenen

ABSTRACT: *Disturbed urban natural areas can benefit from on-going ecological management. In the borough of the Bronx in New York City, an intensive, long-term management project has been underway in a 3 ha woodland since 1980. Located at Wave Hill, a former Hudson River estate that is now a public garden and cultural center, this woodland is under the stewardship of The Forest Project, a managment, education, and applied research program.*

*The vegetation of Wave Hill's natural area reflects not only the vegetation type of the region, Eastern Deciduous Forest, but also its urban location and land-use history. Fragmentation and abundance of nonnative species (46% of 273 vascular species) are two of the most obvious disturbances. Two thirds of the area is urban woodland with 58 species of trees (62% native to the Lower Hudson Valley). Native red oak (*Quercas rubra*), sugar maple (*Acer saccharum*), white ash (*Fraxinus americana*), black cherry (*Prunus serotina*), and hickory (*Carya cordiformis*) show strong regeneration.*

The Forest Project is managing the vegetation to favor natural processes and regeneration of native eastern deciduous forest species. Management practices include closing gaps to unite portions of the woodlands, selective weeding, removing invasive nonnative vegetation, and planting native trees, shrubs, and herbs. The edges, open areas, and gaps, dominated by invasive nonnative herbs, shrubs, and vines, require the most intervention.

A permanent grid of more than 300 10 m x 10 m quadrats is established throughout the natural area to facilitate vegetation analysis. We are analyzing the vegetation at five-year intervals to monitor the progress of our management and to document change.

Wave Hill's woodland is a living laboratory for observing and teaching about human influence on our environment and ecological restoration. For the public, walks and trailside interpretive signs describe the work of The Forest Project. In the summer Bronx teenagers participate in our Field Ecology Work/Study Program. The youth assist with the physical work of management and field research.

KEY WORDS: *urban woodland, permanent quadrats, invasive, regeneration, vegetation analysis, management.*

Susan Antenen (Goddard College, Vermont) has worked as a naturalist in Kenya, Vermont, and now New York City. She is the founder and director of The Forest Project at Wave Hill, 675 W. 252nd St., Bronx, NY 10471.

Mined Land

Revegetation of Abandoned Acid Coal Mine Spoil in South Central Iowa

Chester J. Covert

Abstract: *Direct revegetation of abandoned, acid coal mine spoils was performed in lieu of covering graded spoil with "borrow" material. Spoils samples were collected and analyzed to determine levels of spoil toxicity and to develop a spoil amendment program. Acid-base accounting was used to quantify acid conditions. A strong relationship between electrical conductivity and the ratio of magnesium to calcium was noted. Amendments to the graded spoil included non-dolomitic agricultural limestone and incorporation of large quantities of straw. Plant species selection was based on acid resistance and anticipated droughty site conditions.*

Key words: *direct revegetation, acid-base accounting, electrical conductivity, magnesium-dominated soils, and native warm season grasses.*

Introduction

Reclamation of abandoned coal mine sites was made possible by the Surface Mining Control and Reclamation Act (SMCRA) of 1977. In addition to setting new standards for the reclamation of active coal mines, Title III of SMCRA levied a tax of $0.35 per ton of coal mined to be used for abandoned mine land (AML) reclamation. In the ten years of full-scale reclamation occurring nationwide, natural systems and processes have increasingly been employed in reclamation to increase the long-term success of reclamation and to reduce costs and off-site impacts.

Initially, many AML sites were "reclaimed" by use of "borrow" material from an undisturbed site to cover acid coal refuse and acid spoil before revegetation was attempted. This technique expanded the area of mining impact to the borrow site and was only partially successful due to hydrological and engineering problems which were not fully understood at the time. However, as research and experience in reclamation evolved, new approaches were developed which significantly reduced or eliminated the need for disturbance of surrounding unmined land. The Schmidt site, located in south central Iowa, was typical of midwestern AML sites. Sheer highwalls, acid impoundments, and barren, acid spoil piles comprised the 49 ha site. After a thorough spoil chemistry investigation, it was determined that direct revegetation of the treated spoil would be possible.

Spoil Investigation

Methodology

Eleven borings and 80 test pits were excavated at the Schmidt site. The primary purpose of the borings was to locate existing groundwater levels in the spoil and surrounding undisturbed areas. Logs made of the borings helped identify materials that could be expected in the spoil. Test pits were excavated using backhoes to characterize the spoil by physical and chemical means. Materials were identified and described in the test pit logs. Major transitions between disturbed and undisturbed or subsoil and bedrock were measured and noted. The vertical and horizontal location of test pits were transferred to plan sheets for later reference.

Samples were collected every 1.5 m from borings and from significant material types in test pits. These samples were then field tested for pH as an indicator of acidity. Generally, a pH of less than 4.0 indicates the presence of sulfur oxides which result in chemical imbalances toxic to plants (Sutton 1970). Selected samples were further analyzed in a laboratory for acid potential (by the acid-base accounting method); major bases, such as calcium, magnesium and sodium; electrical conductance; and texture.

Acid-base accounting measures acid potential due to the existence of oxidizable FeS_2 (pyrite). PH measures "free" acidity, i.e., that which is readily exchangeable on the surface of the soil colloid (Barnhisel et al. 1984). Acid-base accounting is a combination of two measurements: neutralization potential, and acid potential in terms of calcium carbonate ($CaCO_3$) equivalence. Acid potential is most accurate when based on the highly oxidizable sulfur component pyrite (Freeman et al. 1984). The neutralization potential analysis can be misleading due to the fact that chemical species of calcium are more stable and therefore do not readily release their bases into solution to counteract acid formation (Barnhisel et al. 1982). In addition, materials dominated by pyrite are inhabited by the bacteria *Thiobaccillus ferrooxidans* which accelerate pyrite oxidation and maintain extreme levels of acidity (Myerson 1981). Even with these drawbacks, this method was selected over the more accurate column leaching method due to budget and time constraints.

High acidity causes the structural breakdown of clays and shales. This breakdown releases large quantities of metals into the soil solution, increases acidity, and produces sulfate salts (Evangelou and Thom 1984). The presence of these salts and their concentration can be identified by measuring the electrical conductivity (E.C.) of the mine spoil. Generally, an E.C. of 4.0 to 8.0 is considered moderately saline and restricts the growth of most plants (Donahue et al. 1976). An E.C. greater than 8.0 is strongly saline and is uninhabitable to all but a few plant species due to greater osmotic potentials required to remove water from a saline soil matrix. Salinity also affects soil structure by causing clay particles to disperse, thereby restricting infiltration of water (Evangelou et al. 1982).

The primary sulfate salts that form in acid spoils are species of calcium, magnesium, aluminum, iron, and manganese (Evangelou and Thom 1984). Magnesium-to-calcium ratios approaching a factor of 4 are toxic to plants and severely restrict plant uptake of potassium (Evangelou and Thom 1983). Even at near neutral pH levels, magnesium-dominated soils severely limit or prevent plant growth.

Physical Properties

The dominant upland soils surrounding the Schmidt site are Clinton silt loam and Gosport silt loam, according to the Marion County Soil Survey. Both series have brown to grayish-brown silty clay loams with maximum depths of 81 cm for the Gosport and 165 cm for the Clinton Series. Both soils are naturally acidic (pH 5.0 to 5.5).

Underlying the silty clay loam are firm to very firm clays followed by a zone of glacial sand to sandy clay and a glacial till zone grading into mottled, very firm clays. The till zone is characterized by small pebbles and gravel in a matrix of clay.

A "typical" spoil cut profile was dominated by a mixture of clay and shale. Layers of sand or sand pockets were observed in approximately one third of all spoil test pits. Occasional pockets of silty clay loam topsoil were also noted. When located on the surface, these topsoil pockets were generally associated with a vigorous vegetative cover. Gypsum crystals were commonly observed on shale fragment surfaces indicating direct precipitation from saline water through weathering.

Chemical Properties

The results of the laboratory analyses performed on spoil samples from the Schmidt site indicate a wide variation in chemical properties. Acid-base accounting tests showed the greatest range followed by E.C. PH showed the least fluctuation. A very strong relationship exists between the ratio of magnesium to calcium and E.C. (Figure 1).

The mean pH of 34 spoil samples was 4.34 with a standard deviation of 1.31 (Table 1). The calcium carbonate equivalent deficiency averaged −10 metric tons per 1,000 metric tons and had a standard deviation of 19.47. Contrary to anticipated results, very little correlation is evident between these two factors. Samples collected from two bore holes in undisturbed shales

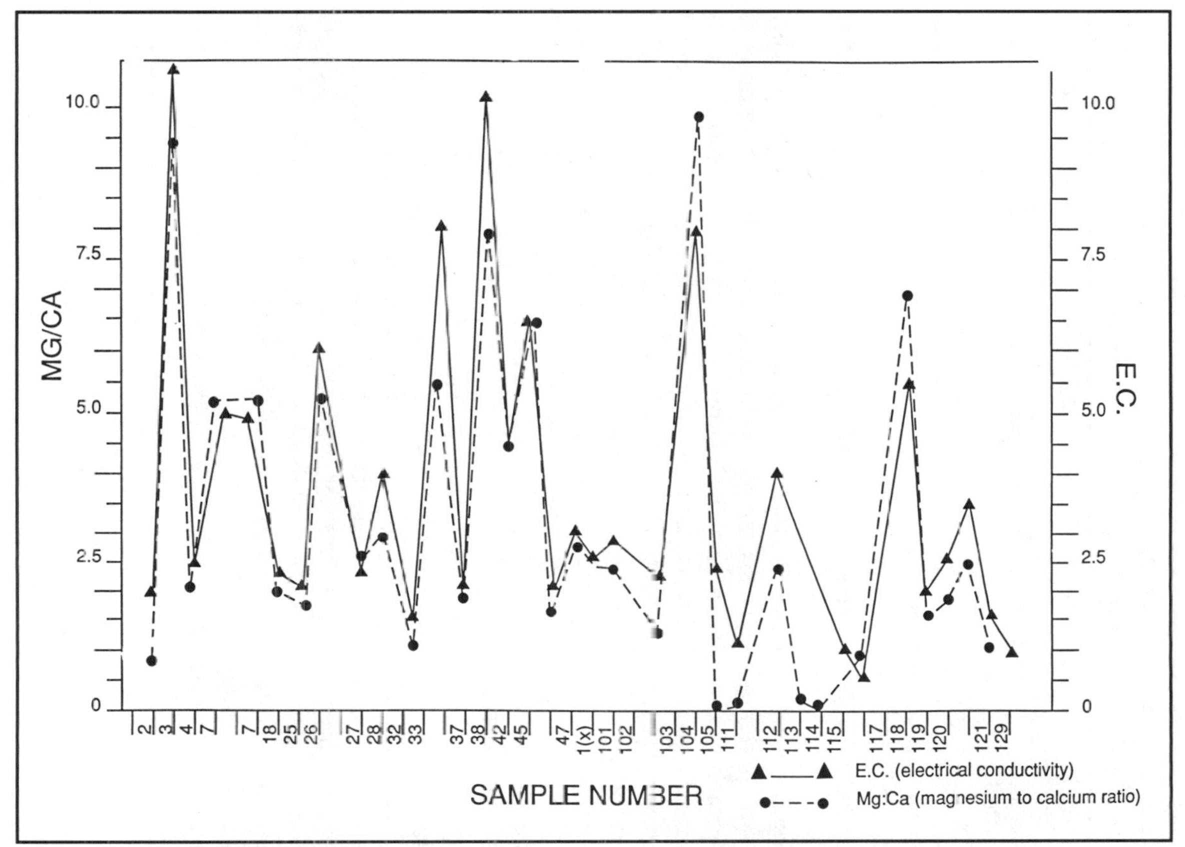

FIGURE 1. *Relationship between the ratio of Mg to Ca and electrical conductivity. Mg and Ca are a ratio of soil levels measured in a meq./liter.*

TABLE 1
INITIAL SPOIL SAMPLES RESULTS (N=34)

	Paste pH	$CaCO_3$* +/−	E.C. (mmhos/cm)
Mean	4.34	−9.87	7.22
Std. Dev.	1.31	19.47	5.15
Range	2.7–7.1	+19.6 to −114.0	0.99–21.36

*Metric tons of $CaCO_3$ equivalence/1,000 tons.

indicate $CaCO_3$ deficiencies of −49 and −24 metric tons per 1,000 metric tons.

The mean E.C. for the 34 spoil samples was 7.22, which indicates very high salt content (Table 1). Six samples exhibited extremely high readings of 12 or greater. Five of these six high E.C. samples were obtained from materials dominated by shales and clays. In these samples (Table 2), it is suspected that magnesium is being released from the slowly decaying shales and combining with sulfate, also in the shale, to form magnesium sulfate salts. Due to the location of these materials within spoil piles, leaching of these salts is a very slow process. The presence of clay in low-pH environments causes low hydraulic conductivity and low dispersion of salts, thereby magnifying the problem of slow leaching.

Table 2 also includes results from two spoil samples collected from the rooting zone of switchgrass and bluestem bunchgrasses located along an adjacent highway. The primary difference between these vegetated and non-vegetated spoil samples is their respective E.C. and the amount of magnesium present. The favorable pH values (for spoil) suggest the site received agricultural lime treatments in order to establish grass along the right-of-way. Once vegetation was established in this material, infiltration has allowed leaching of magnesium sulfate salts.

Ideally, treated spoil requires reaction times of two to three years to pro-

TABLE 2
COMPARISON OF TOXIC VS. PLANT SUPPORTING SPOIL

Sample	pH	$CaCO_3$ +/−	E.C.	Ca	Mg	% Sand
3	2.4	−4.98	21.36	24.20	217.00	17.5
26	3.0	+0.37	12.27	23.20	101.00	12.5
33	3.1	−5.61	15.7	23.20	134.00	17.5
45	3.2	−12.48	12.61	22.00	143.00	27.5
104	3.3	−4.38	16.33	25.60	246.00	35.0
Bluestem	4.3	−2.79	2.89	26.40	6.80	45.0
Switchgrass	5.7	−3.89	2.56	29.40	2.06	27.5

mote leaching of deleterious salts from the rooting zone. Reaction times may vary due to fluctuating field conditions. Unfavorable conditions are due primarily to drought. Lack of moisture can significantly slow chemical reactions and leaching. Below-freezing temperatures also virtually halt reactions; therefore days of 0° C or less should not be counted as part of a reaction period. Excessive rainfall during the reaction period can result in erosion of materials and treatment bases. An established nurse crop of salt-tolerant species can help ameliorate this condition.

The incorporation of straw mulch into the spoil rooting zone was also included in the treatment plan. Mulch adds needed organic matter and is intended to assist infiltration and leaching prior to plant root development. Decaying organic compounds also act as chelating agents which provide a partial sink for the toxic metals entering solution upon liming.

Treatment Program

Construction scheduling and the timing of grants prevented implementation of treatment programs as recommended. The administering state agency, the Iowa Division of Soil Conservation, was under pressure to complete project construction by the fall of 1987, and as a result it was not possible to allow the desired two to three year reaction period or to use a temporary cover crop.

Treatment commenced during the summer of 1986 and was nearly completed by the fall. The treatment program involved application and incorporation of five metric tons of mulch and 18,144 kg of effective calcium carbonate equivalence (E.C.C.E.) per acre into the upper one foot of spoil over the entire 49 ha site. Initiation of the spoil treatment program was critical to allow as much reaction period as possible prior to seeding. The longer the reaction period following treatment, the better the results.

Mild weather during the winter of 1986–1987 allowed completion of the initial spoil treatment program by the end of February 1987. Spoil samples were collected in April 1987 to determine the effects of the initial treatment to neutralize the spoil and the acid-base balance. In all, 48 samples (consisting of 10 subsamples each) were collected from areas of approximately 929 m^2 each.

Plant Selection

Plant species were selected that could survive the harsh conditions at the site. Native warm season grasses are naturally adapted for drought tolerance and low fertility soils. Therefore, switchgrass (*Panicum virgatum*) at 6.7 kg of pure live seed (PLS)/ha; big bluestem (*Andropogon gerardii*) at 5.6 kg PLS/ha; indiangrass (*Sorghastrum nutans*) at 4.5 kg PLS/ha; and sideoats grama (*Bouteloua curtipendula*) at 2.2 kg PLS/ha were seeded. The Kanlow variety of switchgrass was specified at 4.5 kg PLS/ha for drainages and other wet areas due to favorable past performance.

Because of the slowness with which warm season grasses become established (generally over a two-year period), cool season grasses were included in the seed mixture. Salt tolerance was a factor in selecting tall fescue (*Festuca arundinacea*, variety Ky-31) at 9.0 kg PLS/ha, and varieties Barton and Pubescent western wheatgrass (*Agropyron smithii*), a native cool season grass, at 13.5 kg PLS/ha. Reed canarygrass (*Phalaris arundinacea*, variety Ioreed) at 3.4 kg PLS/ha was selected as a cool season species for wet areas.

Legumes were specified for their nitrogen-fixing abilities and included one warm season species: Illinois bundleflower (*Desmanthus illinoensis*) at 1.1 kg PLS/ha; cool season alfalfa (*Medicago sativa*) at 1.7 kg PLS/ha; birdsfoot trefoil (*Lotus corniculatus*, variety Empire) at 4.5 kg PLS/ha; white clover (*Trifolium repens*, variety Ladino) at 3.4 kg PLS/ha; and red clover (*Trifolium pratense*, variety Redbud) at 4.5 kg PLS/ha.

Results

The results (Table 3) indicated an increase in mean pH of 1.77 units (from 4.34 to 6.11); an excess of $CaCO_3$ relative to pyritic sulfur of 17.93 metric tons of equivalence per 1,000 metric tons of material (an overall increase of 27.80 metric tons/1,000 metric tons); and a decrease in electrical conductivity from 7.22 to 2.01 mmhos/cm.

Realistically, these analyses were premature from a reaction period standpoint, since chemical equilibrium had not yet been reached. Salinity can be expected to increase over the short term as weathering occurs and massive amounts of metal bases are released into solution. However, the initial results are encouraging. A significant amount of excess calcium has been introduced into the system to counteract acid generation. Spoil pH has increased and E.C. has decreased to levels more favorable to plant establishment. Upon establishment, the plants will further modify the microsite conditions by lowering oxidation rates and promoting leaching of sulfate salts in the rooting zone.

The result of this project is 49 ha of native warm-season pasture and/or wildlife habitat with two nonacid ponds in place of 49 ha of barren, eroding acid spoil and discharging acid ponds. And this was accomplished without the destruction of adjacent unmined lands.

TABLE 3
SPOIL SAMPLES RESULTS FOLLOWING TREATMENT (N=48)

	pH	$CaCO_3$* +/−	E.C. (mmhos/cm)
Mean	4.34	−9.87	7.22
Std. Dev.	1.31	19.47	5.15
Range	2.7–7.1	+19.6 to −114.0	0.99–21.36

*Metric tons of $CaCO_3$ equivalence/1,000 tons.

The site will be monitored over the next several years to determine if maintenance amendments of lime and/or fertilizer are required.

Conclusions

Spoil Treatment Program

The mean $CaCO_3$ deficiency of -10 metric tons/1,000 metric tons at the Schmidt site is well within the treatable range. However, the wide variance of 19.47 indicates that hot spots of acid salt generation will persist after treatment. It has been generally recognized that liming rates of 30 metric tons/1,000 metric tons are the maximum feasible on soils with moderate cation exchange capacity. Due to a lack of exchange sites on the soil colloids, liming at rates of greater than 30 metric tons/1,000 metric tons would not produce additional benefits. Hot spots beyond the mean deficiencies may require treatment or replacement with suitable material if located on critical grades.

To avoid compounding magnesium imbalances in the spoil, additional magnesium should not be introduced in the form of dolomitic limestones. The magnesium content of agricultural limestones should be kept below 1% when possible.

References

Barnhisel, R. I., J. L. Powell, G. W. Akin, M. W. Ebelhar. 1982. *Characteristics and reclamation of 'acid sulfate' mine spoils.* Soil Science Society of America.

Barnhisel, R. I., J. L. Powell, J. S. Osborne, S. Lakhakul. 1984. Estimating Agricultural Limestone Needs and Observed Responses for Surface Mined Coal Spoils. In *Proceedings: Symposium on the reclamation of lands disturbed by surface mining.* Owensboro, KY.

Donahue, R. L., R. H. Follett, R. W. Tulloch. 1976. *Our soils and their management.* Danville, IL: The Interstate Printers & Publishers, Inc.

Evangelou, V. P., R. E. Phillips, J. S. Shepard. 1982. Salt generation in pyritic coal spoils and its effect on saturated hydraulic conductivity. *Soil Science Society of America Journal* 46(3):457–460.

Evangelou, V. P., and W. O. Thom. 1983. Coal spoil chemistry interpretation and the effect of spoil chemical changes in the continuity of nutrient availability. In *Proceedings: Better reclamation with trees.* Terre Haute, IN.

Evangelou, V. P., and W. O. Thom. 1984. Sampling, testing and lime/fertilizer requirements of acid spoils. In *Proceedings: Conference on reclamation of abandoned acid spoils.* Osage Beech, MO.

Freeman, J. R., J. W. Sturm, and R. W. Smith. 1984. Acid base accounting of acid spoils. In *Proceedings: Conference on reclamation of abandoned acid spoils.* Osage Beech, MO.

Myerson, A. S. 1981. Oxygen mass transfer requirements during the growth of *Thiobacillus ferrooxidans* on iron pyrite. *Biotechnical Engineering* 23:1413–16.

Sutton, P. 1970. Reclamation of toxic coal mine spoilbanks. *Ohio Report* 55(5):99–101.

U.S.D.A. Soil Conservation Service. 1980. *Soil survey of Marion County, Iowa.* Washington, DC.

Chester J. Covert is an Environmental Scientist with George Butler Associates, Inc., 8207 Melrose Dr., Lenexa, MO 66214.

Restoration Symbiosis: Integrating Environmental Programs into Industrial Operations for Quarry Reclamation Success at Coaldale, Colorado

Kenneth S. Klco

ABSTRACT: *Successful, low-cost mined land restoration has been achieved in the Sangre de Cristo Mountains of south-central Colorado by utilizing the expertise of Colorado Mountain College's Environmental Technology Program. Human labor provided by well motivated, well trained and reasonably paid students has proven to be the key resource in attaining reclamation goals on a site with a harsh environment.*

KEY WORDS: *dozer rilling, contour rilling, dormant bare root transplants, mined land reclamation.*

Introduction

SUCCESSFUL MINED LAND reclamation has been achieved in the Sangre de Cristo Mountains of south-central Colorado, notwithstanding severe climatic conditions, limited use of irrigation, and steep slopes. The arid Transition Zone mountain environment (2,100 m elevation, Upper Sonoran Zone), receives an average of 30 cm of precipitation annually, mostly in the form of late-winter snow. The area averages 100 frost-free growing days with a typical hiatus of precipitation of up to eight weeks during the months of May, June, and July. Temperature extremes and irregular precipitation limit natural revegetation. Seventy years of pre-reclamation law surface mining resulted in more than 60 ha of highly disturbed land which, prior to this project, had not changed appreciably since disturbance, which started in the early 1900s.

Reclamation Efforts

A long-range reclamation plan was promulgated by the writer in 1976 in response to the State of Colorado Mined Land Law. Since the reclamation program's inception, it has undergone constant modification, addition, reevaluation, and improvement, due largely to the close association of mine management and the Environmental Technology Program of Colorado Mountain College's Timberline Campus in Leadville, Colorado.

Colorado Mountain College confers an Associate degree on students completing an intensive, two-year curriculum emphasizing land restoration, water quality, waste management, and pollution control. At present, it is the only program of its kind in the western United States. By placing Environmental Technology students in summer practicum programs and in short-term field teams, a number of significant reclamation goals have been achieved.

Over the past eleven years, more than 16 ha of severely disturbed mined land have been restored to productive wildlife habitat in Fremont County, Colorado. State-of-the-art reclamation and revegetation techniques have been developed, such as use of natural moisture-retention devices and steep-slope erosion control measures constructed during contouring operations. Phase One—slope configuration and contour "rilling"—is performed with bulldozers. Phase Two—planting—and Phase Three—maintenance—are performed by hand, including seeding, fertilization, seedling planting, mulching, and watering. Several native-species propagation methods are practiced, including planting of potted-stock tree seedlings; early season planting of bare-root, second-year shrubs while dormant; transplanting of first-year tubeling seedlings from on-site progenitors grown in a nearby nursery greenhouse; and propagating cuttings of certain highly adapted species, such as mountain snowberry (*Symphoricarpos oreophilus*).

Procedures for Successful Wildlife Habitat Restoration

Moisture-Retention/Control

1. *Constructing moisture-collection devices by bulldozer "rilling" along the contour of the slope.* Rilling controls excess run-off during times of extreme moisture by directing water into ditches along the slope contour. (It is a low-cost [less than $200/ha], effective method of collecting and retaining scarce precipitation while minimizing sheet erosion and gullying during times of heavy rainfall.)
2. *Copious application of straw mulch after broadcast seeding of grass-species mixture.* Seeding rate is 45 kg/ha and mulch rate is up to 4,500 kg/ha. Straw mulch may be up to 10 cm thick in a 1-m-diameter zone surrounding each tree and shrub seedling.

3. *Use of water-absorbing or water-retaining media (i.e., an organic polymer) applied to the root zone during seedling installation.* Biodegradable organic starches in granular or powder form are marketed for this purpose.

Timing of Planting

1. Native grass mixture is broadcast in late fall or early spring.
2. Planting of seedlings must be undertaken as early as possible in spring to take advantage of natural precipitation patterns. Propagation method and success rates are related to time of planting. After several years of experimentation and monitoring more than 50% of the 1,600 seedlings to be planted in 1988 will be dormant, potted-shrub species or dormant, bare-root, second-year shrubs.
3. Some native species propagate best via active tubeling seedling plantings in June, after killing frost. This is especially the case for mountain mahogany (*Cercocarpus montanus*), crown vetch (*Coronilla varia*), and ground plum (*Astagalus crassicarpus*). (Tubelings encase a plant's root system in a 3 cm by 10 cm earth-filled tube.)

Planting Technique

1. Planting sites are located at the bottom of the rill trench.
2. A hole is dug with a mattock and shovel, and the seedling is planted using 5 g, slow-release, nitrogen-potassium-phosphate tablets and a moisture-retaining medium. Then up to 10 cm of mulch is applied, surrounding the seedling. Care is taken not to cover up the seedling by rocking the mulch down.
3. The seedling is given 18–24 liters of water, and the seedling root system is never allowed to dry for the first growing season up to and including the plant's fall dormancy period, from late November to early April. Mulch can extend watering intervals for up to five weeks. Moisture conservation and maintenance (i.e., watering, control of weeds, competing vegetation, and animal damage) are critical factors for seedling survival.

Species List by Planting Method for Coaldale, Colorado

Potted Tree Species

These are planted in early April while dormant, and are acquired from the Colorado State Forest Service.

Pinyon pine	*Pinus edulis*
Ponderosa pine	*Pinus ponderosa*

Colorado blue spruce	*Picea pungens*
Rocky Mountain juniper	*Juniperus scopulorum*

Bare-root Dormant Species

Seeds are obtained from on-site progenitors and are planted in early April while dormant.

wax currant	*Ribes cereum*
antelope bitterbrush	*Purshia tridentata*
four wing saltbush	*Atriplex canescens*
winterfat	*Ceratoides lanata*
golden currant	*Ribes aureum*
three leaf sumac	*Rhus trilobata*
mountain snowberry	*Symphoricarpus oreophilus*
Russian olive	*Elaeagnus angustifolia*

Tubeling Seedling Species

Seedlings are grown from seed or obtained by cutting from progenitors on site. Active seedlings are planted in June after killing frosts.

four wing saltbush	*Atriplex canescens*
golden currant	*Ribes aureum*
winterfat	*Ceratoides lanata*
three leaf sumac	*Rhus trilobata*
mountain mahogany	*Cercocarpus montanus*
fringed sage	*Artemesia frigida*
crown vetch	*Coronilla varia*
ground plum	*Astragalus crassicarpus*
mountain snowberry	*Symphoricarpus oreophilus*

Grass Seed Mixture

Grasses are broadcast seeded at 45 kg/ha after fertilization with 338 kg/ha nitrogen and 197 kg/ha phosphorus prior to mulching with straw at 4,500 kg/ha.

20% crested wheatgrass	*Agropyron cristatum*
20% slender wheatgrass	*Agropyron trachycaulum*
15% streambank wheatgrass	*Agropyron riparium*
10% blue grama grass	*Bouteloua gracilis*
10% sand dropseed	*Sporobolus cryptandrus*
10% cicer milkvetch	*Astragalus cicer*
5% wild ryegrass	*Lolium perenne*
5% Indian ricegrass	*Oryzopsis hymenoides*
5% sideouts grama	*Bouteloua curtipendula*

A total of 1,500–1,600 seedlings and 180 kg of seed of all species are planted each season by Environmental Technology students with hand tools on the roughest of terrain. Soil of any kind is scarce and little used. Nevertheless, slopes up to 1:1 have been stabilized and successfully revegetated at low cost. Although greatly dependent on moisture conditions, overall survival rates are above 75%.

Conclusion

The reclamation of the Coaldale Quarry in Fremont County, Colorado has shown that a reclamation program can be successfully integrated into daily production operations of a small mine at a reasonable cost. Effective erosion control and reclamation of arid mined lands can be achieved at relatively low cost and with minimal equipment needs and water consumption. Overall survival rates monitored for the past 12 years are greater than 75%. Human labor is the key resource in restoration efforts, exemplified by the unique and mutually beneficial working relationship fostered by the mining company and Colorado Mountain College.

Kenneth S. Klco is Quarry Superintendent at the Coaldale Gypsum Quarry, Domtar Gypsum, Coaldale, CO 81222.

Ecological Mining

Brian Hill

Abstract: *"Ecological mining" is the newly evolving theory and practice of synthesizing contemporary mining technology with the timely florescence of environmental consciousness to create a new mining industry which complements the stability of the bioregions which provide the minerals we need today for our high-technology civilization.*

This new alliance between environmental and mining communities is finding ways to mine and to create new arable land and/or healthy water systems in the process of mineral extraction.

Key words: *dredging, electrolytic leaching, river restoration, solid waste recovery.*

Brian Hill is an ecological miner and President of Trinity Alps Mining Company, Weaverville, CA 96093.

Constructed Wetlands for Wastewater Treatment— An Overview of Applications and Potentials

Gregory A. Brodie, Donald A. Hammer, Gerald R. Steiner, David A. Tomljanovich, and James T. Watson

ABSTRACT: *Suitably designed and operated wetlands—marshes and swamps—have considerable potential for low-cost, efficient, and self-maintaining wastewater treatment systems. Though not well understood, the complex of wetland vegetation, aerobic and anaerobic substrate, and meandering water column has demonstrated capability to remove nutrients, organic compounds, and metallic ions and to increase oxygen and pH levels in wastewaters from domestic, municipal, mining, industrial, and agricultural sources.*

In addition to its widespread applicability, capital and operating costs of wetlands waste treatment (WWT) systems range from one tenth to one half as costly as conventional systems and require minimal operator training. For small existing public treatment plants alone, potential nationwide savings are greater than $3 billion. Additional billions can be saved by private developers, and Appalachian coal companies could save over $1 million per day currently spent treating acid mine drainage (AMD). WWT systems also require less land area than lagoons or land application flow; typically only 0.8–2 ha per 100,000 gallons per day of flow.

During the last two years, Tennessee Valley Authority (TVA) has successfully established WWT systems that turn AMD into clean water at strip-mined areas, a coal preparation facility, and coal-ash storage ponds. Four types of systems for treating domestic waste are in operation at communities of 500 to 5,000 residents and at a housing development in Kentucky and Tennessee.

TVA's monitoring program at these sites will provide a database for basic design and operating criteria, evaluating cost effectiveness, optimizing design and operating guidelines, and providing information for technology transfer

to the private sector. The monitoring program will investigate standard National Pollutant Discharge Elimination Standards parameters: loading rates, hydraulic conductivity, length-to-width ratios, water levels, substrates, vegetation, selected animal populations, seasonal and annual variability, and changes in effectiveness for three years of operation.

In September 1986, TVA constructed the Acid Drainage Wetlands Research Facility to evaluate efficiencies of various substrate types, then macrophyte types and lastly microbial populations. Carefully designed sampling formats will concurrently provide information on location and capacities of metallic ion sinks, application rates, construction, and operating methods.

Livestock operations and industries also contribute significant pollutants to water bodies. TVA is developing an inexpensive WWT system to demonstrate practical application to hard pressed livestock producers, small industries, and urban stormwater runoff.

Monitoring results at each site will produce the requisite database for development of design guidelines and improved recommendations. Widespread application of the technology provides an economically feasible method to improve the quality of waters nationally and internationally without crippling the economic base of small communities, industries, and farms. Demonstrations of the functional values of wetlands will improve citizen appreciation for the value of wetlands resources.

KEY WORDS: *wetlands, constructed wetlands, created wetlands, acid mine drainage, mine wastewater treatment.*

Gregory A. Brodie is an Environmental Scientist with the Division of Fossil Hydro Power, Tennessee Valley Authority, Chattanooga, TN 37402. Donald A. Hammer is Senior Wetlands Ecologist with the Division of Land and Economic Resources, Tennessee Valley Authority, Norris, TN 37828. Gerald R. Steiner is Program Manager for the Water and Waste Engineering Program, Tennessee Valley Authority, Chattanooga, TN 37401. David A. Tomljanovich is a Biologist with the Division of Air and Water Resources, Tennessee Valley Authority, Knoxville, TN 37902. James T. Watson is Program Manager for the Water Quality Branch, Tennessee Valley Authority, Norris, TN 37401.

Soil Bioengineering and Revegetation

Introduction to Soil Bioengineering Restoration

Robbin B. Sotir

ABSTRACT: *Soil bioengineering is an applied science that uses living plant material as a main structural component. It is useful in controlling problems of land instability, where erosion and sedimentation are occurring. This technology offers a distinctively different approach to land stabilization as it relates to restoration of the land to its natural, stable state. Recognizing land as a living, changing system, these units increase land stability by using living structures. The goal of soil bioengineering is the reestablishment of a native plant community capable of self-repair as it adapts to the land's stresses.*

Initially, unrooted live plant materials are installed in various stabilizing systems to protect the soil mechanically. As they root and grow, the plants further stabilize the site through the development of a vegetative cover and a reinforcing root matrix within the soil mantle. Selected native pioneer species typical of the surrounding area are initially established as a foundation for natural development. Ultimately, additional species invade, creating a species-rich community for long-term natural site protection.

Soil bioengineering goes a step beyond conventional hard engineering in that it works with nature to restore the land to a self-supporting, naturally beautiful state. This living system gives land an opportunity to restabilize more rapidly, more completely, and more permanently. Unlike conventional structures, which deteriorate with age, this system provides an enduring, attractive structure that becomes stronger with time. It is an alternative to conventional engineering where traditional methods might not be economically feasible, technically effective, or suitable because of other constraints (social, environmental, political, etc.). As an addition to conventional approaches, soil bioengineering methods can reduce costs while increasing overall project effectiveness, permanence, and aesthetic appeal.

KEY WORDS: *soil bioengineering, erosion, land stability, sedimentation.*

Introduction

It makes sense to respond to land instability problems with soil bioengineering, a system that is able to work with the principles of nature. Soil bioengineering responds to the ever-changing, fragile balances of land and water with live, dynamic, flexible, self-repairing, natural units that are themselves part of nature.

Soil bioengineering is a true land-reinforcing system, not just a surface revegetation system. It is a highly developed technology that offers an alternative solution to many shallow, mass-waste erosion and sedimentation problems, a solution often more immediate and permanent than conventional stabilization methods. It may be used, however, in conjuction with conventional engineering to provide a more permanent, aesthetic, and environmentally responsible product.

Soil Bioengineering

Live systems are not usable for all land instability problems, but are very useful in areas where conventional engineering may be *uneconomical, unsuitable* or *ineffective*. All the conventional data, such as slope stability, compaction rates, soil nutrient levels, moisture contents, soil composition, etc., must be carefully assessed. Then soil bioengineering goes a step beyond conventional engineering to investigate the live qualities of land and respond to them with living structures. Proper execution, from the site assessment/design stage to the final installation and follow-up, is of extreme importance.

Unlike conventional structures, bioengineered systems grow stronger with age. These are the only systems I know in which this is the case. Time is on our side, so to say. An anchor bolt in conventional engineering acts similarly to an anchor root in soil bioengineering—both exhibit shearing and tensile strength.

Living systems are bioadaptive and can remove excess moisture from land sites. Many conventional or quasi-conventional systems are copies of nature; geogrid, reinforced earth, polyfiber systems, anchor bolts, etc., simulate reinforced root matrixes, both lateral and vertical. Lateral roots tie or consolidate soil masses together, while vertical roots create anchors and buttressing units as they become part of the land.

Soil bioengineering uses largely native plants collected in the immediate vicinity of a project site. This assures that the plant material will be well adapted to site conditions. While a few selected species may be installed for immediate protection, the ultimate goal is for the natural invasion of a diverse plant community. Typically, pioneer plants are installed that will meet the needs of the land and act to stabilize and improve soils to prepare the site for natural invasion or for specific cash crops.

Plant parts are used as the major structural component in soil bioengineering construction. Unrooted live vegetation is installed as structural members, which provide immediate stabilization, while shoots and roots develop to form a permanent vegetative cover and root-reinforcing matrix. Soil bioengineering is in many ways a more complicated and sophisticated approach than conventional engineering. Soil bioengineering does not directly repair a site as does conventional engineering; instead, it sets in place a mechanical and living foundation on which the land is intended to recover.

Soil bioengineering systems have been developed worldwide for centuries, and are based on a long history of research and success. Public agencies in the United States, such as the Soil Conservation Service, practiced an abbreviated form of soil bioengineering briefly in the 1930s and 1940s. After World War II, these systems were abandoned as high-technology solutions were developed for land instability problems. In today's era of high energy costs and depleted resources, soil bioengineering is a practical, cost-effective alternative to many land stabilization problems. It is labor and skill intensive rather than capital/energy intensive.

Soil bioengineering is useful in a wide range of situations, including the revegetation of highway roadsides, cut-and-fill slopes, dredge disposal sites, mined land, streambanks, floodplains, shorelines, bluffs, and spawning streams. It is also helpful in the repair of gullies, earthslides, slumps, stream bank erosion, streambed degradation, and on transportation and transmission corridors, park and recreation land, forests, commercial and agricultural land, as well as on wildlife and wetland habitats.

Soil bioengineering works closely with nature to cause land to become its own self-supporting structure. By offering a low maintenance, rapid recovery system, soil bioegineering produces structures that grow stronger and more beautiful with age and become part of the land.

Ms. Robbin B. Sotir is President of Robbin B. Sotir & Associates, Inc., 627 Cherokee Street NE, Suite 11, Marietta, GA 30060.

Erosion Control Effectiveness: Comparative Studies of Alternative Mulching Techniques

Michael V. Harding

Abstract: *The wide range of revegetation and erosion control problems is paralleled by a diversity of products and techniques employed to address them. Current state specifications for erosion control products are as varied as the products themselves. Consequently, no uniform standard exists which can be used to select materials based on their erosion control effectiveness. Utilizing a rainfall simulator device, tests were conducted to determine the erosion control effectiveness of alternative materials and practices. Treatments were applied to a soil bed containing a Blount silt loam inclined to a 9% slope and subjected to rainfall intensities of 10.66 cm and 14.73 cm for one hour. Runoff water (kg) and sediment (g) were collected and weighed for each erosion control practice tested. Sediment concentration in the runoff water, water velocity reduction and soil loss from each practice was compared against values obtained from a bare soil control. The results appear to indicate a difference in the erosion control effectiveness of the alternative mulching techniques tested. Where water velocity was reduced by the mulch matrix, sediment concentrations in the runoff water and, ultimately, soil losses appeared to be correspondingly reduced. Overall, nylon monofilament materials did not appear to be as effective as natural organic materials, possibly due to their lack of moisture-holding capacity and general inability to trap soil particles as efficiently. Although laboratory evaluation of erosion control products may not be the best approach to determine performance and effectiveness in the field, the purpose of the rainfall simulation test was to demonstrate a method for evaluating alternative mulching techniques within the framework of erosion control parameters they were designed to address.*

Key words: *erosion control effectiveness, rainfall simulation, sediment concentration.*

Introduction

LAND DISTURBING ACTIVITIES have the potential to develop severe erosion and sedimentation problems due to massive disturbance of the landscape and exposure of bare soil to weathering by wind and water. Soil erosion and the resultant sedimentation contributes to the overall cost of a project as slopes must be reworked to fill in gullies and deposited sediment must be removed from catch basins and traps for disposal or redistribution.

Short-term erosion and sedimentation problems can become long-term revegetation and maintenance problems. An eroding seedbed not only loses soil, but also nutrients and seed applied for revegetation. The resulting effects of erosion are not only aesthetically displeasing: sites can be left with infertile, droughty subsoil or with rills and gullies which limit accessibility to maintenance crews and delay permanent vegetation establishment and, hence, permanent erosion control.

The wide range of erosion control and revegetation problems is paralleled by a diversity of products and techniques employed to address them. If optimum growing conditions occurred after final seed preparation, many erosion control products and practices would be unnecessary. But the fact is that, in most parts of the country, bare soil is exposed to weathering for varying time periods ranging from weeks to months. Various mulching techniques and materials have been developed that provide temporary ground cover until mature, permanent vegetative ground cover is established.

Blown straw or hay mulch, mulch grown in-situ, hydroseeding, hydromulching, soil sealants, blankets of degradable organic materials, man-made fabrics, and matrices of nondegradable synthetic materials all have their specific applications.

Current state specifications and use of erosion control products and techniques are as varied as the products themselves. In most cases, final revegetation and erosion control is based on relative cost, on the assumption that the public interest is best served by providing the least costly alternative. Consequently, no uniform standard exists which selects materials based on their erosion control *effectiveness*. An exclusively economic orientation to erosion control is acceptable on large, flat areas where conventional seeding and mulching is effective in establishing permanent vegetation and thereby permanent erosion control. On steep slopes, critical areas, and under unusual circumstances, these lower-cost techniques typically are not effective.

In recent years, materials have been developed which, though costing more initially, are correspondingly more effective in controlling soil erosion and promoting permanent vegetation establishment. The purpose of this study was to quantify the performance of some of these materials under similar conditions in order to show justification for their use on the basis of erosion control effectiveness and to demonstrate rainfall simulation as a potential technique for the standardization and evaluation of erosion control materials.

METHOD

The use of rainfall simulators has become an accepted practice in determining the erodibility of soils. Factors which influence soil erosion have been defined for undisturbed soils (Wischmeier and Smith 1978) and, in recent years, for reclaimed strip mine soils (Stein et al. 1983; Barfield et al. 1983). The amount of ground cover, either in the form of permanent vegetation or temporary ground litter/mulch directly affects the rate of soil erosion (Lee and Skogerboe 1984). Many studies have demonstrated the benefits of agricultural crop residues in the reduction of soil losses. Some work has been done to evaluate the effectiveness of manufactured erosion control materials. Ingold and Thomson (1986) quantified the relative performance of manufactured synthetic and natural fiber erosion control systems using rainfall simulation to simulate various storm events.

For the purpose of this study, a rainfall simulation device was used which produced storm intensities of 10.66 cm and 14.73 cm for one hour. Twelve manufactured erosion control materials were selected for testing based on their prior specification for critical area erosion control. In addition, conventional applications of 2,240 kg/ha and 4,480 kg/ha blown straw mulch were included in the evaluations. The straw was not crimped or tacked to the soil surface. Table 1 lists the materials tested.

A 1.2 m by 3.04 m soil bed containing a Blount silt loam was constructed beneath the rainfall simulator and inclined to a 9% slope. The soil bed was wetted before each test until field capacity was reached, at which time the material to be tested was placed on the bed according to the manufacturer's

TABLE 1
MATERIALS LIST FOR RAINFALL SIMULATION TESTS

Mulch Material	Characteristics
1.	100% wheat straw/top net
2.	100% wheat straw/two nets
3.	70% wheat straw/30% coconut fiber
4.	70% wheat straw/30% coconut fiber
5.	100% coconut fiber
6.	Nylon monofilament/two nets
7.	Nylon monofilament/rigid/bonded
8.	Vinyl monofilament/flexible/bonded
9.	Curled wood fibers/top net
10.	Curled wood fibers/two nets
11.	Anti-wash netting (jute)
12.	Interwoven paper and thread
13.	Uncrimped wheat straw – 2,242 kg/ha
14.	Uncrimped wheat straw – 4,484 kg/ha

instructions. Each treatment was subjected to rainfall intensities of 10.66 cm or 14.73 cm for one hour. Runoff water (kg) and sediment (g) were collected and sediment was filtered from the runoff water, weighed and compared against bare soil (control) losses for the respective rainfall intensity. Dye tests were conducted to determine the relative velocity of the runoff water from each test.

Results

Water Velocity Reduction

All mulch treatments were compared to the bare soil (control) plots. To determine the decrease in water velocity afforded by each mulch material, the values obtained for each treatment were compared to those from the bare soil to derive percent water velocity reduction, as reported in Figure 1.

Treatments which involved the use of organic materials in the form of straw, wood shavings, paper, and coconut or jute fiber exhibited a higher level of water velocity reduction ranging from 45%–78%. In contrast, synthetic nylon and vinyl monofilament materials reduced velocities by 24%–32% respectively; an exception to this trend was the netted nylon blanket, no. 6, which achieved a 74% reduction.

Sediment Concentration in Runoff Water

Concentration of sediment in the runoff water (reported as grams of soil per kilogram of runoff water) illustrates the effectiveness of the various treatments in reducing sediment delivery. Table 2 reports the materials tested under the 10.66 cm/hr and 14.73 cm/hr rainfall intensities.

Sediment concentration data parallel the relative order of the water velocity reduction results, but with a wider separation in the values. When comparing runoff values for each material to the values of the respective control, organic materials show a propensity to reduce sediment concentrations at a rate greater than the synthetic products. An exception is the netted nylon blanket, no. 6 (0.046 g/kg), for which sediment concentration was less than 1% of the control value, and the vinyl/monofilament, no. 8 (1.78 g/kg), which exhibited concentrations which were 7% of the control (22.09 g/kg). By contrast, no. 4 (0.38 g/kg), no. 5 (0.514 g/kg), and no. 1 (0.944 g/kg) reduced concentrations to less than 1 kg of soil per kg of runoff water.

Soil Loss

As might be expected, a measurement of the actual soil loss occurring from each test plot mirrors the results of the sediment concentration tests (Table 3).

Reductions in soil loss were determined by subtracting the soil loss for each material tested from the soil loss for the control. The difference in soil

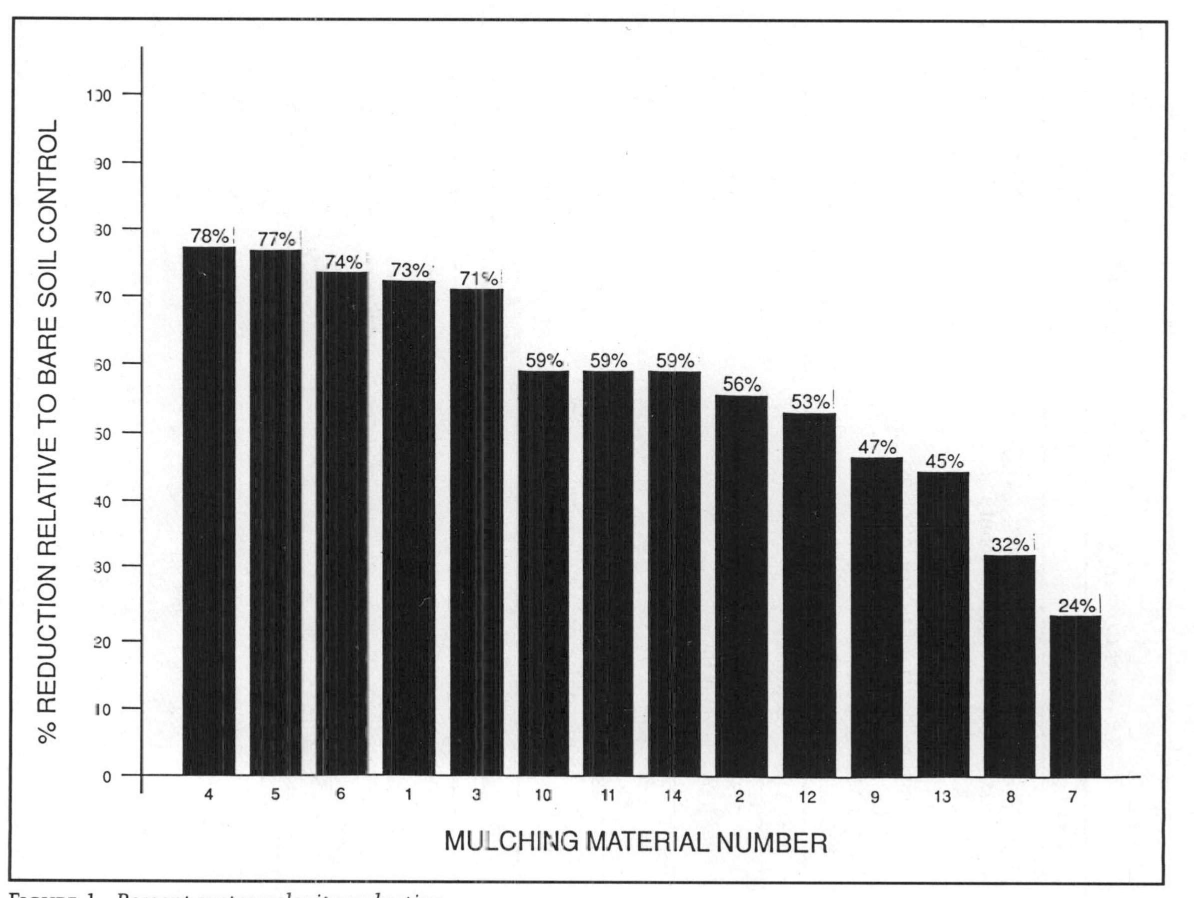

FIGURE 1. *Percent water velocity reduction.*

TABLE 2
SEDIMENT CONCENTRATIONS (G SOIL/KG RUNOFF WATER)

Test 1 (10.66 cm rain)	g/kg
6. Nylon monofilament/two nets	0.046
4. 70% wheat straw/30% coconut fiber	0.113
2. 100% wheat straw/two nets	0.326
11. Anti-wash netting (jute)	1.50
8. Vinyl monofilament/flexible/bonded	1.78
9. Curled wood fiber/top net	1.91
13. 2,242 kg/ha straw	2.81
Bare soil control	22.09
Test 2 (14.73 cm rain)	**g/kg**
3. 70% wheat straw/30% coconut fiber	0.38
5. 100% coconut fiber	0.514
1. 100% wheat straw/top net	0.944
10. Curled wood fiber/two nets	1.99
12. Interwoven paper and thread	2.24
14. 4,484 kg/ha straw	3.14
7. Nylon monofilament/rigid/bonded	15.18
Bare soil control	30.06

TABLE 3
ACTUAL SOIL LOSS

Test 1 (10.66 cm rain)			Test 2 (14.73 cm rain)		
Material	Soil loss (g)	% Reduction	Material	Soil loss (g)	% Reduction
6.	4.9	99.8	3.	57.3	98.7
4.	10.1	99.5	5.	70.9	98.4
2.	29.0	98.6	1.	110.4	97.5
11.	168.0	91.8	10.	278.3	93.5
9.	195.4	90.4	12.	303.0	93.0
8.	213.2	89.6	14.	458.9	89.3
13.	325.4	84.0	7.	2,033.6	53.0
Bare soil	2,032.3	–	Bare soil	4,268.5	–

loss was then divided by the control value and subtracted from 100 to determine percent soil-loss reduction. No. 6 (99.8%), no. 4 (99.5%), no. 2 (98.6%), no. 3 (98.7%), no. 5 (98.4%) and no. 1 (97.5%) exhibited reductions greater than 97.5%. All other materials exhibited a high degree of soil loss reduction (84%-93.5%) with the exception of no. 7 (53%).

Discussion

The results of the tests appear to indicate a significant difference in the erosion control effectiveness of alternative mulching techniques. Materials which are fibrous in nature and interwoven so as to achieve a matrix high in surface area are more efficient in reducing rainfall impact, dislodgement and transportation of soil particles. This conclusion is illustrated by the fact that 1, 2, 3, 4, 5, and 6 are manufactured under the same process which produces an extended matrix and therefore rank as the top six materials for each test.

These materials showed a clear superiority over alternative mulching techniques in the reduction of water flow velocities. Lower water velocities carry less sediment and scouring of the soil surface decreases as water velocity decreases. It follows that if soil particles are not dislodged through rainfall impact and overland water flow, their concentration in runoff water will also be reduced. Sediment concentration data (Table 3) show mulch matrices also appear to trap soil particles carried by runoff water, further reducing sediment delivery, an important water quality consideration.

Finally, extended matrix materials such as those previously mentioned ultimately reduce soil losses; in this case up to 99.8%. But the key to their effectiveness is that they stay in place and maintain a high level of soil-fiber contact. The relatively poor performance of the conventional 2,240 kg/ha and 4,480 kg/ha straw mulching can be attributed to movement of the mulch downslope during the rainfall events, lack of contact with the soil surface, and dislodgement of particles by rainfall impact on exposed soil.

Conclusion

A variety of materials and techniques is available to control soil erosion. Manufactured materials vary in their physical composition, the manner in which they are applied and their cost. Low cost alternatives are not always effective on critical areas, and some higher cost products do not guarantee a correspondingly high level of erosion protection. With increased public awareness of erosion, regulations enforcing its control and the development of new products designed to provide solutions, a uniform method for evaluating erosion control materials should be established.

Standard American Society for the Testing of Materials (ASTM) specifications which evaluate geotextiles as to their grab strength, flexibility, tensile strength or elongation do not apply to the effectiveness of a material to control erosion. Yet, state departments of transportation frequently use ASTM parameters to compare dissimilar erosion control products, such as nylon monofilament and straw.

Rainfall simulation is one technique which can be used to evaluate materials within the framework of conditions they were developed to address.

Although laboratory evaluation of erosion control products may not be the best approach to determine performance and effectiveness in the field, rainfall simulation is an appropriate method for comparative evaluations of alternative mulching techniques under exact, controlled conditions. A general ranking can be inferred from such tests, permitting a design engineer or contractor to evaluate cost-effectiveness prior to installation.

Finally, uniform standards should be established which specify the use of erosion control materials based upon their effectiveness and appropriateness for a given set of conditions. It is incumbent upon members of the erosion control industry to work with organizations such as the Transportation Research Board and the International Erosion Control Association to develop guidelines for the establishment of uniform standards.

Acknowledgment

The author would like to acknowledge the work of Stephanie Schroeder, graduate student at Purdue University, for the collection and measurement of runoff and sediment for the test plots.

References

Barfield, B. J., R. I. Barnhisel, M. C. Hirschi, I. D. Moore, J. L. Powell. 1983. *Erodibility and eroded size distribution of western Kentucky spoil and reconstructed topsoil.* CRIS Project no. 907-15-2. Lexington, KY: College of Agriculture, University of Kentucky.

Ingold, T. S., and J. C. Thompson. 1986. Result of current research of synthetic and natural fiber erosion control systems. In *Erosion control: Protecting our future. Proceedings of Conference XVII of the International Erosion Control Association.* Steamboat Springs, CO: International Erosion Control Association.

Lee, C. R. and J. G. Skogerboe. 1984. *Restoration of pyritic soils and the quantification of erosion control.* Miscellaneous Paper EL-84-7, (October). Vicksburg, MS: U.S. Army Engineer Waterways Experiment Station, CE.

Stein, O. R., C. B. Roth, W. C. Modenhauer, and D. T. Hahn. 1983. Erodibility of selected Indiana reclaimed strip mined soils. In *Proceedings 1983 symposium on surface mine hydrology, sedimentology and reclamation.* Lexington, KY: College of Engineering, University of Kentucky.

Wischmeier, W. H. and D. D. Smith. 1978. Predicting rainfall erosion losses, a guide to conservation planning. U.S. Department of Agriculture, Science and Education Administration, Agricultural Handbook no. 537. Washington, DC: U.S. Department of Agriculture.

Michael V. Harding is President of Aspen Environmental Consultants, Inc. of Fenton (St. Louis), MO 63026.

Use of Erosion Control Blankets on Harsh Sites

Michael V. Harding, Christopher D. Betts, Alan W. Juncker

ABSTRACT: *Case studies are presented which illustrate the ability of erosion control blankets to promote permanent vegetation by moderating moisture and temperature, and in stability problems on unusual sites. In the first study, a coconut fiber blanket was used in conjunction with American beachgrass (*Ammophila breviligulata*) to stabilize and build sand dunes. The beachgrass was planted through slits cut in the blanket. The dune is then built as the plants grow and wind-blown sand is deposited around them, vegetatively anchoring the sand and protecting the beaches and homes from erosion.*

Another study demonstrates the ability of straw and coconut fiber blankets to moderate moisture and temperature extremes in the reclamation of mining waste (coal slurry) impoundments in the midwestern United States. The technique represents a cost-effective vegetative alternative to covering the waste materials with topsoil and is applicable to a wide range of abandoned refuse areas. Finally, protection of areas disturbed during extended winter construction periods is addressed through the use of fiber blankets for erosion control and the application of dormant seed. A gravel pit reclamation site in southern New England was seeded, fertilized, and protected with erosion control blankets in early December of 1986. The slopes were protected over the winter by the anchored mulch and showed a vigorous stand of vegetation the following spring, with no offsite movement of sediment.

KEY WORDS: *stabilize, moderate, revegetation, coal slurry, dormant seeding.*

Michael V. Harding is President of Aspen Environmental Consultants, Inc. of Fenton (St. Louis), MO 63026. Christopher D. Betts is Regional Sales Manager for North American Green, Inc., in Charlotte, NC. Alan W. Juncker is Regional Sales Manager for North American Green, Inc., in Evansville, IN 47711.

URBAN ISSUES, PARKS, AND WASTE TREATMENT

Habitat Restoration on a Central Texas Office Building Site

David Mahler and Judy Walther

Abstract: *In November 1986, habitat restoration was begun on 0.8 ha of steep, bare limestone slopes surrounding a newly completed building. The slopes were restructured to mimic the original stairstep typography; naturalistic drainage channels were added. Sod of perennial grasses and forbs was salvaged from nearby construction areas and planted into the slopes. Approximately 98% of the transplants survived. Seed was harvested from native habitats using a "Grin Reaper" (invented by the author), which facilitated the harvest of large quantities of various seed reflecting the diversity of the habitat. By October 1987, over 120 native species were identified in the restoration area.*

Key words: *seed harvest, Grin Reaper, harvesting equipment, grasslands, commercial restoration, and landscaping.*

Introduction

In mid-1986, Environmental Survey Consulting (ESC) entered into contract with Yancey Hausman Interests, the owner of II Wild Basin Office Building in Austin, Texas, to provide a site analysis with design recommendations for habitat restoration of the disturbed construction areas. This contract was carried out under the supervision of landscape architect Steven Domigan. Recommendations were incorporated into the final landscape design, and restoration of approximately 0.8 ha was contracted to begin December 1986. The site is located in the central Texas hill country in Upper Glen Rose limestone habitat.

Analysis and Design

The analysis and design stage of work included: (1) analysis of the undisturbed natural areas of the site, (2) comparison of the site to the almost pristine and similar habitat of nearby Wild Basin Preserve, (3) analysis of physical conditions of the construction areas, (4) recommendations for structural modifications, (5) recommendations of more than 200 species which were probably native to the site and possibly available for utilization, and (6) participation in negotiations with City of Austin officials to accept this design as an alternative to existing regulations developed primarily for nonnative revegetation.

The 2.8 ha building site consists of a ridge top and the upper half of a valley (244–287 m elevation) carved from Cretaceous limestone, marl, and dolomite. The plant community on the undisturbed areas of this site is a high quality remnant of the woodland/grassland mosaic typical of this habitat. The woody component is dominated by juniper (*Juniperus ashei*), Texas oak (*Quercus buckleyi*), and live oak (*Quercus fusiformis*), with a diversity of shrubs. Little bluestem (*Schizachyrium scoparium*) and several species each of *Bouteloua, Muhlenbergia,* and *Sporobolus* characterize the grasses. There is a large assortment of native perennial wildflowers, mostly in the 15.25–61 cm height category. Studies of a similar area in Wild Basin Preserve indicate that infrequent patchy fires, several decades apart, shift the habitat toward the grasses and wildflowers with significant islands of shrubs which resprout strongly after fire. Fire suppression leads to development of a closed canopy woodland. There also are frequent vegetational variations which correspond to the stairstep topography produced by the alternating hard and soft layers of rock that determine the variation in topsoil depth and root penetration depths. These hillsides also have significant species variation between the water-shedding and water-gathering areas. These natural "microniche" variations in Wild Basin Preserve provided models which approximately correspond to the physical variations in the disturbed site. This allowed for some aesthetic choices for reestablishing the appearance of natural open grass and wildflower areas, the screening effect of evergreen shrubs, or the shade of trees.

The development of this site consisted of the construction of a large office building surrounded by parking on the ridge top, with an entrance drive and utility corridor cut from the bottom of the site and leading up to the building. Part way up the hillside, an evapotranspiration (ET) bed was constructed for effluent. The ridge top leveling created fill slopes at the edge of the parking areas consisting of limestone and dolomite rubble at a slope of between a 2:1 or a 3:1 angle. The entrance drive produced exposed rock cuts on the uphill edge and steep fill slopes on the downhill edge. The ET bed created a cut rock face, a flat unnatural 0.9 m deep pit of sandy loam, and a front rubble slope. Additionally, a spring and seep area had been altered, and in places, covered with fill and sediment from uphill erosion. All construction areas had high erosion potential.

Recommendations were developed to physically restore the slopes using frequent dry stack rock walls to mimic the stairstep terracing of natural slopes. Water gathering channels were designed with natural rock sides and bottoms to resemble the natural steep drainage channels. Additionally, a design was developed to uncover the seep area at the site entrance and develop a rock stream channel. A list of approximately 200 species which would be appropriate for use in the different niches within this site was provided, along with a one year schedule for transplanting, seeding, and maintenance. These recommendations were adopted for all construction areas except for those immediately adjacent to the building where a more formal xeriscape landscape design was used. The City of Austin, which has extensive landscape requirements in its development ordinances, accepted the restoration design. However, the uniqueness of this design necessitated more complicated negotiations with city staff and a longer time period during which a performance bond was required until the establishment of vegetation was proven.

Restoration

Erosion control and restructuring of part of the construction areas began in November 1986. Construction areas ready for restoration were released to ESC in sections with the last area being released in March 1987. Structural work was followed by the transplanting of native sod into the area, followed by the first seeding in the spring of 1987. All areas received a second seeding in August 1987.

Transplant stock was primarily sod of native grasses and flowers (Table 1). Approximately 125 trees and shrubs were transplanted, most being less than 0.3 m in height. Transplant stock was salvaged from the right-of-way of a road scheduled for construction in high quality examples of similar habitat. The sod was placed into 1,858 m^2 restaurant bussing pans, watered immediately during dry periods, and trucked to the restoration site for planting within several days. Conditions during transplanting varied from cold (1.7°C) wet days during plant dormancy to a period in April of no significant rain for one month and temperatures near 32°C. During times of high moisture content in the soil, no supplementary watering was utilized. During the dry spell, transplant areas were soaked before and after transplanting and every several days for approximately two weeks. Over 98% of the grass and flower transplants survived. Almost all loss of plants occurred within the first two weeks, with adequate soil moisture from the moment of digging being the crucial factor. Approximately half of the woody stock survived, with the survival rate directly correlating to the amount of root ball dug from the salvage area.

Seed for this site was harvested from wild stock in high quality habitat similar to the restoration site. The Grin Reaper, a box-like attachment for string line trimmers (Figure 1), was used for all of the seed harvesting except for small amounts of woody species, which were hand harvested. The Grin Reaper allowed harvest of seed mixes that included both the common species in large quantities as well as the less common or less prolific species in smaller amounts. Seed for the spring seeding was collected primarily during October

FIGURE 1. *Harvesting a wild seed assortment with a Grin Reaper.*

through December 1986. Additionally, about 10% was from the previous harvest in the spring of 1986. The seed mix included approximately 100 species with large quantities of the perennial grasses and composites, most of which produce their seed in the late fall (Table 1). Seed heads were stored at outside temperatures to mimic natural conditions. This material was shredded by a gas-powered grinder to release the seed from the seed heads. Approximations of seed quantities were made for selected species. The August 1987 seeding was a diverse mix of species harvested primarily during April through June 1987, but also including some fall harvested species. Seed material was spread by hand and raked into the rough surfaces. Some species were utilized throughout most of the site. Moisture-tolerant species were specially harvested for the wet areas. Other separately harvested species were utilized only in appropriate niches.

Results

Analysis of the restoration area during October through December 1987 documented the establishment of over 120 species (Table 1). Because of the inexactness of the data both on the amount of seed utilized and on the quantities of each species transplanted and our need to monitor eight additional restoration sites, we utilize broad observational estimates of success for tabulating results. The observations, while lacking precision, are still very useful in evaluating our seed harvest efforts and for planning future restorations. In summary, a very diverse association of species was established with the perennial grasses and composites dominating. In October 1987, the site met the City of Austin's revegetation requirements. In November 1987 this site, including both the restored and landscaped areas, was awarded Second Place in the City of Austin Commercial Xeriscape Contest.

References

Correll, D. S., and M. C. Johnston. 1979. *Manual of the vascular plants of Texas.* Dallas, TX: University of Texas.

Bureau of Economic Geology. 1979. Geologic quadrangle map no. 38: Austin West. Austin, TX: University of Texas.

Garner, L. E., and K. P. Young, Bureau of Economic Geology. 1976. *Environmental geology of the Austin area: An aid to urban planning.* Austin, TX: University of Texas.

Mahler, D., and J. Walther. 1986. *The importance of wild seed: A new harvesting tool.* Austin, TX: University of Texas.

Mahler, D., and J. Walther. 1987. *The process of habitat restoration: With specific application to the upper Glen Rose geologic formation of central Texas.* Austin, TX: University of Texas.

Walther, J., and D. Mahler. *High diversity restoration of a central Texas grassland.* Austin, TX: University of Texas.

David Mahler and Judy Walther are co-partners of Environmental Survey Consulting, 4602 Placid Place, Austin, TX 78731.

Key for Table 1

Column 1: Seed Use (SU)
3 Large quantities in general site mixes
2 Moderate quantities in general site mixes
1 Small quantities in general site mixes
A Large quantities in selected areas
B Moderate quantities in selected areas
C Small quantities in selected areas
S Not seeded but probably in the soil bank
0 Not seeded

Column 2: Seeding Results (SR)
3 High seedling frequency per quantity seeded
2 Moderate seedling frequency per quantity seeded
1 Low seedling frequency per quantity seeded
0 Seedlings not yet found on the site
* No results available

Column 3: Transplant Use (TU)
3 Large quantities used
2 Moderate quantities used
1 Small quantities used
0 Not transplanted
D Probably in transplant sod as dormant roots or bulbs

Column 4: Transplant Results (TR)
3 High survival rates (95 + %)
2 Moderate survival rates
1 Low survival rates (25 − %)
A Frequently appearing from unseen rootstock
B Moderately appearing from unseen rootstock
C Occasionally appearing from unseen rootstock

Column 5: Site Occurence (SO)
3 Common on much of the site
2 Moderate numbers on much of the site
1 Occasional throughout the site
A Common in restricted areas
B Moderate numbers in restricted areas
C Occasional in restricted areas

TABLE 1
ANNOTATED LIST OF NATIVE SPECIES ESTABLISHED ON THE SITE 7–12 MONTHS AFTER TRANSPLANTING AND SEEDING

	SU	SR	TU	TR	SO
ACANTHACEAE (Acanthus family)					
Ruellia sp. (False petunia)	1	0	1	3	C
ANACARDIACEAE (Sumac family)					
Rhus lanceolata (Lance-leaf sumac)	B	0	1	3	1
R. virens (Evergreen sumac)	C	0	1	2	1
AQUIFOLIACEAE (Holly family)					
Ilex vomitoria (Yaupon)	C	0	1	2	1
BERBERIDACEAE (Barberry family)					
Berberis trifoliolata (Agarita)	0	0	1	2	1
BORAGINACEAE (Borage Family)					
Heliotropium tenellum (Pasture heliotrope)	2	2	D	B	2
CARYOPHYLLACEAE (Pink family)					
Paronychia parksii (Park's nailwort)	C	0	1	3	1
COMMELINACEAE (Spiderwort)					
Commelina erecta (Erect dayflower)	0	1	1	3	C
COMPOSITAE (Sunflower family)					
Ambrosia psilostachya (Western ragweed)	S	*	0	0	1
Baccharis neglecta (False willow)	S	*	0	0	1
Brickellia cylindracea (Gravelbar brickle-bush)	2	2	1	3	2
Chrysactinia mexicana (Damianita)	C	0	1	2	C
Conyza canadensis (Horseweed)	S	*	0	0	1
Coreopsis sp. (Coreopsis)	A	3	0	0	A
Eupatorium havanense (Shrubby boneset)	0	0	1	3	1
Evax prolifera (Bighead rabbit tobacco)	1	2	0	0	1
Gaillardia pulchella (Indian blanket)	3	3	0	0	3
Gutierrezia texana (Texas broomweed)	2	3	0	0	3
Gymnosperma glutinosum (Tatalencho)	1	0	1	3	1
Helenium elegans (Pretty sneezeweed)	3	3	0	0	2
Helianthus maximiliani (Maximilian sunflower)	C	0	1	3	1
Hymenoxys linearifolia (Fineleaf bitterweed)	3	3	2	3	3
H. scaposa (Plains bitterweed)	2	1	2	3	2
Iva angustifolia (Narrowleaf sumpweed)	1	3	0	0	3
Liatris mucronata (Gay feather)	2	2	1	3	2
Marshallia caespitosa (Barbara's button)	2	2	2	3	2
Melampodium leucanthum (Plain's blackfoot)	2	1	3	3	2
Palafoxia callosa (Small palafoxia)	2	2	2	3	3
Ratibida columnaris (Mexican hat)	2	3	0	0	3
Rudbeckia hirta (Brown-eyed susan)	B	1	0	0	C
Simsia calva (Bush sunflower)	2	1	1	3	1

TABLE 1, CONTINUED
ANNOTATED LIST OF NATIVE SPECIES ESTABLISHED ON THE SITE 7–12 MONTHS AFTER TRANSPLANTING AND SEEDING

	SU	SR	TU	TR	SO
Tetragonotheca texana (Plateau nerve ray)	2	1	1	3	2
Thelesperma simplifolium (Slender greenthread)	1	1	2	3	2
Verbesina encelioides (Cowpen daisy)	A	3	0	0	A
V. lindheimeri (Lindheimer crownbeard)	1	1	1	3	1
V. virginica (Frostweed)	C	1	1	3	C
Vernonia lindheimer (Wooly ironweed)	1	2	1	3	1
Viguiera dentata (Plateau goldeneye)	3	3	2	3	3
Wedelia hispida (Hairy zexmenia)	2	3	2	3	3
CORNACEAE (Dogwood family)					
Garrya lindheimer (Lindheimer silktassel)	0	0	1	2	1
CUPRESSACEAE (Cypress family)					
Juniperus ashei (Mountain cedar)	0	0	1	3	1
CYPERACEAE (Sedge family)					
Carex planostachys (Cedar sedge)	0	0	2	3	2
Carex sp. (Sedge)	0	0	1	3	A
EBENACEAE (Ebony family)					
Diospyros texana (Texas persimmon)	0	0	1	2	1
EUPHORBIACEAE (Spurge family)					
Argythamnia simulans (Plateau wild mercury)	1	0	1	3	1
Croton monanthogynus (Prairie tea)	S	*	0	0	2
Euphorbia marginata (Snow-on-the-mountain)	0	0	1	3	1
Euphorbia spp. (Euphorbia)	S	*	D	B	2
Phyllanthus polygonoides (Knotweed leaf-flower)	1	0	1	3	1
Tragia ramosa (Catnip noseburn)	1	0	D	C	1
FAGACEAE (Beech family)					
Quercus buckleyi (Texas oak)	0	0	1	2	1
Q. fusiformis (Plateau live oak)	0	0	1	2	1
Q. sinuata var. *brevilova* (White shin oak)	0	0	1	2	1
GENTIANACEAE (Gentian family)					
Centaurium beyrichii (Mountian pink)	1	1	1	2	1
GERANIACEAE (Geranium family)					
Erodium texanum (Texas filagree)	1	1	1	3	1
GRAMINAE (Grass family)					
Andropogon glomeratus (Bushy bluestem)	B	2	1	3	B
Aristida sp. (Threeawn)	2	3	1	3	3
Bouteloua saccharoides (Silver bluestem)	S	*	1	3	1
B. curtipendula (Side-oats grama)	2	2	1	3	2

TABLE 1, CONTINUED
ANNOTATED LIST OF NATIVE SPECIES ESTABLISHED ON THE SITE 7–12 MONTHS AFTER TRANSPLANTING AND SEEDING

	SU	SR	TU	TR	SO
B. pectinata (Tall grama)	3	2	3	3	3
B. rigidiseta (Texas grama)	B	1	1	3	B
Buchloe dactyloides (Buffalo grass)	S	*	0	0	C
Cenchrus incertus (Stickergrass)	S	*	0	0	C
Eragrostis sp. (Lovegrass)	1	2	1	3	2
Erioneuron pilosum (Hairy tridens)	B	1	1	3	C
Leptochloa dubia (Green sprangletop)	0	0	1	3	1
Leptoloma cognatum (Fall witchgrass)	1	1	1	3	1
Muhlenbergia involuta (Canyon muhly)	0	0	1	2	C
M. lindheimeri (Lindheimer muhly)	B	0	2	3	B
M. reverchonii (Seep muhly)	3	0	3	2	2
Panicum capillare (Witchgrass)	1	0	2	3	2
P. hallii (Hall's panicum)	1	0	1	3	1
P. virgatum (Switchgrass)	B	3	0	0	A
Schizachyrium scoparium (Little bluestem)	3	3	3	3	3
Sorghastrum avenaceum (Yellow indian grass)	C	2	0	0	B
Sporobolus asper (Tall dropseed)	3	2	2	3	2
S. vaginaeflorus (Poverty dropseed)	3	3	0	0	3
Tridens buckleyanus (Endemic tridens)	1	0	D	C	C
T. muticus (Slim tridens)	3	2	2	3	3
IRIDACEAE (Iris family)					
Sisyrinchium sp. (Blue-eyed grass)	1	0	1	3	2
KRAMERIACEAE (Ratany family)					
Krameria lanceolata (Trailing ratany)	1	0	1	3	1
LABIATAE (Mint family)					
Hedeoma acinoides (Slender hedeoma)	2	1	0	0	1
H. drummondii (Limoncillo)	3	1	2	3	3
Monarda citriodora (Horsemint)	3	2	0	0	3
Salvia roemeriana (Cedar sage)	C	0	1	3	C
S. texana (Texas sage)	3	1	2	3	2
Scutellaria drummondii (Drummond skullcap)	2	0	2	3	2
LEGUMINOSAE (Legume family)					
Cercis canadensis (Redbud)	C	0	1	2	1
Desmanthes sp. (Bundleflower)	1	0	1	3	1
Eysenhardtia texana (Texas kidneywood)	C	3	1	3	2
Indigofera miniata (Scarlet pea)	1	0	1	2	1
Lupinus texensis (Texas bluebonnet)	B	3	1	3	A
Mimosa borealis (Pink mimosa)	0	0	1	3	1
Senna lindheimeriana (Lindheimer senna)	B	3	0	0	2
S. roemeriana (Two-leaf senna)	1	2	D	C	1

TABLE 1, CONTINUED
ANNOTATED LIST OF NATIVE SPECIES ESTABLISHED ON THE SITE 7–12 MONTHS AFTER TRANSPLANTING AND SEEDING

	SU	SR	TU	TR	SO
LILIACEAE (Lily family)					
Allium drummondii (Drummond onion)	0	0	D	A	A
Cooperia drummondii (Rainlily)	1	0	D	2	2
Nolina texana (Sacahuista)	C	0	1	2	B
Yucca rupicola (Twisted-leaf yucca)	1	0	2	3	2
LINACEAE (Flax family)					
Linum rupestre (Flax)	1	2	1	3	1
MALVACEAE (Mallow family)					
Abutilon incanum (Indian mallow)	C	1	1	3	1
Sida abutifolia (Spreading sida)	1	0	D	C	1
OLEACEAE (Olive family)					
Forestiera pubescens (Elbow-bush)	0	0	1	3	1
Fraxinus texensis (Texas ash)	0	0	1	3	C
ONAGRACEAE (Evening primrose family)					
Calylophus drummondianus (Drummond sundrops)	2	3	2	3	2
Oenothera missouriensis (Flutter-mill)	C	0	1	2	1
ORCHIDACEAE (Orchid family)					
Spiranthes cernua var. *odorata* (Ladies' tresses)	0	0	D	C	1
POLYGALACEAE (Milkwort family)					
Polygala alba (White milkwort)	0	0	1	3	1
P. lindheimeri (Shrubby milkwort)	1	0	1	3	1
POLYPODIACEAE (True fern family)					
Adiantum capillus-veneris (Southern maidenhair)	0	0	1	3	B
RANUNCULACEAE (Crowfoot family)					
Anemone heterophylla (Tenpetal anemone)	0	0	D	A	2
RUBIACEAE (Madder family)					
Hedyotis nigricans (Bluets)	2	2	1	3	2
RUTACEAE (Citrus family)					
Thamnosma texana (Dutchman's britches)	1	0	1	3	1
SALICACEAE (Willow family)					
Salix nigra (Black willow)	0	0	1	3	C

TABLE 1, CONTINUED
ANNOTATED LIST OF NATIVE SPECIES ESTABLISHED ON THE SITE 7–12 MONTHS AFTER TRANSPLANTING AND SEEDING

	SU	SR	TU	TR	SO
SCROPHULARIACEAE (Figwort family)					
Agalinis edwardsiana (Plateau gerardia)	1	0	D	C	1
SOLANACEAE (Nightshade Family)					
Chamaesaracha coronopus (Green false-nightshade)	1	0	1	3	1
Solanum elaeagnifolium (Trompillo)	0	0	D	C	1
TYPHACEAE (Cat-tail family)					
Typha domingensis (Cat-tail)	0	0	1	3	B
VERBENACEAE (Verbena family)					
Phyla incisa (Texas frog-fruit)	C	3	1	3	C
Verbena bipinnatifida (Dakota vervain)	2	2	2	3	2
V. canescens (Grey vervain)	2	2	1	3	2
VITACEAE (Grape family)					
Cissus incisa (Cow-itch vine)	0	0	1	3	1
Vitis sp. (Grape)	0	0	1	3	1

Management Strategies for Increasing Habitat and Species Diversity in an Urban National Park

Robert P. Cook and John T. Tanacredi

Abstract: *Gateway National Recreation Area consists of 10,522 hectares located within and adjacent to New York City. Gateway lands range from relatively intact natural landforms to dredge spoil and landfill that underline most of the upland habitat. Since its establishment in 1972, the National Park Service has been implementing a policy of restoring and enhancing these altered and once degraded habitats, and the flora and fauna which utilize them. Specific activities include active and passive revegetation of impacted sites, restoration of grassland habitat utilized by regionally rare grassland-dependent bird species, creation of freshwater habitat, and a transplant program to restore populations of native amphibians and reptiles. Through these activities, the diversity of native biota will be preserved and made accessible to the millions of people in the New York Metropolitan area who visit this unit of the National Park System.*

Key words: *national parks, species diversity, habitat management, grasslands restoration, coastal ecosystems.*

Introduction

Gateway National Recreation Area consists of 10,522 ha distributed among four counties in the states of New York and New Jersey. Located within and adjacent to New York City, Gateway lands range from relatively intact natural landforms to the dredge spoil and landfills that underlie most of the upland habitats. Typical upland habitats include dunes dominated by beachgrass (*Ammophila breviligulata*), mixed grasslands, thickets of bayberry (*Myrica pennsylvanica*), early successional woodlands of grey birch (*Betula populifolia*), black cherry (*Prunus serotina*), poplars (*Populus* spp.) and American holly (*Ilex opaca*) forest. On the more disturbed sites, generally associated with landfill, monocultures of common reed (*Phragmites australis*) and mugwort (*Artemisia vulgaris*) predominate.

With the establishment of Gateway in 1972, the National Park Service adopted a policy of "restoring" and enhancing the altered and heavily impacted habitats it had acquired. To this end a number of strategies are being pursued, depending on the specifics of the site. Guiding these strategies have been two developments: the establishment of a greenhouse and nursery at Floyd Bennett Field, Gateway's headquarters, enabling the propagation of native plant species not available commercially, and a habitat maintenance plan which defines the habitat management goal for a particular site, and in turn guides its restoration. At some sites, restoration will proceed naturally once disturbance factors are eliminated. At other sites, barriers to exclude vehicles are erected, and native grass, shrubs, and trees are planted. To date, major plantings have included the planting of 300,000 culms of American beachgrass to revegetate a beach nourishment project; the planting of 10,000 *Spartina alterniflora* to restore a salt marsh and control erosion; and the planting of several hundred trees and shrubs in the past five years. Over 30 typical coastal species have been planted. Other species, such as butterflyweed (*Asclepias tuberosa*) and butterflybush (*Buddleia davidii*) have been planted to improve habitat for lepidopterans.

Coastal Ecosystem Restoration

The dynamic geomorphological processes that shape the coastline are testimony to nature's landscaping abilities. Erosion and accretion of our coastal barriers has both plagued and nurtured man's desire to settle closer and closer to our rising seas. At the Sandy Hook Unit of the park, a natural sand spit exists where the shoreline dynamics have separated the tip area (a "spit" in geological terms) from the mainland time after time over the past several hundred years.

Considerable effort on the part of many federal and state agencies has gone into evaluating alternatives for solving the potential breaching of the "critical zone," a narrow strip where erosion and overwash threatened to separate the spit from the mainland. A single access route to the tip is the only access for over two million beach users annually and is the only way the U.S. Coast Guard, the National Oceanographic and Atmospheric Administration, the New Jersey Marine Science Consortium, and the American Littoral Society have continued to operate their facilities. A special Congressional appropriation of $12 million, to be reimbursed to the Federal Government by charging beach user fees, was allotted to help restore the original 1954 Sandy Hook shoreline profile.

The plan called for some 3 million m^3 of sand material dredged from

Sandy Hook and Ambrose Channels to be hydraulically pumped into the critical zone, beginning in March 1983. In addition to the sand material, an attempt at dune construction and stabilization would be made along a 2.5 km strip of beach once the sand had been placed, establishing the original shoreline. Three hundred thousand culms of dune grass, *Ammophila breviligulata,* were purchased, planted, and fertilized over a two-month period beginning in November 1983. Aerial photographs of the entire planted area were made and were coupled with an on-site ground evaluation of the survivorship of the culms. The choice of plant species depends upon the site's location along the coastline. Only a handful of species are tolerant of the stresses associated with the beach environment. The plants must be able to survive sand blasting, burial by sand, salt spray, saltwater flooding, drought, heat, and low nutrient supply. A perennial grass such as *Ammophila breviligulata* is most appropriate at Sandy Hook. This species is a vigorous cool-season dune grass which grows in dense clumps spreading laterally by rhizomes; it is easy to propagate, harvest, and store, and is readily available from commercial nurseries. The grass is extremely easy to transplant; it establishes and grows rapidly, and begins trapping sand by the middle of the first growing season. The rhizomes may spread as much as 3–4 m/yr while accumulating as much as 1.2 m of sand in one growing season.

The workforce for this project consisted of the Sandy Hook maintenance staff equipped with a tractor, an agricultural transplanter, and 50 volunteers. The volunteers included representatives of the National Park Service, National Oceanographic and Atmospheric Administration, United States Coast Guard, local Sierra, Audubon, and Garden Clubs, Boy Scouts, camping clubs, and others. The planting went quickly to total approximately 150 rows, each with nearly 1,200 plants. To encourage rapid establishment of the beach grass, a complete fertilization was applied shortly after planting in the fall of 1983 and then re-applied in the spring of 1984.

As the new dune system takes shape, powered by natural processes, other plant species will be introduced both naturally and with the help of the park personnel. To do this, Gateway has begun a nursery system with the prime objective of growing coastal plant species that often are unavailable from commercial sources. Along with the nursery sites at Sandy Hook and Floyd Bennett Field in Brooklyn, a 7.3 x 12.2 m, free-standing greenhouse will provide transplants of such plants as dusty miller (*Artemisia stellaria*), seaside goldenrod (*Solidago sempervirens*), northern bayberry (*Myrica pennsylvanica*), lance-leaved coreopsis (*Coreopsis lanceolanta*), prickly pear (*Opuntia compressa*), beach plum (*Prunus maritima*), and eastern red cedar (*Juniperus virginiana*).

The dune planting exercise at Sandy Hook is by no means intended to stop erosional processes there. We do anticipate, however, that the grass will aid in stabilizing a new dune system; reduce the rate of erosion at the critical zone as the dunes become stabilized; increase coastal habitat diversity, thereby providing for additional passive recreational opportunities; and continue to maintain the connection that provides access to Sandy Hook.

Grasslands Management

While the majority of habitat restoration efforts have focused on plantings, a number of other projects are also underway. At Floyd Bennett Field, 56.6 ha of the site's 579 ha have been designated as managed grasslands. The Park Service, working cooperatively with the New York City Audubon Society and the Seatuck Research Program of Cornell Laboratory of Ornithology, has begun restoring this grassland habitat. The site originally was a series of salt marsh islands, filled in by dredge spoil and rubble to create New York City's first municipal airport and later a naval air station (Blakemore 1979; Black 1981). Through plantings and natural colonization, communities of mixed grassland, giant reed (*Phragmites australis*), shrub thickets of bayberry (*Myrica pennsylvanica*) and winged sumac (*Rhus copallinum*), and scattered stands of black cherry (*Prunus serotina*), grey birch (*Betula populifolia*), and cottonwood (*Populus deltoides*) have developed (Lent et al. 1985).

From Gateway's beginnings in 1972, Floyd Bennett Field had become known for supporting populations of breeding and wintering grassland birds. Since grassland habitat in the northeastern United States has declined dramatically, as evidenced locally by the loss of the Hempstead Plains (Stalter 1981), many of these grassland birds are listed as endangered, threatened, or of special concern. The possibility of losing these regionally unique grassland birds at Floyd Bennett Field due to successional changes led to research into bird-habitat relationships. As a result of this research, a "grassland indicator species," the Grasshopper Sparrow (*Ammodramus savannarum*), was identified, and prime habitat was delineated (Lent et al. 1985). The result is a habitat-management scheme for Floyd Bennett Field that provides for a number of successional-stage habitats, thereby maintaining a greater overall species diversity on site, and also contributing to greater species diversity regionally. Grassland restoration began in 1985, with Audubon Society volunteers and Park Service staff removing trees and shrubs from the designated grasslands. Pending studies on other techniques, such as fire, these grasslands will be maintained through biennial mowing. Monitoring of bird usage indicates a slight increase in grasshopper sparrow abundance, as well as a return to restored areas that had been recently abandoned. Other New York State-listed grassland species that utilize these grasslands for feeding, winter roosts, and migratory stopovers include northern harrier (*Circus cyanus*), common barn owl (*Tyto alba*), short-eared owl (*Asio flammeus*), and upland sandpiper (*Bartramia longicauda*).

Freshwater Wetlands Creation

Through the deposition of dredge spoil and other landfilling processes which occurred in the Gateway area, existing wetlands were destroyed, and few were created. Recognizing that wetland habitats are part of the original landscape and are an extremely important component of habitat diversity, and thus species diversity, freshwater wetlands are being created. The first ponds were created in 1953, when the City of New York managed the Jamaica Bay Wildlife Refuge. Two ponds, 18.2 and 40.5 ha, were created by diking off sections of salt marsh with dredge spoil. These two ponds, though much fresher than adjacent Jamaica Bay, are still slightly brackish, and do not support true freshwater systems.

More recently, at the Jamaica Bay Wildlife Refuge, a 0.16 ha pond and several small ones were created by bulldozer and with hand tools. Upon excavation, these ponds are planted with emergent and submergent aquatic vegetation, obtained locally and through commercial sources. This not only accelerates the development of vegetation in the pond, but also provides an infusion of invertebrates into the newly created system. At Floyd Bennett Field, a joint project of the National Park Service and the New York State Department of Environmental Conservation, funded through a "Return-A-Gift" grant for wildlife enhancement, will create a 0.8 ha pond in similar fashion.

Amphibian and Reptile Restoration

In an urban area such as Gateway, habitat restoration and management can only go so far in increasing animal species diversity. Urbanization creates many dispersal barriers to wildlife (Campbell 1974) and while avian and insect diversity at Gateway are high, the non-volant taxa, such as amphibians, reptiles, and mammals, are impoverished relative to the area's habitat diversity and original fauna (Cook and Pinnock 1987). In order both to restore at Gateway an animal community more representative of the original fauna of the New York City region and to help preserve the populations of locally declining species, a program of amphibian and reptile transplantation was begun. Most of the releases have been at the Jamaica Bay Wildlife Refuge, using local animals often collected from sites scheduled for development. Since 1980, individuals of 13 species, collected as eggs, larvae, neonates, or adults, have been released. Results to date indicate that two species are established, and nearly all are surviving and beginning to reproduce (Table 1). Based on the experiences of this pilot program, amphibian and reptile population restoration has begun Gateway-wide.

Conclusion

As a result of these habitat restoration activities, the National Park Service is enhancing the altered and once-degraded habitat at Gateway National Rec-

TABLE 1
POPULATION STATUS OF AMPHIBIANS AND REPTILES RELEASED ON RULER'S BAR, HASSOCK ISLAND, JAMAICA BAY WILDLIFE REFUGE, NEW YORK, NEW YORK 1980–1987

Species Released	Year	No. of Individuals	Overwinter Survival	Breeding Records	Estab-lished
Spring peeper	80-83	58 adult 3600 larvae	yes	Innumerable	yes
Grey treefrog	1987	1000 larvae	a	a	a
Green frog	85–87	130 adult 212 larvae	yes	a	a
Spotted salamander	1987	14,000 embryos	a	a	a
Redback salamander	83–86	361 juvenile 1443 adult	yes	12 offspring recorded	a
Northern brown snake	80–84	23 juvenile 49 adult	yes	42 offspring recorded 83–84	yes
Smooth green snake	81–86	17 juvenile 64 adult	yes	10 offspring recorded	a
Eastern hognose snake	84–85	21 hatchling 4 adult	yes	a	a
Eastern milk snake	84–87	19 juvenile 13 adult	yes	1 offspring recorded	a
Black racer	85–87	6 juvenile 18 adult	yes	25 offspring recorded	a
Snapping turtle	83–87	320 hatchling 12 juvenile 38 adult	yes	3 offspring recorded	a
Eastern painted turtle	82–87	28 juvenile 361 adult	yes	8 offspring recorded	a
Eastern box turtle	80–86	12 juvenile 183 adult	yes	6 offspring recorded	a

a = insufficient elapsed time or data to determine

reation Area (Tanacredi 1987). By restoring habitats and maintaining habitat diversity, greater wildlife utilization is foreseen. In the heavily urbanized New York City area, the existence of such wildlife habitat is important not only for the wildlife it supports, but also for the millions of people who have an opportunity to recreate in this part of the National Park system.

References

Black, F. R. 1981. Jamaica Bay: A history of cultural resource management. Study 3, Washington, DC: U.S. Department of the Interior National Park Service.

Blakemore, P. R. 1979. Historic structure report–historic data. Floyd Bennett Field, Gateway National Recreation Area. Denver, CO: U.S. Department of the Interior, National Park Service. Denver Service Center, Branch of Historic Preservation. Package 109.

Campbell, C. A. 1974. Survival of reptiles and amphibians in urban environments. In *Wildlife in an Urbanizing Environment*, ed. J. H. Noyes and D. R. Progulske. Amherst, MA: University of Massachusetts, Cooperative Extension Service.

Cook, R. P., and C. A. Pinnock. 1987. Recreating a herpetofaunal community at Gateway National Recreation Area. In *Integrating man and nature in the metropolitan environment.* Chevy Chase, MD: National Symposium on Urban Wildlife, National Institute of Urban Wildlife, 4–7 November 1986.

Lent, R. A., T. S. Litwin, and N. Giffen. 1985. Bird-habitat relationships as a guide to ecologically-based management at Floyd Bennett Field, Gateway National Recreation Area. Islip, NY: Seatuck Research Program, Cornell Laboratory of Ornithology.

Stalter, R. 1981. Some ecological observations of Hempstead Plains, Long Island, New York. In *Proceedings of Northeast Weed Science Society.*

Tanacredi, J. T. 1987. Natural resource management policy constraints and trade-offs in an urban national recreational area. In *Integrating man and nature in the metropolitan environment.* Chevy Chase, MD: National Institute of Urban Wildlife, 4–7 November 1986.

Robert P. Cook is a Natural Resource Management Specialist, and John T. Tanacredi is Chief of the Division of Natural Resource Management and Compliance, National Park Service, Gateway National Recreation Area, Brooklyn, NY 11234.

Fact or Fantasy: From Landfill to National Park

(A Summary of Investigations Regarding the Natural Resources of Jamaica Bay and Their Relationships to the Future Management of an Abandoned Hazardous Waste Landfill Site)

John T. Tanacredi

ABSTRACT: *Since 1972, the National Park Service (NPS) has been charged with the responsibility to conserve Jamaica Bay, primarily as a wildlife refuge. Jamaica Bay represents some of the last remnant portions of the Hudson-Raritan estuarine ecosystem. As such, it is an important habitat type and laboratory for the restoration of abused natural resources. In the legislation that established Gateway National Recreation Area and in the policies of the NPS, our goal is to restore and maintain the general environment in its natural condition.*

Gateway National Recreation Area inherited not only the Jamaica Bay Wildlife Refuge, but some highly stressed adjacent land areas, such as the Pennsylvania/Fountain Avenue Landfills (PAL/FAL). In addition, pollution from the atmosphere, combined sewer overflows, John F. Kennedy Airport (JFK), wastewater treatment facilities, etc., have all contributed to the pollutant load of the bay and estuary. Nevertheless, the bay has endured surprisingly well, and preliminary inventories of both vertebrate and invertebrate species—their numbers, abundance and population stability—reveal a truly diverse ecosystem.

The most challenging long-term threat to the bay system are impacts from two large landfills acquired by NPS from New York City. The last deeded right for use of the Fountain Avenue Landfill (at 5,442–7,256 tonnes of solid refuse per day) expired in December, 1985. The city fought hard to retain its use of the Fountain Avenue landfill, but the NPS closed both landfills and discontinued garbage disposal, citing "bird strikes" at JFK, environmental impacts from xenobiotic leachates on the bay, and operating violations by the city as overriding concerns.

A major concern now is that the leachate associated with these old landfills contains toxic wastes. Testimony before Congress by illegal ("midnight") toxic waste haulers indicated that over 30 million gallons of such wastes were illegally dumped at these two landfill sites. PCB's have been detected leaching from both landfill sites. The NPS is conducting research efforts to define the boundary of these pollutants within Jamaica Bay and, more importantly, their potential long-term ecological health risks. This is not an easy task since it is expensive, complex, and requires a network of specialists in the realm of hazardous materials impact analysis. Currently a 30-year time period is projected before the landfills can be "used" in some as yet undefined manner. However, Jamaica Bay continues to perplex us with both its problems and diversity of wildlife.

KEY WORDS: *National Parks, landfills, estuary, hazardous wastes.*

John T. Tanacredi is Chief of the Division of Natural Resource Management and Compliance, National Park Service, Gateway National Recreation Area, Brooklyn, NY 11234.

WASTEWATER DISPOSAL IN A FOREST EVAPOTRANSPIRATION SYSTEM

BEVERLY B. JAMES

ABSTRACT: *The operation since 1978 of a prototype marsh forest evapotranspiration (ETA) system at the Mt. View Sanitary District has demonstrated that one can take advantage of the high water demand of selected trees to maximize the disposal rate of treated wastewater through irrigation of trees without adverse effects on the soil. Since evergreen trees such as redwood (*Sequoia spp.*) and Douglas-fir (*Pseudotsuga menziesii*), which are adapted to the mild wet winters of the Pacific Coast, continue to take up water in the winter months, a forest ETA system can function as a year-round treatment and disposal system in many mild climates.*

The results of the study have particular relevance to California communities faced with discharge prohibitions and a shrinking supply of agricultural land to accommodate land disposal. A forest ETA system has several advantages for wastewater disposal:

- *Less land is required per gallon of water.*
- *Wastewater treatment and disposal can continue in the fall and winter months when other irrigation disposal systems are shut down.*
- *The irrigation water is underground, reducing the chance of public contact.*
- *Maintenance requirements are minimal.*
- *The trees provide wildlife habitat and other environmental benefits.*
- *In the long term, the trees can be harvested for offsetting revenue.*
- *Seasonal discharges can be circulated through the forest for polishing before discharge.*

KEY WORDS: *wastewater treatment, stormwater treatment, evapotranspiration, urban forest restoration, marsh restoration, water reuse.*

Beverly B. James is a principal in James Engineering, Inc., El Cerrito, CA 94530.

Solid, Toxic, and Radioactive Waste Management

Composting and Recycling Organic Wastes: The Environmental Imperative

Barton Blum

Abstract: *Composting and recycling the organic waste stream have significant potential in reducing the amount of landfilled wastes and restoring nutrients to the soil. At the municipal and commercial level, these processes require a high level of procedural input and expertise for practical gain to be realized, and the information needed must be derived from several different sources.*

There are three interrelated levels of information which affect organic waste reuse: (1) the analysis of the waste stream and knowledge of potential sources for materials; (2) field and laboratory testing of materials; and (3) knowledge of potential markets for component wastes. The judicious use of this information results in a higher quality product, a better match of product and material, and a lower resource cost.

Waste-stream analysis involves a broad assessment of wastes generated by commercial and industrial sources. Food processing, pharmaceutical, and other industries produce a wealth of clean, nontoxic wastes which have potential uses as amendments, provided they are thoroughly tested.

Field and laboratory testing is necessary to determine physical and chemical characteristics of the components and appropriate rates of incorporation into the soil. The size of particles, fertilizer value, total nutrient content, carbon/nitrogen ratio, temperature, and aeration are some of the important factors that need to be considered.

Well-developed markets are vital to the success of a municipal organic waste recycling program. Finding the appropriate applications for materials is a multifaceted process that necessitates knowledge of the needs of different clientele, including landscapers, farmers, gardeners, nursery stock growers, public works and reclamation project managers. Developing a marketing strategy entails inventorying existing and potential markets, based on knowledge of the product, product use, constraints on use, and the value of the product to the user.

The problems of solid-waste management are critical, and solutions will be derived from more process-oriented thinking that links economic, sci-

entific, and social knowledge in a communicative model. The recycling of organic wastes can be effectively achieved by utilizing information in the most efficient and creative manner possible.

KEY WORDS: *composting, recycling, solid waste, organic waste, marketing, soil science.*

Barton Blum is General Manager and Director of Recycling, American Soil Products, 2222 Third St., Berkeley, CA 94710.

The United States of America's Future in Toxic Waste Management: Lessons from Europe

BRUCE W. PIASECKI

ABSTRACT: *The search for superior toxic waste management systems in the United States is reminiscent of the nation's response to the energy crisis in the 1970s. We were forced then to acknowledge that reducing our dependence on imported oil was only half a solution. The more difficult part of the solution involved establishing entirely new energy conserving technologies and adjusting a massive infrastructure of existing energy using technologies to make it more energy conserving.*

Similarly, today, Americans must restructure our response to the toxics challenge. The major remedy to the toxics challenge traditionally has been a legal response: but it can only go so far. In addition to enforcing and improving our end-of-the-pipe environmental regulations, we must mobilize a probusiness response which both reduces and treats the nation's toxic waste load at its source. This takes a fully structured industrial response: blending the Resource, Conservation and Recovery Amendments with building an industrial infrastructure that renders alternatives to land disposal profitable and reliable. The goal is to reform America's industrial mechanism, not just the law. The vehicle we employ to extinguish our toxic waste liabilities is like a decrepit automobile. Replacing bald tires helps the car run smoother but it does not revitalize American transport. What is needed is a new, revitalized business outlook, a new manufacturing "design triangle" to propel the great change. It is not only our legal system that must undergo repair, but the very design of our industrial processes, which emit large volumes of liabilities.

Herein rests the promise of the new design triangle. In the face of mounting environmental liabilities, many industrial firms now find that when they reassess waste as indicative of inefficiencies in their processes, they need a new approach based on pollution prevention. The yield from this is twofold: the economics of production can be improved at the same time the environment is protected. Whether it is called pollution prevention, waste re-

duction, waste minimization, low- and non-waste technology, or clean technology, the essence of the concept is the same. Pollution, or waste, is eliminated or reduced at the source within industrial processes rather than "controlled" at the end of the pipe or stack. The terms "low and non-waste technology" and "clean technology" are the most widely used in Europe and are broader than "waste minimization" or "waste reduction" as presently used in U.S. programs.

The force of these broad definitions rests in their advocacy of a new design triangle, whereby the desires to conserve energy, save materials, and reduce waste join in a shared dynamic. "Clean technology" is one based on constantly made decisions about what products a nation produces and how they are produced.

KEY WORDS: *toxics, waste reduction, environmental management, industrial design.*

Bruce W. Piasecki is Associate Director of the Hazardous Waste and Toxic Substances Research and Management Center at Clarkson University, Potsdam, NY 13676, where he is also a professor in the Center for Liberal Studies.

Environmental Assessment, Site Selection and Characterization for the California Low Level Radioactive Waste Disposal Facility

Mark F. Winsor

Abstract: *The three-year process of selecting a site for California's required low-level radioactive waste (LLRW) disposal facility was unique because it gave equal consideration in decision-making to technical suitability, environmental issues, and public acceptance. Ward Valley is the proposed site for the facility because of its superior geologic, surface runoff, and deep groundwater conditions; its remote location in an unpopulated area 40 km west of Needles, California; and general acceptance by the local populace.*

*The State's approval of the site may be determined largely on the basis of the potential impact on the desert tortoise (*Gopherus agassizii*), the only special-status species at Ward Valley. Between 5 and 50 individuals may be affected by facility development. Following selection of the Ward Valley site, the tortoise was classified as a candidate for California's endangered species list. That action by the Department of Fish and Game was precipitated by a petition of the Desert Tortoise Council (DTC), a conservation group seeking to prevent further impact on the species. The tortoise has experienced major losses of population and habitat throughout the Mojave Desert.*

The determination of the significance of impact of the LLRW facility on the tortoise centers on two issues: (1) minimizing disturbance to tortoises and their habitat, and (2) effective habitat restoration at the time of facility closure in about 30 years. DTC asserts that the facility contributes to significant cumulative impacts on the tortoise and that habitat restoration should not be viewed as a negotiable part of a licensing agreement because preservation offers higher potential for protecting tortoises. Some researchers agree that restoration of fragile, native, desert vegetation and highly sensitive animal species like the tortoise may not be possible.

In contrast, the project sponsor, US Ecology, Inc. (USE), asserts that impacts can be minimized and has carried out a careful program of impact avoidance to date. USE points out that long-term impacts will be minimized because all but 36 ha of the 1.67 km² site will be undisturbed; access to off-road recreation vehicles (a major source of tortoise kills and habitat destruction) will be prohibited; and the facility will likely repel other development in the vicinity. Habitat restoration is assumed a priori *to be an integral part of facility development and a license requirement. USE appears committed to making restoration work.*

In sum, the fundamental issue dividing DTC and USE is the viability of restoration compared to preservation. DTC appears to have little faith in the ability of restoration techniques to reestablish destroyed tortoise habitat and, therefore, views preservation of the site (by rejection of the facility) as the most effective approach. In contrast, USE is optimistic that habitat restoration can be successful and, therefore, believes that the State should not turn back a lengthy and costly model-site-selection program on grounds of assumed infeasibility of restoration.

KEY WORDS: *radioactive waste, desert tortoise, Mojave Desert.*

Mark F. Winsor is supervisor of the Earth and Water Sciences Division of Environmental Science Associates, Inc., 760 Harrison St., Oakland, CA 94107, a private environmental consulting firm.

THE DECOMMISSIONING OF NUCLEAR POWER PLANTS

MICHAEL J. MANETAS

ABSTRACT: *Each of the 400 nuclear reactors currently operating worldwide will reach the end of their useful production life within the next thirty years, and will have to be decommissioned. This decommissioning process involves the cleanup and removal of all radioactive materials, and restoration of the site. Several major problems which may prevent the achievement of this goal are discussed along with decommissioning techniques and costs.*

Current decommissioning plans call for postponement of dismantling nuclear facilities for several decades. Lack of federal policy and technology to accept highly radioactive materials, and the tardiness of states to construct low-level dumps contribute to this delay, which may result in hundreds of "permanent" nuclear repositories scattered throughout the landscape.

Decommissioning costs have been greatly underestimated, and when dismantling begins, complete restoration may not be achievable. Also, this process may drain dollars from other needed projects.

The high-level repository, as well as the low-level dumps, will create serious new concerns and impacts on the environment.

The Shippingport Reactor, which is currently being decommissioned, serves as an example of the above concerns.

KEY WORDS: *decommissioning, radioactivity, high-level waste, low-level waste, repository, dump, nuclear power plant.*

Michael J. Manetas is a Lecturer in the Department of Environmental Resources Engineering, Humboldt State University, Arcata, CA 95521.

PART II

Restoration of Aquatic Systems

Estuaries, Rivers, and Lakes

Biscayne Bay: A Decade of Restoration Progress

Anitra Thorhaug, Eugene Man, and Harvey Ruvin

Abstract: *A major multijurisdictional restoration program for Biscayne Bay, Florida is described.*

Key words: *estuarine restoration, Biscayne Bay.*

Introduction

A rapidly growing population of about 2.5 million persons—with tens of millions of tourists visiting each year—occupies Dade County, Florida. A major Florida estuary, Biscayne Bay, occurs throughout the center of this community.

The shallow subtropical/tropical estuary is 61 km long and 18 km wide at its broadest. It drains the Everglades and farmlands through several rivers and a flood canal system. The Atlantic Ocean has a marked tidal effect through a chain of barrier islands to the east. The bay is shallow; the shores and bottom were previously heavily vegetated, sustaining fish nurseries.

About half of the population lives within 1 km of the bay. Most tourists choose to stay around the bay. It is heavily used for commercial and sports fishing, sailing, windsurfing, hobycatting, bathing, snorkeling, scuba diving, motor boating, jet ski and parasailing, beaches, and wading. The upper half is occupied by a large number of residential water front homes, apartments and condominiums. The Port of Miami, the largest cruise-ship port in the world and a major world cargo port, is in the mid-bay. The lower bay area includes a large federal preserve: Biscayne Monument, one of the nation's first federal estuarine preserves, regulated by the Department of Interior. Eleven municipalities have various jurisdiction around the shoreline. The State of Florida has declared a portion of the north bay an aquatic preserve.

Restoration Efforts

In 1974 the University of Miami Research Council (Dr. Eugene Man, chair) formed a Biscayne Bay Committee to focus the community for the first time on the problems of the bay as a whole. Prior to this, activities had been piecemeal. The Biscayne Bay Committee (Anitra Thorhaug, chair) organized a symposium on Biscayne Bay in October 1976 to review how the Bay originally functioned, what impacts had occurred, and what was feasible in the future. The proceedings were published. Participants included citizens, environmentalists, taxpayer groups, industry, public works (local, state, federal), utilities, government decision-makers and managers (local, state, federal), scientists (natural and social), urban planners, lawyers, engineers, and media. Information on what was in and around the bay, how it originally functioned, and what had impacted it was presented. All sectors presented "where should we go from here" ideas. Heated discussions between various interests and positive recommendations occurred. The very badly abused central and north bay sections, and future action for increased citizen use were discussed.

A second workshop, based on an "American assembly" organizational model was held with key selected representatives from all the above sectors participating. (The group divides itself into four or five working groups who meet to come to consensus on a definite set of agenda items. The leaders of the groups report and a total group consensus is reached.) In three days one hundred twenty-five persons came to a consensus that the north and central portions, previously considered hopeless to enhance, could be restored given new technologies available, but that a large integrated effort would be necessary. Dade County Commissioner Harvey Ruvin made it known he would personally commit himself to such a program. The appropriate body to manage the bay was considered by the group to be Dade County. The combined efforts of science, industry, government, commerce and citizens would be necessary. Large-scale media coverage of this whole event occurred. The National Oceanographic and Atmospheric Administration Sea Grant Program sponsored this initial phase. Many individuals were helpful in completing this vision of where to proceed.

The University Biscayne Bay committee (Dr. Thorhaug, Dr. Man, and other members) proceeded immediately with state senator Robert McKnight to formulate state appropriations and proposals. Federal sources of funds and proposals for ameliorating conditions in the bay as well as private participation were sought.

Commissioner Ruvin organized a Dade County "Biscayne Bay Committee," including citizen groups, government agencies and scientists, who began to prepare county programs.

A second symposium, "Wither Biscayne Bay?" occurred in 1977 to review progress on master plans, funding, ongoing research, and implementation programs and other progress.

During the succeeding four to five years, state and federal regulatory agencies cooperated in viewing regulatory permits as a portion of the restoration effort, so that mitigation for necessarily permitted public works, such as roads, causeways, ports, sewer and water pipelines, and other facilities, contributed to restoring portions of north and central Biscayne Bay.

During this same period, federal, state, and local regulation of discharges helped eliminate many of the pollution sources to Biscayne Bay, such as sewage, industrial effluents, and municipal waste elimination, which declined. Consequently, when dredging was done, far less turbidity occurred.

The Dade County Environmental Regulation Department and Dade County Planning Department commenced a series of enhancement activities, which included:

1. Riprapping vertical seawalls using private funds through permit procedures;
2. Enhancing causeways with parking and beach access wherever possible for citizen use;
3. Creating "user" areas for zoned nonconflicting bay and shoreline use, such as bathing, boating, commercial activities, and marina access;
4. Providing facilities, such as bike paths, walkways, parking, picnic tables, and toilets;
5. Making a master map of bay resources;
6. Monitoring bay water and biota for a baseline;
7. Scientific studies of water circulation patterns, bottom sediments, fisheries, and functional ecosystems;
8. Construction of artificial reefs.

A completely unexpected event that greatly enhanced the community and global image of the bay resulted from the New World Art Festival. The Bulgarian artist Christo chose Biscayne Bay for an art event, "The Surrounded Islands." This event took three years of preparation, employed hundreds for construction, and resulted in a film and exhibit on the bay and a book. All this drew the attention of global media, and millions of visitors who had not appreciated the northern bay were introduced to it by this event.

State government participated in the bay enhancement from 1978 onward by a multi-million dollar funding program directly for bay enhancement; by cooperation of the Department of Transportation in planting mangroves on causeways; by water management districts' increasing concern about run-off in canals and rivers; by dredge-and-fill permits, and storm water run-off regulations. A joint county-state effort to "clean up Miami River" was launched during Baynanza, an annual bay celebration, and cleanup by Dade County and its citizens.

A third public symposium, organized by the authors, was held in October 1987, to review a decade of progress and to allow public, scientific, and environmental groups to participate in directing and focusing the restoration effort for the next decade. There was support for continuing and intensifying ecosystem restoration, public access, cleaning and maintaining shorelines and facilities, constructing artificial reefs, and public education about the bay through the school system and not-for-profit groups. Enthusiasm of fishermen for the "hatcheries" release program of federal government and private parties was evident.

Other estuaries in Florida that used our restoration model included Tampa Bay and Indian River Lagoon.

A major symposium reviewing all Florida's efforts at estuarine restoration, as well as those on Chesapeake Bay, was held in conjunction with the Decade of Progress symposium in October 1986.

Conclusion

Conflicts still remain between various participants. About 14 million dollars of state, county, and private money has contributed to this implementation thus far. At least $10 million more is needed to complete the restoration.

Our present goal is to restore water quality and circulation, and abundant biota while preserving use for citizens and tourists. This means partially, not completely, restoring the bay to its original condition. The amount of dredging and filling for land and the use patterns of the causeways, shoreline reconstruction, artificial island residence and port would not allow complete restoration. Conflict between environmental interests and developers, especially concerning marinas, remains intense in this estuary. Use has improved in quality and quantity due to improved biota and water quality.

Anitra Thorhaug, Dade County Cultural Affairs Council, Science Committee, and Research Professor of Biological Sciences, Florida International University, 555 NE 15th St., #PH-H, Miami, FL 33132. Eugene Man, University of Miami and Dade County Cultural Affairs Council, Science Committee. Harvey Ruvin, Commissioner, Dade County.

Willow Ecophysiology: Implications for Riparian Restoration

John G. Williams and Graham Matthews

Abstract: *Significant physiological and phenological differences exist among and probably within species of the cottonwoods and willows, as illustrated with data on water potential and stomatal conductance of arroyo willow (*Salix lasiolepis*) and red willow (*S. laevigata*) along the Carmel River. These indicate that red willow has a more effective root system, explaining its greater tolerance for water table lowering. This suggests that red willow are better for replanting in restoration efforts. However, arroyo willow leafs out earlier in the year than does red willow, and probably provides an important food resource for insects that are in turn an important resource for steelhead smolts and other insectivors.*

Key words: *willow, water potential, stomatal conductivity, ecophysiology.*

Introduction

Since 1981, the Monterey Peninsula Water Management District (MPWMD) has been studying riparian vegetation along the Carmel River, in coastal central California. Diversions for municipal water supply have caused the lower Carmel River to go dry in the summer for many years, and, in recent decades, ground water pumping for the same purpose has so lowered the water table that riparian vegetation has been devastated.

As has been described elsewhere (Kondolf and Curry 1986; Groeneveld and Griepentrog 1985), the loss of vegetation initiated a major episode of erosion in the unconsolidated sandy deposits that make up the riverbanks. The MPWMD is now involved in the effort to restore the riparian corridor of the Carmel River by replanting; restoring perennial flow where that is possible; and irrigation where it is not.

The California American Water Company (Cal-Am), the local water purveyor, did not and does not agree that its pumping has affected the vegetation, so when Cal-Am applied for permits to put in new wells in a previously unaffected reach of the river, the MPWMD began studies to clarify this issue. Since 1981, this has involved thousands of individual measurements of leaf water potential, hundreds of measurements of stomatal conductance, and many measurements of environmental variables, including soil moisture profiles, atmospheric humidity, radiation, and so on.

Monitoring, Findings, and Discussion

The main results of that work will be reported elsewhere (Williams, Woodhouse and McNeish in preparation), and are available in reports to the MPWMD (Woodhouse 1983; McNeish 1986). For now, suffice it to say we have proved the obvious, although small-scale variation in soils and other complications made that surprisingly difficult. We also learned some unexpected things, one of which is described below.

When the new wells began pumping, the effect on the vegetation was dramatic. Many of the leaves of nearby cottonwood and willows soon turned yellow and then fell off. However, it quickly became clear that the red willow (*Salix laevigata*) tolerated the drop in the water table better than the arroyo willow (*S. lasiolepis*).

As part of monitoring the effects of wells on the vegetation, McNeish (1986) rated the conditions of plant canopies up and down the river from one of the wells, using eyeball estimates of the extent of leaf yellowing and defoliation on a four point scale: (1) full, green canopy; (2) some leaf yellowing to 10% defoliation; (3) 30%–50% defoliation; and (4) greater than 50% defoliation. Figure 1 shows average ratings for 61 m reaches upstream from one of the wells on September 30, 1985. Ratings for red willow averaged 2.27 and ratings for arroyo willow averaged 2.83. Ratings in a control reach at this time were 1.5 to 1.8, with both species counted together.

To explore why red willow tolerated the drop in the water table better than arroyo willow, we measured water potential and stomatal conductance of adjacent trees 161 km up the channel from the well, through the morning of August 19, 1985. (Stomatal conductance describes the ease with which carbon dioxide and water vapor can diffuse into and out of a leaf. Water potential describes the water status of the leaf: the drier the leaf, the more negative the water potential.)

The results showed that stomatal conductance was substantially higher in the red willow although the water potential was only marginally lower (Figure 2). This means that the rates of both evapotranspiration and photosynthesis will be higher in the red willow in these conditions. (The values reported here for conductance seem suspiciously high. The data came from a Delta-T porometer, with which we had some difficulty. We have more confidence in the relative values than in the absolute magnitudes.) We in-

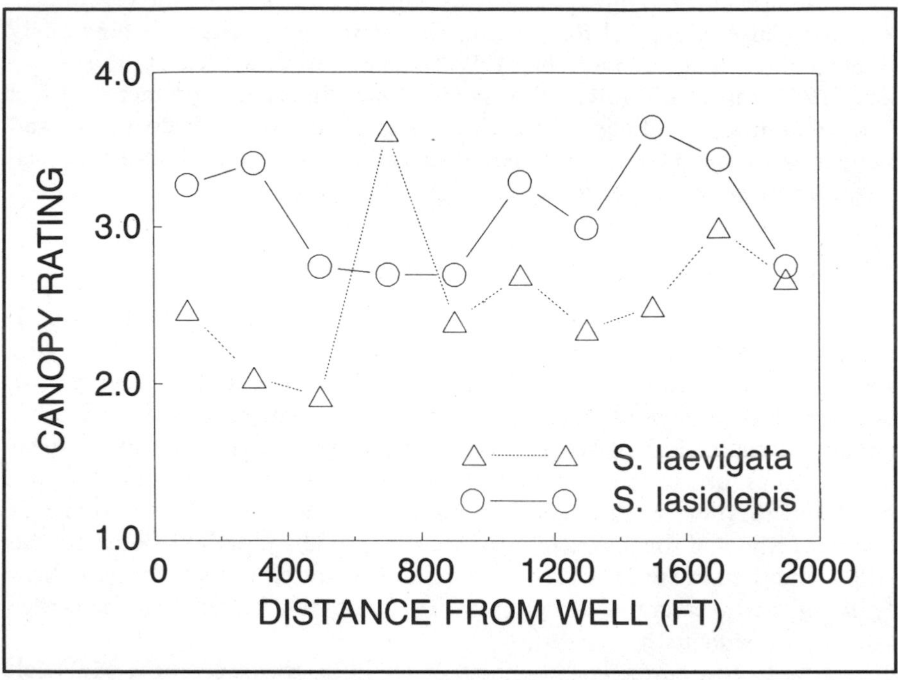

FIGURE 1. *Canopy ratings versus distance from well. Canopy ratings on September 30, 1985. Weighted averages over 200 feet reaches upstream from a productive well. The scale is: full green canopy (1); some yellowing to 10% defoliation (2); 30%–50% defoliation (3); and more than 50% defoliation (4).*

terpret the data as showing that red willow is better able to maintain a supply of water to the leaves than is arroyo willow. Red willow seems to have a more effective root system than arroyo willow; its greater tolerance of the drop in the water table does not come from restricting water loss from the leaves in the fashion of, for example, a broad-leaved chaparral plant.

There can also be a similar difference between adjacent plants of the same species. Figure 3 shows the results of measurements on three adjacent trees farther up the river on September 20, 1987, in an area that has long been affected by pumping. This was later in the year, and the defoliation of an arroyo willow was nearly complete. One of the two red willow was partially defoliated, and the other seemed relatively unaffected. This kind of within-species variation needs to be taken into account in the design of programs to monitor or evaluate restoration efforts. Whole stands need to be evaluated as well as individual plants.

Within the red willow pair, the differences in water potential were again minor compared with the differences in stomatal conductance; and once again the differences in canopy conduction reflected differences in water supply to the leaves, rather than greater control of water supply from the leaves. The

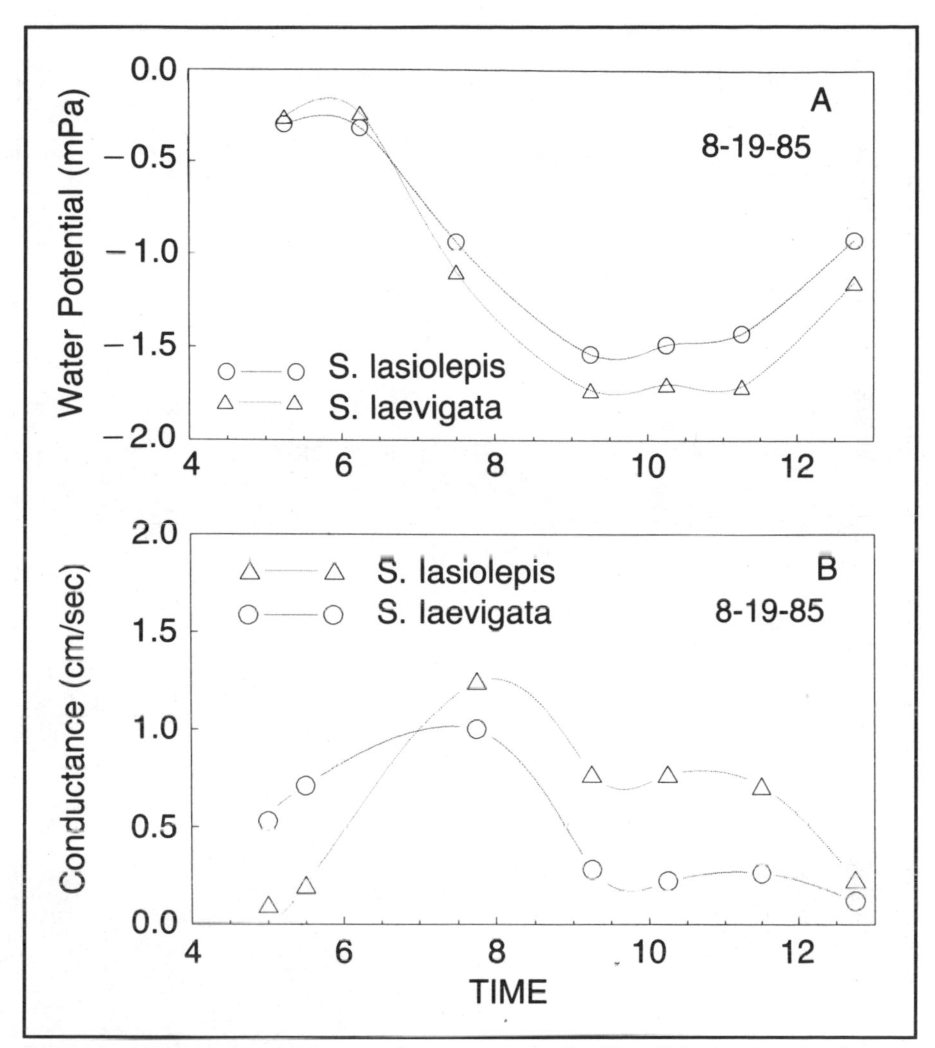

FIGURE 2. *(A) Water potential in an adjacent pair of red willow (*Salix laevigata*) and arroyo willow (*S. lasiolepis*). (B) Stomatal conductance in the same pair.*

high water potential of both trees early in the morning (dawn water potential) shows that both had access to moist soil somewhere in their root zones, but the divergence of the data later in the day showed that one had better access. In contrast, the arroyo willow did not have any roots in this moist soil, as shown by low water supply potential early in the morning, and was unable to supply water even to the few leaves remaining.

It is tempting to conclude that red willow is a better choice for the riparian restoration projects now in progress or in prospect for the Carmel River, or in other areas where drawdown of the water table is a problem. However this

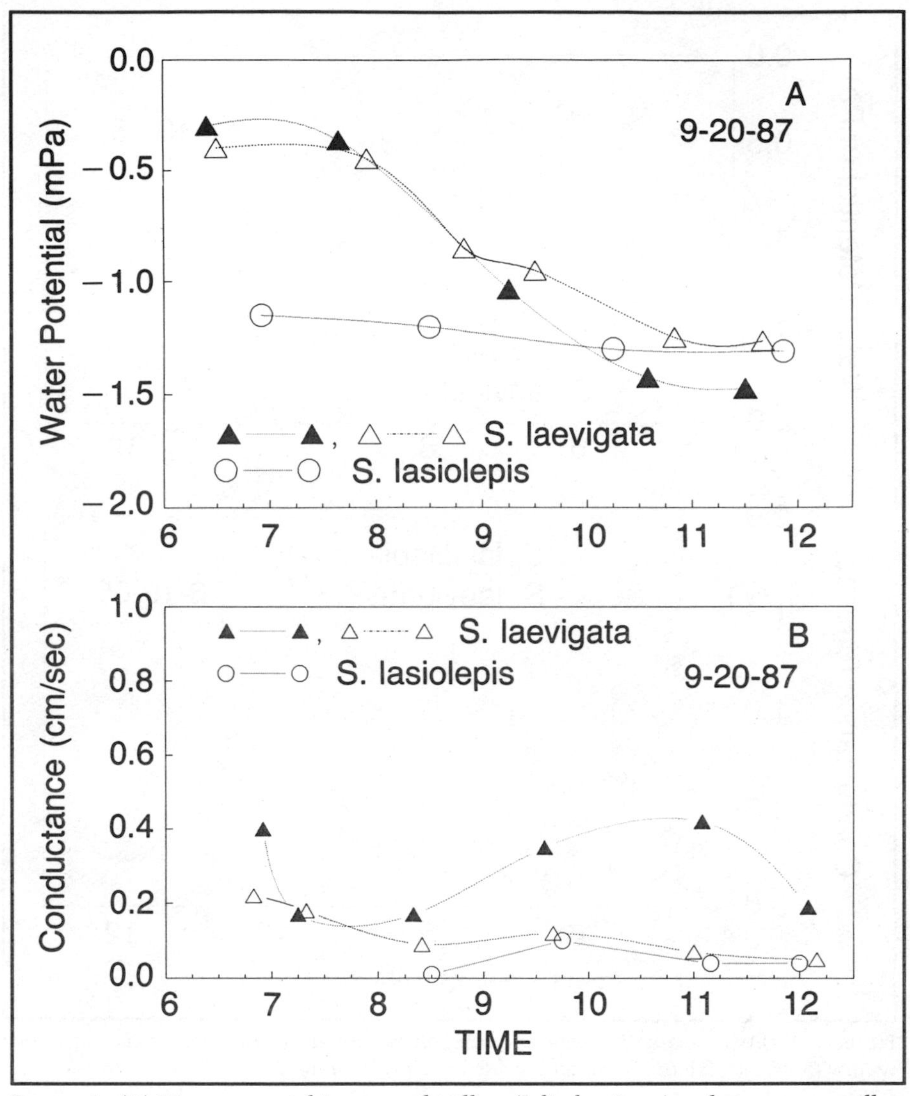

FIGURE 3. *(A) Water potential in two red willow (*Salix laevigata*) and one arroyo willow (*S. lasiolepis*). (B) Stomatal conductance in the same three.*

ignores other differences between the species, which appear to be important for their roles in the food web of the riparian system.

One easily visible difference is phenological. The leaves of arroyo willow develop earlier in the year than those of red willow. We have some data on arroyo willow: trees sampled in 1981 began to leaf out in February, and reached 60–100% of maximum leaf number by March (Woodhouse 1983). We do not have comparable data for red willow, but we know from causal information that it begins to leaf out several months later.

We do not have any data on the insect fauna of the willows along the Carmel River, but it seems safe to say that their numbers are mostly lower in the winter when the trees are mostly bare, and increase with time after the trees leaf out. Young leaves provide a particularly important type of food resource, because they have not yet synthesized the various compounds that plants have evolved to make their leaves less palatable. Replacing arroyo willow with red willow would delay the appearance of this food resource in the riparian system, and it does not seem too speculative to argue this would result in a significant decrease in insect numbers during the spring – the period of downstream migration of the steelhead smolts.

According to Fields (1984), "Terrestrial organisms, mostly insects, are an important component of food of the steelhead trout. In the lower Carmel River, terrestrials comprised 17.8% of the total number and 42% of the total volume of the food organisms consumed by smolts collected in early May, 1982. Where streamside vegetation was sparse, terrestrials comprised 2.8% by number and 11.4% by volume of the total diet. Where vegetation was abundant, however, they made up 18.7%–38.9% of the total number and 50%–59.1% of the total volume."

Because of the loss of vegetation and bank erosion along much of the Carmel River, it was possible for Fields to compare the input of terrestrial invertebrates to the stream in paired vegetated and unvegetated reaches in late spring. He did this by setting up two fine mesh nets, 100 m apart. The upstream net blocked off invertebrates drifting down from upstream, and the downstream net sampled the surface of the water to catch organisms falling into the stream between the two nets.

Based on five paired samples, the input of organisms in vegetated reaches was more than three times greater by volume, and six times greater by number. Moreover Fields thinks that these numbers underestimate the importance of riparian vegetation, because differences in wind and sunlight biased results in two of the five pairs. Accordingly, selecting willows for "drought hardiness" alone would delay the appearance of a major food resource for phytophageous insects, and, by reducing the insect populations on vegetation overhanging the stream during the downstream migration of smolts, could degrade the river as a steelhead habitat.

The steelhead (*Salmo gairdnerii*) is unusual in that its diet has been studied. There are insectivorous birds and amphibians in the riparian corridor, and it seems also likely that a decrease in insect numbers during the spring would affect them, too. It seems better to try to maintain the riparian forest, even if that makes migration or restoration somewhat more expensive.

The real point here is that although we are starting to learn about riparian systems, we still do not know very much, and, given that, we will probably do better to maintain or mimic natural riparian systems, rather than try to use the knowledge we have to design new systems to match altered environmental conditions. Many times we will have to try to design new systems because environmental conditions will be changed so much that we have no choice, and in these situations we should use all the knowledge that we have. But we should do so with a sense of humility, remembering that our knowledge is still limited, and that a little of it is well known to be a dangerous thing.

References

Fields, W. C., Jr. 1984. The invertebrate fauna of the Carmel River system and food habits of fish in the Carmel River system. In Appendix C of *Assessment of Carmel River steelhead resource: Its relationship to streamflow and to water supply alternatives,* ed. D. W. Kelley and Associates. Unpublished report to the Monterey Peninsula Water Management District.

Groeneveld, D. P., and T. E. Griepentrog. 1985. Interdependence of groundwater, riparian vegetation, and streambank stability: A case study. In *U.S. Department of Agriculture Forest Service General Technical Report RM 120,* 44–48.

Kondolf, G. M., and R. C. Curry. 1986. Channel erosion along the Carmel River, Monterey County, California. In *Earth Surface Processes and Landforms* 11:307–19.

McNeish, C. M. 1986. *Effects of production well pumping on plant water stress in the riparian corridor of the lower Carmel valley.* Unpublished report to the Monterey Peninsula Water Management District.

Williams, J. G., R. M. Woodhouse, and C. M. McNeish. The effects of groundwater pumping on riparian vegetation: Carmel Valley. To be submitted to *Environmental Management.*

Woodhouse, R. M. 1983. *Baseline analysis of the riparian vegetation in the lower Carmel Valley.* Unpublished report to the Monterey Peninsula Water Management District.

John G. Williams is senior associate, Philip Williams and Associates, Pier 33 North, The Embarcadero, San Francisco, CA 94111. Graham Matthews is hydrologist, Monterey Peninsula Water Management District, Monterey, CA.

Aquatic Weed Removal as a Nutrient Export Mechanism in Lake Okeechobee, Florida

Dean Mericas, Paul Gremillion, and Ed Terczak

ABSTRACT: *Eutrophication in Lake Okeechobee, Florida, has occurred as a result of nutrient loading from surrounding agricultural watersheds. A demonstration was undertaken to assess the efficacy of operational-scale macrophyte harvesting as a nutrient-export strategy. Harvesting operations are being conducted in a 202-ha study area using specially modified harvesting machinery. Harvesting strategies emphasized selective harvesting to maximize nutrient yield per unit effort. Preliminary estimates of daily exports are 12 kg of phosphorus and 80 kg of nitrogen per harvester. Phosphorus data from the literature suggests that rates in excess of 20 kg may be attainable.*

KEY WORDS: *macrophytes, harvesting, hydrilla, nutrients, Lake Okeechobee, eutrophication, lake restoration.*

Introduction

LAKE OKEECHOBEE, LOCATED in south-central Florida, has a surface area of over 1,800 km² and a mean depth of 3 m. The lake is the heart of south Florida's water supply and flood control system, and the second largest freshwater lake in the United States.

There has been increasing concern in recent years over the degradation of water quality in the lake. The watershed surrounding Lake Okeechobee is dominated by agricultural uses, principally beef cattle, dairy cattle, and sugar cane. Nutrient loading from nonpoint sources has been implicated in the eutrophication of the lake. Phosphorous concentrations have been observed to double in the past 12 years (Jones 1987), and hypereutrophy (the most extreme stage of eutrophication) appears to be imminent. Algae blooms have become common with an apparent shift in species from green to blue-green in recent years (Center for Aquatic Plants 1987). The blue-green algae *Ana-*

baena circinalis was observed in a massive bloom in August 1986, covering approximately 333 km^2 of the lake. This noxious species is frequently dominant in highly eutrophic lakes, but had never been previously observed in a bloom in Lake Okeechobee.

The results of modeling studies on Lake Okeechobee indicate that a reduction in nutrient loading is required to obtain a stable and desirable trophic level (Davis and Reel 1983; Kratzer and Brezonik 1984). A 40% reduction in phosphorus loading has been defined as the management target. Both external and internal strategies for accomplishing this are being examined by the South Florida Water Management District (SFWMD). External reduction strategies include implementation of best management practices (BMPs) and diversion of nutrient-rich runoff away from the lake. Internal lake nutrient-reduction techniques include sediment removal and macrophyte harvesting.

The objective of the present study was to examine the demonstrated performance of large-scale macrophyte harvesting as a mechanism for nutrient removal from Lake Okeechobee.

Methods

The Lake Okeechobee Aquatic Weed Harvesting Demonstration Project was conducted on the northern shore, immediately south of where the Kissimee River discharges into the Lake (Figure 1). The 174-ha study area was defined by eight parallel transects, each 152 m wide and ranging in length from 610 to 3,049 m. The transects were separated by 61 m wide unharvested "wind fences." The primary purpose of the wind fences was to contain sediment and loose plant material that result from harvesting operations. A secondary benefit was the enhancement of the sport fish habitat by providing structural diversity.

Each transect was subdivided into 152 m "blocks." Each block represents an area of 2.3 ha. Transect corners and 152 m intermediate points were marked with white PVC plastic poles. The poles were positioned by a land surveying crew using a Wild T-2 theodolite and a Geodimeter Electronic Distance Measuring Device (EDM). The transect boundaries were precise to within approximately 0.3 m.

The harvesting equipment used for this project consisted of two Aquamarine H10-800 mechanical weed harvesters, supported by an Aquamarine T12-14,000 "Hydrashuttle" high-speed transport, and two Aquamarine TC-800 trailer conveyors pulled by 6x6 (i. e., all-wheel drive) semi-tractors. The harvesters and transport were 3 m-wide machines representing the largest harvesting machinery generally in use in North America.

The data collection program included machinery logs maintained by each operator, load weight estimate records, and periodic sampling of the harvested material for nutrient analysis. Each harvester, transport, and truck operator maintained a daily operations log. Times were recorded at the start and ter-

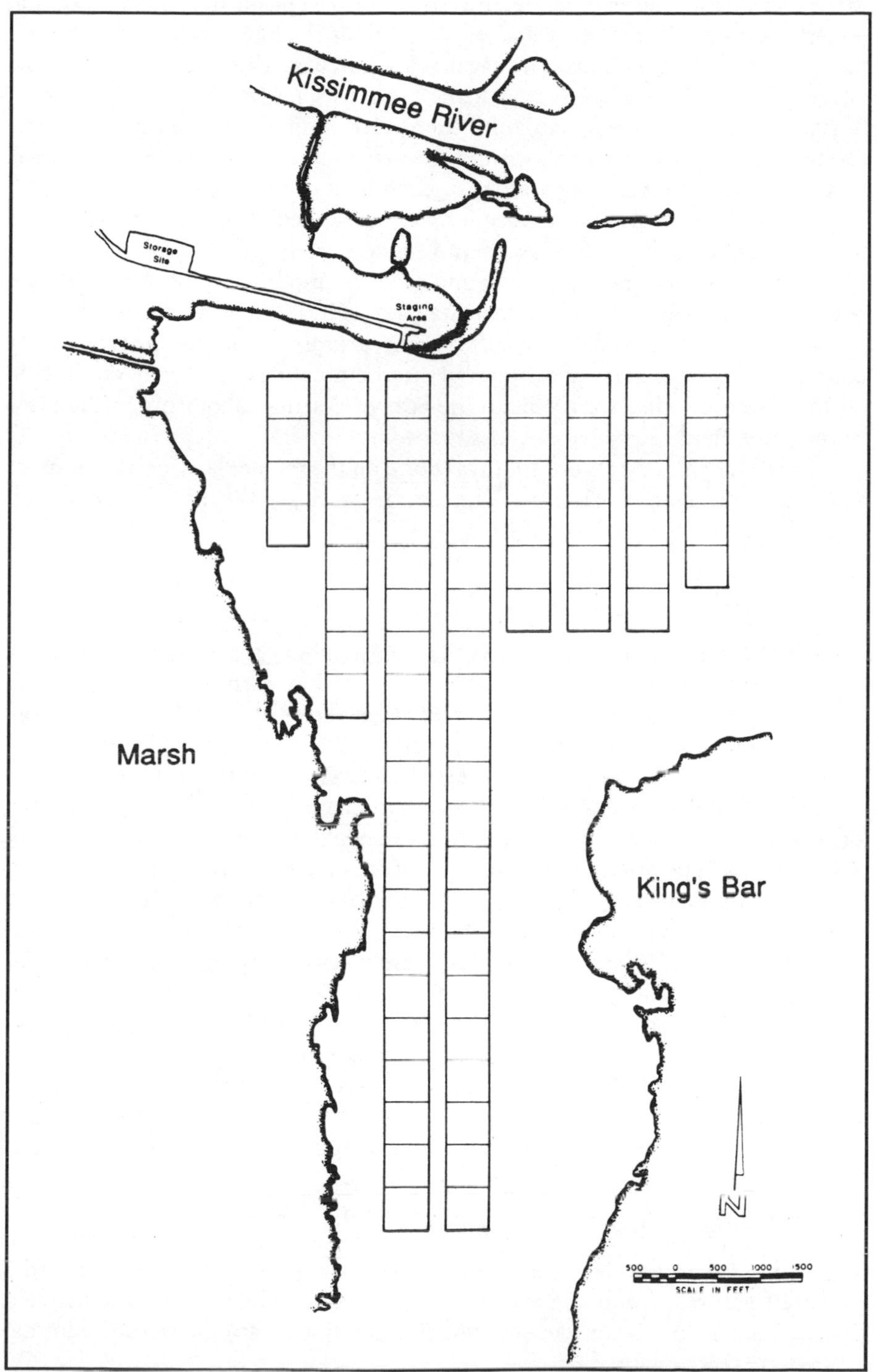

FIGURE 1. *Map of the study area on Lake Okeechobee, including harvesting transect boundaries.*

mination of each operation. Records were also kept on harvesting location (by transect and block) and weather. Weights of randomly selected harvester loads were estimated using draft gauges located at the four corners of the vessels' hulls. For each selected load, draft was recorded with the load of harvested macrophytes aboard, and then with the harvester empty. The difference in draft was used to estimate vessel displacement change due to the load of weeds, and thus the weight of each load of wet weeds.

Approximately five weed samples were collected from randomly selected harvester loads each week. These samples were analyzed for total phosphorus (P) using a wet persulfate digestion and ascorbic-modified phosphomolybdate reduction determination. Chemical analyses were accompanied by blanks and National Bureau of Standards (NBS) audit samples to allow correction for contamination and bias. Additional plant chemical analyses were conducted on three samples that were sent to the Forage Testing Laboratory of the New York Dairy Herd Improvement Center.

This paper presents work that was done on the project between September 4, 1987, and December 31, 1987. The demonstration project is still underway as of this writing in January 1988.

Results

A total of 148 ha of macrophytes were harvested between September 4, 1987, and December 17, 1987. This took 67 days of harvesting with one or two harvesters in operation, for a total of 127 harvester-days. During this period, 1,183 harvester loads of material were removed from the lake. The bulk (i.e., over 95%) of the macrophytes harvested were *Hydrilla verticillata*.

Forty-four load weight estimates were calculated from harvester draft gauge readings. The mean load weight was 3,989 kg (sd = 968) and the median was 3,664 kg. The Shapiro-Wilk test for normality indicates that the data are not normally distributed ($P<0.01$). The distribution of these data appear to be bimodal or complex (Figure 2).

It is suspected that the very high weights observed represent loads that were collected at the beginning of the project. During the first few weeks of operations, it was discovered that the harvesters were capable of safely and effectively handling a substantially greater maximum load than the trailer conveyors. As a result, harvester load sizes were reduced to avoid overloading and damaging the trailer conveyors.

The result of the plant nutrient determinations are presented in Table 1.

Discussion

The nutrient content data are consistent in magnitude with values reported in the literature. Sutton and Portier (1983) reported phosphorus (P), nitrogen (N) and potassium (K) contents of hydrilla from five locations in south Florida,

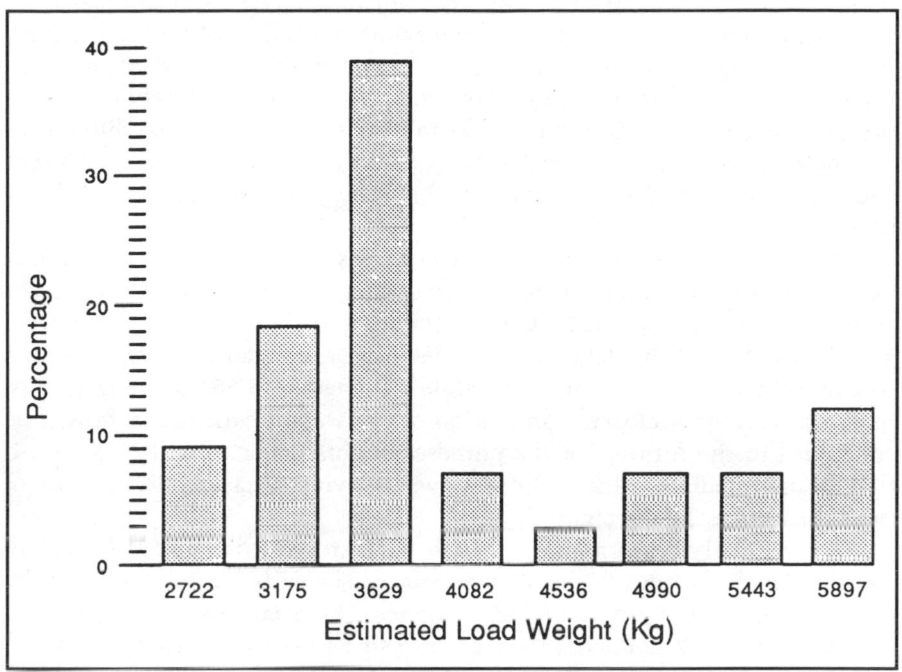

FIGURE 2. *Distribution of observed harvester load weights.*

TABLE 1
SUMMARY OF PLANT NUTRIENT ANALYSES (MG/G)

		Wet Weight		Dry Weight	
Parameter	n	Mean (mg/g)	sd[1]	Mean (mg/g)	sd[1]
Phosphorus	22	0.33	0.11	3.97	1.24
Nitrogen	3	2.29	0.24	34.4	2.61
Potassium	3	3.80	0.53	56.2	4.68
Magnesium	3	0.35	0.04	5.23	0.38
Calcium	3	1.53	0.32	22.3	3.04

[1]sd = standard deviation

including the Harney Ponds area of Lake Okeechobee, within 16 km of the demonstration project study site. Mean dry weight concentrations reported for hydrilla from Lake Okeechobee were 8.26 mg/g P, 25.1 mg/g N and 52.1 mg/g K (n = 10). Sabol (1987) reported P and N dry weight content of hydrilla from Orange Lake in Florida as 5.2 mg/g and 35.9 mg/g, respectively. The plants collected during the demonstration project appear to be lower in P than those in other studies. It is notable that Sutton and Portier's samples from Lake Okeechobee had the highest nutrient content of all the samples collected, P content being four times greater than the minimum observed site average. It appears that the data presented from the present study is likely to present the lower end of hydrilla P that occurs in Lake Okeechobee. The differences may be attributable to site-specific factors, such as sediment nutrient content and availability. Further discussion of this issue is beyond the scope of this paper.

An estimate of the weight of a typical harvester load is required to project nutrient export estimates. There is some uncertainty as to which load weight statistic is more appropriate for use in estimating total mass of weeds removed. The distribution of the data would suggest that the mean 3,989 kg will tend to overestimate true load weight, while the median 3,664 kg may lead to underestimation. A closer examination of the weight estimate data will be conducted in the future. For the purposes of this paper, it will be assumed that an intermediate value of 3,818 kg will provide a reasonable estimate of typical harvester load weight.

Average daily performance for each harvester consisted of 9.3 loads of harvested hydrilla from 1.2 ha of lake surface. The daily biomass removed by each harvester was approximately 35.5 tonnes. These daily estimates are based on harvesting data collected over a range of 213–2,130 m from the shore transfer site, and under a wide variety of environmental conditions. Individual daily performance could be much higher or lower, depending on the specific location of the working area, local plant density and random effects, such as the weather. As an example, the number of loads harvested by an individual machine ranged as high as 17 per day.

An average of 12.2 tonnes of biomass was removed per ha harvested in 1987. Plant density based on total harvester loads required to clear out individual blocks ranged from 5–112 tonnes/ha. It should be noted that these values are based on harvested material, not total standing crop. Total standing crop would be higher, since the harvesters generally cut the plants approximately 0.3 m above the bottom of the lake. Additionally, typical scientific studies of macrophyte density measure the weight of blotted plant samples, while this study considers plant density in terms of weight-as-harvested, which includes some entrained water.

Total phosphorus export rates may be estimated from the average daily biomass removed and the nutrient content data presented in Table 1. Using the observed P content of 0.33 mg/g, the calculated daily P export rate is 11.7 kg/harvester-day. Assuming that typical hydrilla moisture content is 93%, Sutton and Portier's data may be converted to a wet-weight P content of 0.58 mg/g. An estimated daily P export rate of 20.6 kg/harvester-day is obtained using this higher reported value. Thus a 75 % increase in export weight might be obtained by selecting a working area, such as Harney Pond, that supports high P content in the hydrilla.

An estimated total of 4,518 tonnes of harvested weeds were removed from the lake during 1987 harvesting operations. Based on the nutrient content data presented in Table 1, this represents 1.49 tonnes of P, 10.1 tonnes of N, 17.2 tonnes of K, 1.6 tonnes of magnesium, and 6.9 tonnes of calcium. N and K are important and potentially limiting nutrients. One of the benefits of mechanical removal of macrophytes as a P-reduction strategy is that a suite of plant nutrients are exported as harvested biomass. Again, site-specific conditions might increase the P export by as much as 75% above the result presented here.

There may be other significant benefits, beyond the export of various nutrients, associated with harvesting macrophytes. The potential effect of hydrilla on sport fishing has only recently begun to be investigated. Colle et al. (1987) examined the effects of dense hydrilla growth on the sport-fishing industry in Orange Lake, Florida, and found it to be generally deleterious. (Comparisons were made between large-scale presence or absence of hydrilla beds.) The strategy of alternating rows of harvested and unharvested macrophytes is likely to provide habitat enhancement to sport fish (Martin, personal communication). Engel (1987) found that selective channelizing of macrophyte beds increased structural diversity and expanded feeding opportunities for largemouth bass (*Micropterus salmoides*) in Halverson Lake, Wisconsin.

The productive use of the harvested biomass from a macrophyte would provide another benefit. Potential uses include composting, animal feed, and biogas (i.e. methane) production. Greening et al. (1988) present an analysis of the practical alternatives for a resource recovery program using harvested hydrilla from Lake Okeechobee.

Conclusions

This study provides nutrient export estimates that may be practically and routinely attained by large-scale harvesting of hydrilla in Lake Okeechobee. Based on the available nutrient content data from the literature, the observed P-export rate of 11.7 kg/harvester-day may be on the low side of rates attainable in this system. Rates may be as much as 75% higher in other areas of the lake. The export rates estimated for other nutrients are probably representative of what may be expected elsewhere in the lake since P-content in hydrilla appears to be more variable than N or other nutrients.

There are no data available to quantitatively examine the ancillary benefits of harvesting hydrilla in alternating strips. Qualitatively, it may be assumed that some significant enhancement of sport-fish habitat will result from the increased structural complexity. There is a distinct need for an examination of alternative harvesting strategies to optimize habitat enhancement.

The economic practicality of mechanical harvesting as a mitigative strategy for eutrophication is dependent on a variety of situation-specific factors, such as plant nutrient content, lake nutrient budget, machinery and operating costs, economic benefits provided, and available alternatives. Previous investigators (Burton et al. 1978; Conyers and Cooke 1983) have reported on the practical implications of nutrient removal as a benefit of mechanical harvesting. Future analyses of the data collected during this study will quantify the costs involved in implementing a harvesting program at this scale. The potential for cost recovery associated with reuse of the harvested plant material is also being investigated as part of this effort.

Acknowledgments

This research was conducted by International Sciences and Technology, Inc. for the South Florida Water Management District under Contract No. 621-M87-0519. The material has not been reviewed by the district, and does not necessarily reflect the views or policies of the district.

References

Burton, T. M, D. L. King, and H. L. Ervin. 1978. Aquatic plant harvesting as a lake restoration technique. In *Lake Restoration,* 177–185. Washington, DC: U.S. Environmental Protection Agency 400-15-79-001.

Center for Aquatic Plants. 1987. *Aquaphyte* 7(2):2. Gainesville, FL: University of Florida.

Colle, D. E., J. V. Shireman, W. T. Haller, J. C. Joyce, and D. E. Canfield. 1987. Influence of hydrilla on harvestable sport-fish populations, angler use, and angler expeditions on Orange Lake, Florida. *North American Journal of Fisheries Management* 7:410–417.

Conyers, D. L, and G. D. Cooke. 1983. A comparison of the costs of harvesting and herbicides and their effectiveness in nutrient removal and control of macrophyte biomass. In *Lake Restoration, Protection and Management,* 317–21. Washington, DC: U.S. Environmental Protection Agency 440/5-83-001.

Davis, F. E., and J. S. Reel. 1983. Water quality management strategy for Lake Okeechobee, Florida. In *Lake Restoration, Protection and Management,* 313–16. Washington, DC: U.S. Environmental Protection Agency 440/5-83-001.

Engel, S. 1987. The impact of submerged macrophytes on largemouth bass and bluegills. *Lake and Reservoir Management* 3:227–34. North American Lake Management Society.

Greening, H. S., C. E. Mericas, L. K. Featherstone, P. T. Gremillion, and J. Pursley. 1988. *An examination of utilization options for harvested aquatic plants from Lake Okeechobee.* Technical Report for the South Florida Water Management District.

Jones, B. 1987. Lake Okeechobee eutrophication research and management. *Aquatics* 9(2):21–26.

Kratzer, C. R., and P. L. Brezonik. 1984. Application of nutrient loading models to the analysis of trophic conditions in Lake Okeechobee, Florida. *Environmental Management* 8(2):109–20.

Martin, B. Personal communication. Sport-Fishing Institute, Washington, DC.

Sabol, B. M. 1987. Environmental effects of aquatic disposal of chopped hydrilla. *Journal of Aquatic Plant Management* 25:19–23.

Sutton, D. L., and K. H. Portier. 1983. Variation of nitrogen, phosphorus, and potassium contents of hydrilla in South Florida. *Journal of Aquatic Plant Management* 21:87–92.

Dean Mericas is Senior Engineer, International Science & Tecnology, Inc., 11260 Roger Bacon Drive, Suite 204, Reston, VA 22090. Paul Gremillion is Staff Engineer, International Science & Technology, Inc., 11260 Roger Bacon Drive, Suite 204, Reston, VA 22090. Ed Terczak is Aquatic Plant Specialist, South Florida Water Management District, 3301 Gun Club Road, West Palm Beach, FL 33416.

Fisheries and Streams

Hydrologic and Channel Stability Considerations in Stream Habitat Restoration

G. Mathias Kondolf

Abstract: *When restoration projects are planned on streams where habitat has been degraded as a result of water diversions, changes in sediment load, or other such factors, these factors must be studied, understood, and explicitly accounted for in the design of the restoration. If they are not, the forces that destroyed the original channel are likely to undo the restored channel. On the Carmel River, a restoration has been designed with the instability induced by groundwater withdrawal and bank devegetation in mind. On Rush Creek the effects of upstream water diversions on aquatic and riparian habitat must be considered in planning future restoration work.*

Key words: *stream restoration, riparian vegetation, instream flows, trout habitat, bank erosion.*

Introduction

Streams may need restoration because of direct modifications (e.g., channelization, culverting), or they may have become unstable as an indirect result of other events, such as changes in the flow and sediment load delivered to them; destruction of bank-stabilizing vegetation; or incision (downcutting) of the channel. When restoration projects are planned on streams where habitat has been degraded as an indirect result of other factors, these other factors must be studied, understood, and explicitly accounted for in the design of the restoration. If they are not, the forces that destroyed the original channel are likely to undo the restored channel.

A variety of factors can be important in determining channel stability and inducing channel change (Kondolf and Sale 1985); these must be considered in designing a stable channel for restoration. In this paper, we present two case studies that illustrate the importance of understanding the recent history of channel changes and their probable cause; the extent of riparian vegetation and its role in bank stability; and interactions between stream and groundwater.

Design of a restored channel is usually thought of mainly in terms of its physical structure (e.g., cross-sectional geometry and meander pattern). However, on regulated rivers, flow may be a design variable as well. How much water is needed to maintain the riparian vegetation along the newly restored channel? How much water is needed for the fish at their various life stages? How does the visual experience differ at different flows, and what flow is optimal for this resource?

Recent studies have attempted to predict the potential impacts of flow reductions on riparian, recreational, and aesthetic resources by rating the relative quality of these resources at different flows (e.g., FERC 1986). There is a larger body of work in which the habitat available for fish at different flows has been predicted, notably with the computer (see Loar and Sale 1981 for a review). To specify flows for the design channel involves (1) drawing on published sources and the expertise of agency personnel to estimate the quality of each resource at various proposed flows, (2) establishing the cost of the water released from the dam, and (3) negotiations among the parties involved.

Once the design flows are agreed upon, however, downstream changes in flow (resulting from stream-groundwater interactions, evapotranspiration, surface inflow or diversions, etc.) must be taken into account to insure that the minimum design flow is satisfied along the entire reach. In a case study presented here, we measured flow losses of up to 50%; in such a case, the flow release from the dam must be augmented substantially to satisfy minimum flow requirements along the entire reach.

The purpose of this paper is to draw attention to certain aspects of channel form and to the process that must be addressed if the design channel is to function properly and maintain its integrity. The channels described in our case studies both lie downstream of dams that divert water for municipal use, and both have suffered massive dieoff of riparian vegetation. In the Carmel River case, groundwater extraction is the primary cause of channel instability; in the Rush Creek case, devegetation of the channel and incision from base-level lowering are the primary causes of channel instability. Restoration projects are underway on the Carmel River and proposed for Rush Creek on the study reaches.

Case Study: Carmel River

The Carmel River rises in the rugged Santa Lucia Mountains, flows through steep canyons in its upper reaches, and then, downstream of Robles del Rio, flows through an alluvial valley for 24 km before debouching into the Pacific Ocean near Carmel, California (Figure 1). The upper basin receives the most rainfall and produces most of the basin's runoff. Over 90% of the annual

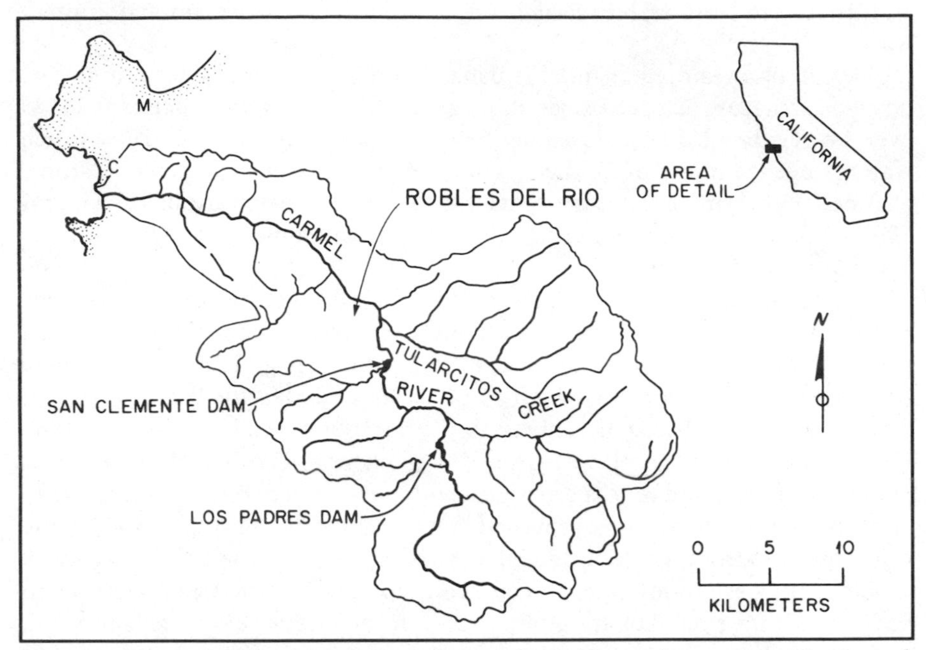

FIGURE 1. *Location and basin map, Carmel River. Capital letters denote towns: C = Carmel, M = Monterey. (From Kondolf et al. 1987, copyright Elsevier, used by permission.)*

precipitation falls from November through April and runoff is likewise seasonal (U.S. Army Corps 1974).

In the lower, alluvial valley the river bed and banks consist of sands and gravels; these sediments are permeable and store a substantial amount of water. The water below the level of the thalweg (the deepest part of the stream) is available to deep-rooted plants or can be extracted by shallow wells. Above the thalweg, the banks store water from the winter high flows for later release as baseflow. This process, termed *bank storage,* is important because it augments base flow in the river in the spring during critical times for willow establishment and seaward migration of steelhead trout smolts.

The bank storage process is illustrated in Figure 2, beginning with equilibrium low-water conditions in the fall (a). During the first high flows of the winter, a hydraulic gradient exists from the river into the banks, and water flows into the banks (b) until the water table in the banks is equal to the stage of the river (c), a condition not usually achieved in nature because of the short duration of flood peaks. As the stage falls, bank-stored water begins to drain into the river when the stage of the river falls below the water table (d). Bank storage has been studied mostly as a short-term phenomenon acting over a period of days to dampen flood hydrographs as they pass through an alluvial valley (Todd 1955; Pinder and Sauer 1971). However, as long as the river stage continues to fall, a hydraulic gradient from the banks to the river

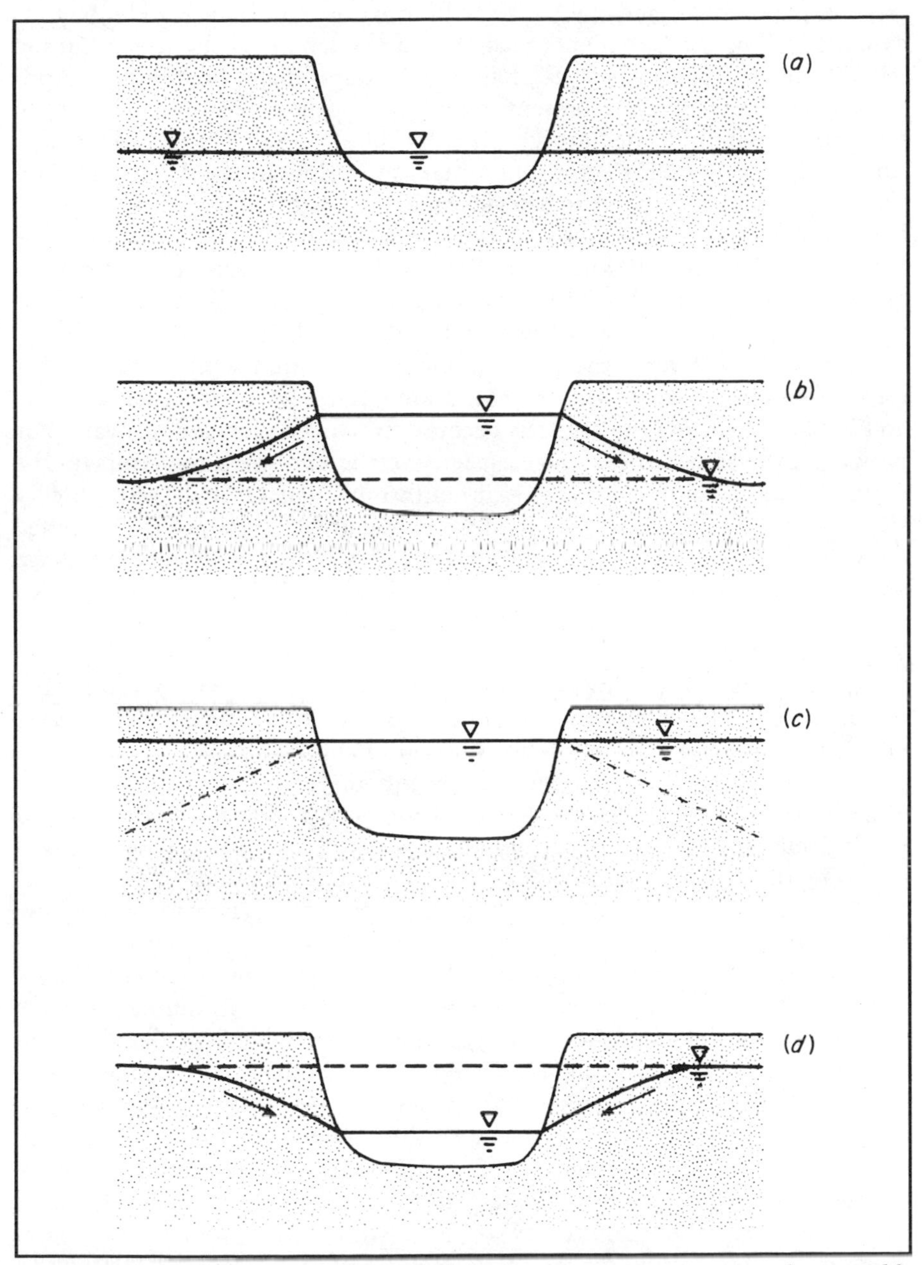

FIGURE 2. *Idealized time series showing relationships between river stage and water table in banks: (a) equilibrium at low flow; (b) recharge to banks during high flow (arrows indicate direction of groundwater flow and dashed lines show water levels from previous frame); (c) equilibrium at high flow, a condition not actually achieved during flood peaks because of their short duration; and (d) discharge from the banks to the stream following high flow. (From Kondolf et al. 1987, copyright Elsevier, used by permission.)*

will be maintained, and the banks will continue to drain. Bank storage is evident in flow measurements made along the length of the lower Carmel River from May to August 1982, following the last major high flow on April 11 (Figure 3). Each point represents a (nearly) simultaneous measurement of flow on the date indicated at right (Kondolf et al. 1987). Through the end of June, the measurements show a net increase in flow in the downstream direction, largely attributable to bank storage. In later measurements, however, bank storage has been exhausted and there is a net loss in flow near the Schulte Road station, 10 km above the mouth. Several major pumping wells are located in this reach, and they lower the water table so that streamflow is locally influent to the groundwater (Kondolf et al. 1987).

The Carmel River is the source of most of the municipal water supplies for the Monterey Peninsula cities of Monterey, Carmel, Seaside, Pacific Grove, and Pebble Beach. As the area has become urbanized, demand for water has grown. At the same time, storage capacity of the two small reservoirs on the Carmel River has been reduced by sedimentation. To make up the water supply shortfall, the water company (a privately owned utility) has increasingly relied on wells in the alluvium of the lower river valley. These wells take advantage of the storage provided by the sand-and-gravel valley fill, locally drawing down the water table, and relying upon high winter flows to recharge the aquifer. Willows, cottonwoods, and other riparian species died off near the wells as pumping increased. During the drought of 1976–1977, heavy pumping and the absence of recharge from river flow combined to produce drawdowns in excess of 10 m over 3 km of the river (Kondolf and Curry 1986), accelerating dieoff of riparian vegetation. During subsequent-year moderate floods (with 6–8 year return period), massive bank erosion occurred locally in the pumped reach. Banks here are composed of unconsolidated sand and gravel; if unprotected by the dense armor of willow roots characteristic of downstream, unpumped reaches, they offer little resistance to erosion even by relatively low-magnitude flows (Kondolf and Curry 1986).

The bank erosion has resulted in loss of valuable property and loss of aquatic habitat by converting a meandering channel, with numerous pools and shaded by well-vegetated banks, into a wide, braided, sand-and-gravel-bedded channel. Although individual property owners have used rubble, riprap, and automobile bodies to protect their banks, these efforts have been uncoordinated and have given the river a trashy, unnatural look. The Monterey Peninsula Water Management District (MPWMD), the government agency charged with regulating the private water company and with developing a coordinated policy for management of the basin's water resources, has begun a program of restoration for the most severely eroded reach with assistance from Urban Stream Restoration Program of the California Department of Water Resources. The restoration design, based in part on Curry and Kondolf (1983), emphasizes revegetation in bank stabilization and river training, as advocated by Acheson (1968), Nevins (1969), and Gray and Leiser (1982). A

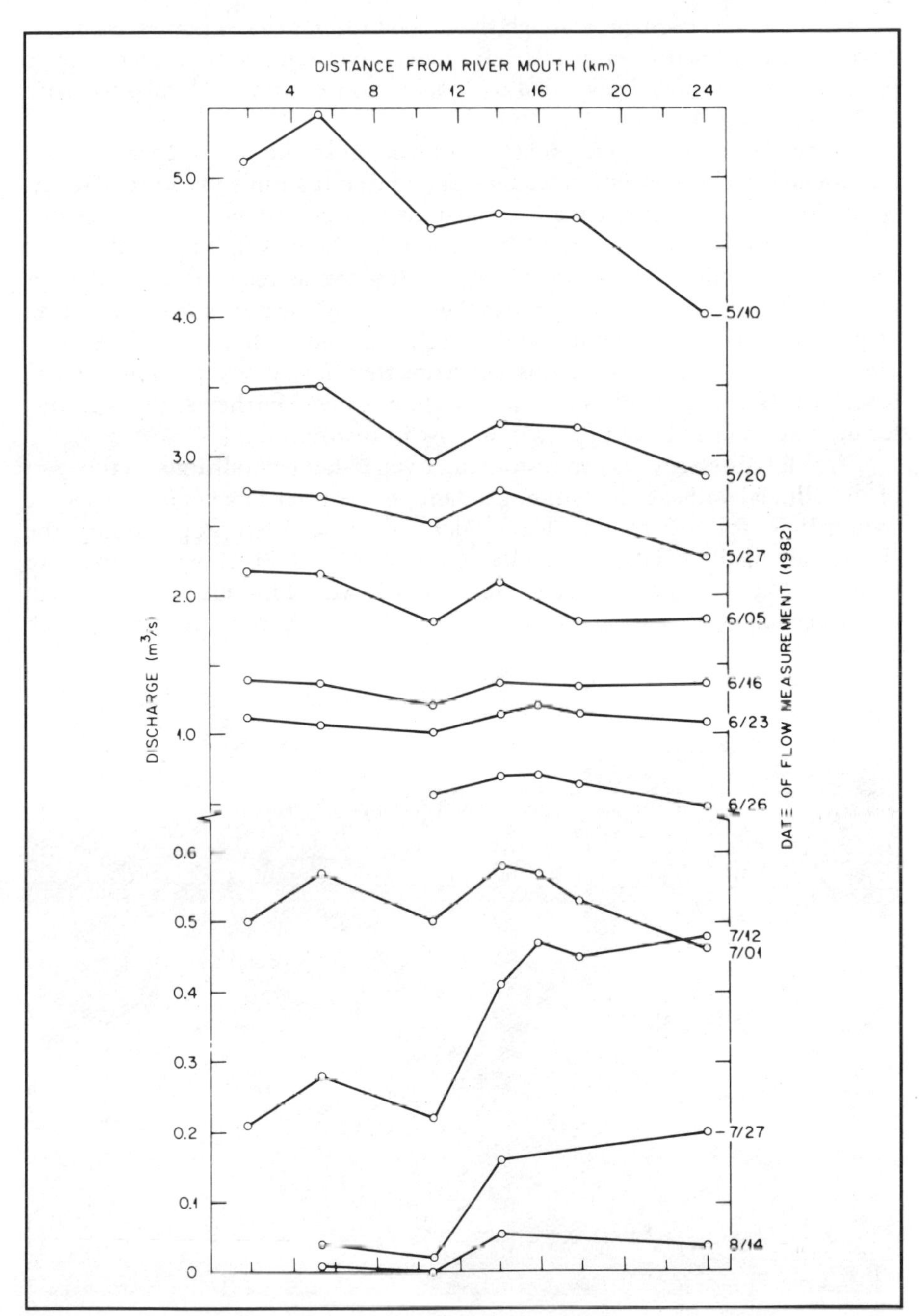

FIGURE 3. *Flow measurements along the Carmel River, May to August 1982. Downstream is from right to left; each point represents a (nearly) simultaneous measurement of flow on the date indicated at right. Note change in scale on ordinate above 0.6 m³/s. The net increase in flow downstream in measurements through June 5 can be attributed primarily to bank storage. All measurements show a net decline in flow about 10 km above the mouth, where major pumping wells have drawn down the water table. (From Kondolf et al. 1987, copyright Elsevier, used by permission.)*

pilot channel has been cut through the eroded reach. At key points, its banks have been lined with permeable wire fence (extending several feet deep to protect against scour); back-filled with cobble; and extensively planted with willows (Figure 4).

Because the underlying problem, drawdown of the alluvial water table, has not ended, the restoration would clearly fail if it simply involved planting trees. Irrigation systems have been installed to permit willows to become established and to flourish above the water table. However, it is not clear that the willow stands produced in this way will prove as resistant to erosion as have the dense, natural stands in reaches with high water tables, unaffected by pumping. Moreover, it is unlikely that these stands will develop the species diversity characteristic of the riparian strips that have developed naturally in reaches with a high water table. (See Williams and Matthews, this volume, for discussion of red versus arroyo willow in restoration.)

Probably the only way to restore high-water-table conditions to this part of the alluvial valley is by building a dam on the river large enough to store winter flows for year-round release. Such a dam has been proposed near the site of one of the existing small dams (MPWMD 1987). However, the dam would be expensive, and there are fears that it would induce further growth in the region, exacerbating future problems. License conditions for any such

FIGURE 4. *Installation of permeable wire fence along the Carmel River. This fence was backfilled with cobbles; it and the reconstructed floodplain behind it were extensively planted with willows. (Photo by W. V. G. Matthews.)*

dam would include minimum flow releases for maintenance of fish and riparian resources (MPWMD 1987). In specifying these flow releases, stream-groundwater interactions must be considered so that minimum flows are met along the entire reach.

Case Study: Rush Creek

Rush Creek, the largest tributary to Mono Lake, drains the eastern front of the Sierra Nevada and flows from the mountain front across alluvium and deposits of a large Pleistocene lake into the modern lake (Figure 5). Its natural flow is dominated by snowmelt runoff in the spring and early summer. From the 1860s until 1941, irrigation diversions from Rush Creek reduced summer flow in the lower reaches below Grant Lake, although a considerable amount of the diverted water returned to the channel through springs and channel seepage. Since completion of the Mono Basin extension of the Los Angeles aqueduct in 1941, increasing amounts of water have been diverted for export to Los Angeles. Flow in the creek was not significantly affected until dry years in 1947–1951, when releases to the channel from Grant Lake Reservoir were eliminated and irrigation was reduced. As a result, baseflow in the "meadows" section of lower Rush Creek declined from 0.68 m^3/s in 1947 to 0.06 m^3/s in 1949 (Vestal 1954). From 1952 to 1969, occasional wet-year releases from Grant Lake Reservoir combined with enough return flow to maintain a riparian corridor in the meadows. In 1970, however, the "second barrel" of the aqueduct became operational and made more exports from Mono Basin possible. From 1970–1978, there were no releases into the Rush Creek channel and irrigated acreage was reduced, thereby causing a decline in the return flow. As a result, the riparian corridor in the meadows was severely degraded (Vorster 1985). The skeletons of thousands of riparian trees (mostly willows and cottonwoods) killed during this period are still visible in the meadows.

The years 1982 and 1983 were unusually wet in the Sierra Nevada. Excess water in the aqueduct system forced the Los Angeles Department of Water and Power to release water into the Rush Creek channel beginning in 1982, and a naturally reproducing population of brown trout became reestablished. With the advent of drier conditions in late 1984, the city announced the shutoff of the releases into Rush Creek; fishery and wildlife groups filed suit to prevent the shutoff and consequent destruction of the fishery. The court granted a preliminary injunction and required that a 19 cfs release be maintained (*Dahlgren et al. versus City of Los Angeles*). Pending trial, the parties were directed to conduct an instream flow study to determine flow releases needed to maintain the fishery. As part of this study, the senior author has (1) conducted flow measurements along Rush Creek to determine flow losses and gains under various conditions so that these downstream changes can be accounted for in specifying releases from the dam, and (2) examined recent historical

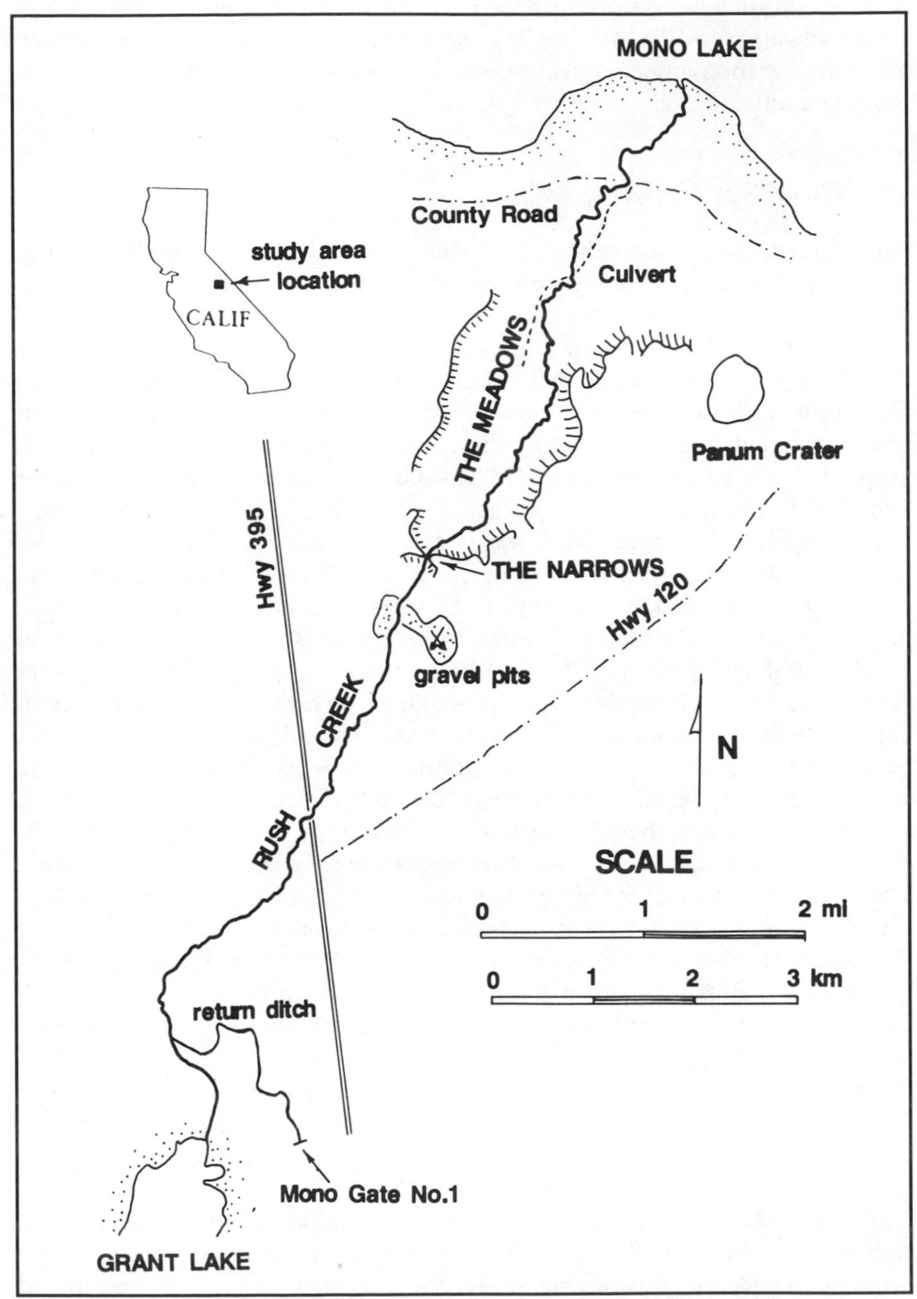

FIGURE 5. *Location map, lower Rush Creek (downstream of Grant Lake Reservoir). The Narrows is a constriction in the valley produced by the outcrop of a granodiorite dike. From Narrows downstream to the County Road is the "Meadows" reach, so named because of the riparian vegetation that flourished there in earlier decades.*

channel changes to evaluate the stability of the channel under present conditions.

Results of flow measurements along Rush Creek show some interesting patterns (Figures 6 and 7). First, flow losses are concentrated at two points in the channel, both at transitions (contacts) in the underlying geological units. The greatest flow loss, averaging 0.2 m^3/s, occurs over a 2 km reach, from about 3.5–5.8 km downstream of the dam, near the U.S. Highway 395 bridge. It is at about this point that the creek, which had been flowing on silty lacustrine deposits, cuts through those deposits and flows over gravels (deposited by the ancient Rush Creek) with a deep water table (Lajoie 1968). The high permeability of the gravels and the deep water table result in high rates of infiltration. Downstream, the stream continues to flow mostly over gravels through the meadow reach, but here the water table is relatively shallow and the streamflow is relatively constant as groundwater and tributary inflow are balanced by evapotranspiration and losses to groundwater. However, about 11.5 km downstream of the dam, the stream begins to flow over volcanic blast deposits from a nearby crater (Bailey 1987). These deposits are very

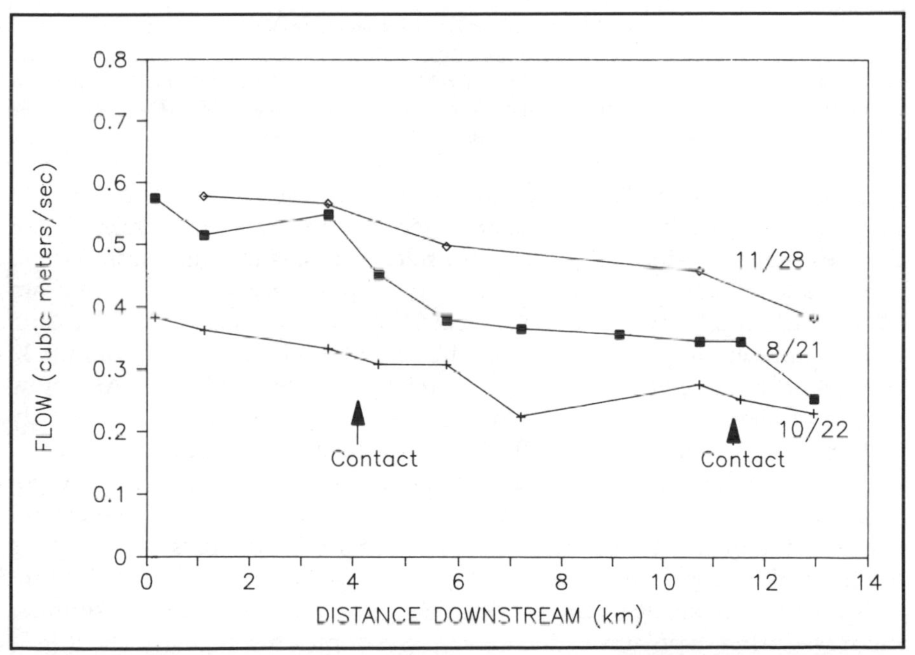

FIGURE 6. *Flow measurements along Rush Creek at releases 0.54^3/s and loss. Downstream is from left to right; each point represents a nearly simultaneous measurement of flow on the date indicated at right. Losses were concentrated at two geologic contacts: between lake silts and gravels (at about 2.5–3.0 km), and between granitic gravels and volcanic blast deposits (at about 7 km). The summer measurement August 21 showed a much greater net loss than did the fall-winter measurements of October 22 and November 28.*

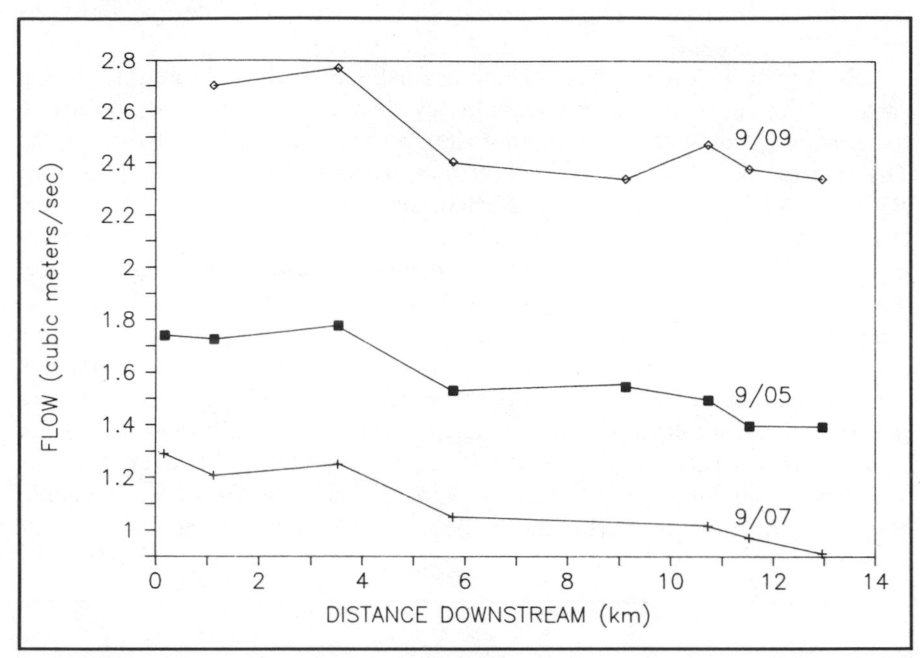

FIGURE 7. *Flow measurements along Rush Creek at releases of 1.3, 1.7, and 2.8 m³/s. All these releases were made between September 5 and September 9. Net flow losses were similar in all three measurements, over a large range in releases.*

permeable and, evidently, well-drained so that an average of 0.1 m³/s is lost into them over a 2 km reach. Second, losses can be quite high. Over 50% of the presently prescribed minimum flow release of 0.54 m³/s is lost into them over a 2 km reach. Third, for a given flow, losses are very different from summer to winter, but are quite similar at very different releases in one season. In the summer, net flow losses over the entire reach varied only from 0.32 to 0.38 m³/s for releases ranging from 0.54 to nearly 2.88 m³/s. At a flow release of about 0.54 m³/s, flow loss in the winter measurement was only 0.20 m³/s compared to a loss of 0.32 m³/s in the summer measurement.

One is inclined to attribute seasonable differences in loss rates to evapotranspirative demand (ET). In this case, however, the differences are far greater than can reasonably be explained by ET. Estimates of riparian acreage along Rush Creek and probable rates of ET (Patten, personal communication) indicate that only about 0.023 m³/s can be attributed to ET in the summer months. Earlier estimates of ET losses were somewhat higher (NAS 1987), and much of this vegetation is newly emergent so its ET demand may be higher than expected. Nonetheless, the difference between the summer and winter loss rates is unexplained. It may be that freezing temperatures affect the rate of groundwater infiltration, but a satisfactory mechanism for this is lacking. Even if we do not have an explanation for the seasonal pattern, the empirical observations can, nonetheless, be profitably applied to specifying flow releases required from the reservoir to satisfy minimum flows along the stream.

Since 1941, the channel of Rush Creek has been unstable, especially in its lower reaches because of incision and devegetation of the channel. Because tributaries to Mono Lake have been diverted for export, the lake level dropped about 14 m from 1941 to 1981, but has risen about 2 m since because of greater inflow in the wet years 1982–1986 (NAS 1987). Mono Lake serves as base level for Rush Creek, and as the lake level has dropped, it has triggered episodes of channel incision during the brief floods of 1967, 1969, 1980, 1982, 1983, and 1986 that came down the otherwise dry streambed. The destruction of potentially stabilizing bank vegetation was a contributing factor to the channel instability.

Restoration of the Rush Creek channel would be doomed if it proceeded without recognition of the potential for further instability if the level of Mono Lake continues to fall. There is growing concern, however, over the possible consequences of any further drop in lake level, especially in the wake of a National Academy of Sciences report that concluded that many lake-dependent resources (e.g., brine shrimp, alkali fly, numerous bird species, and shoreline vegetation) would be severely affected if the lake were to drop further (NAS 1987). With increasing support for maintenance of Mono Lake at its present or a higher level, there may be cause for optimism that Rush Creek can stabilize in its present profile.

Revegetation of the Rush Creek riparian corridor would be an important component in any restoration plan. Although volunteer willows (most dating from 1982 and 1983) are now flourishing along the stream edges, deliberate planting of willows in strategic points would serve to encourage stabilization of the channel. Because perennial flows are once again assured in the stream, plantings are likely to be successful unless scoured by large floods in their first few years.

Conclusions

Many habitat restoration projects are undertaken on streams where habitat has been degraded by indirect effects of other events. Restoration projects on these channels must take into account the recent history of channel changes and their probable cause; otherwise, the forces that caused the original problems may undo the newly designed channel. The case studies presented here illustrate the importance of understanding the extent of riparian vegetation and its role in bank stability, and interactions between stream and groundwater.

On regulated rivers, channel design must also specify the flow needed to maintain riparian, aquatic, aesthetic, and recreational resources. Stream-groundwater interactions must be considered when specifying releases from the upstream dam needed to satisfy minimum flow requirements along the entire reach.

References

Acheson, A. R. 1968. *River control and drainage in New Zealand.* Christchurch, New Zealand: Ministry of Works.

Bailey, R. 1987. Unpublished map. U.S. Geological Survey. Menlo Park, CA.

Curry, R. R., and G. M. Kondolf. 1983. *Sediment transport and channel stability, Carmel River, California.* Report to the Monterey Peninsula Water Management District, Monterey, CA.

FERC (Federal Energy Regulatory Commission). 1986. *Final environmental impact statement, Owens River basin: Seven hydroelectric projects.* Docket No. EL85-19-102. Washington, DC: FERC.

Gray, D. H., and A. T. Leiser. 1982. *Biotechnical slope protection and erosion control.* Englewood Cliffs, NJ: Van Nostrand Reinhold Co.

Kondolf, G. M, and R. R. Curry. 1986. Channel erosion along the Carmel River, Monterey County, California. *Earth Surface Processes* 11:307–19.

Kondolf, G. M., and M. J. Sale. 1985. Application of historical channel stability and analysis (HCSA) to instream flow studies. In *Proceedings of the symposium on small hydropower fisheries,* 184–94. Denver, CO: American Fisheries Society, Bio-Engineering Section and Western Division.

Kondolf, G. M., L. M. Maloney, and J. G. Williams. 1987. Effects of bank storage and well pumping on base flow, Carmel River, Monterey County, California. *Journal of Hydrology* 91:351–69.

Lajoie, K. R. 1968. Late quaternary stratigraphy and geologic history of Mono Basin, Eastern California. Ph.D. thesis, University of California, Berkeley.

Loar, J. M., and M. J. Sale. 1981. *Analysis of environmental issues related to small scale hydroelectric development: V. Instream flow needs for fishery resources.* Oak Ridge, TN: Oak Ridge National Laboratory, Environmental Science Division, ORNL/TM-7861.

MPWMD (Monterey Peninsula Water Management District). 1987. *Draft environmental impact report/environmental impact statement, New San Clemente Project, Monterey County, California.* Corps Permit Application no. 16516509.

NAS (National Academy of Sciences). 1987. *The Mono Basin ecosystem, 2 effects of changing lake level.* Washington, DC: National Academy Press.

Nevins, R. H. F. 1969. River training—the single thread channel. *New Zealand Engineering* (December):367–73.

Patten, D. 1987. Unpublished data. Arizona State University.

Pinder, G. F., and S. P. Sauer. 1971. Numerical simulation of flood wave modification due to bank storage effects. *Water Resources Research* 7:63–70.

Todd, D. K. 1955. Groundwater flow in relation to a flooding stream. *Proceedings of the American Society of Civil Engineers* 81:628. New York: ASCE.

U.S. Army Corps. 1974. *Hydrologic engineering office report, Carmel River basin,*

Monterey County, California. U.S. Army Engineer District, San Francisco, CA.
Vestal, E. H. 1954. Creel returns from Rush Creek test stream, Mono County, California, 1947–1951. *California Fish and Game,* April 1954:89–104.
Vorster, P. 1985. A water balance forecast model for Mono Lake, California. Masters thesis. Hayward, CA: California State University.
Williams, J. G., and W. V. G. Matthews. 1988. Willow ecophysiology: implications for riparian restoration. (This volume.)

G. Mathias Kondolf, a hydrologist and geomorphologist, is an Assistant Professor of Environmental Planning at the University of California, Berkeley, CA 94720.

Volunteer Stream Restoration on an Agricultural Watershed in Northwest Ohio

Justine Magsig

Abstract: *Obstructed rivers in Ohio are often channelized to provide agricultural drainage. In 1973 the Sugar Creek Protection Society was formed to provide an alternative to channelization and a means of continuous monitoring and improvement for a rural stream in northwest Ohio. Obstructions were removed with borrowed equipment and volunteer labor, and a plan for annual care of the creek was developed and adopted by county and state agencies. These efforts saved the river from modification and provided a basis for preservation of the river which is still in use today. The procedure may be easily replicated in other streams.*

Key words: *assessments, channelization, drainage, farmland, Ohio, obstructions, rivers.*

Introduction

The year 1988 is the twentieth year of the Ohio Scenic Rivers program. This river protection law predated the national adoption of a wild and scenic river designation. Ten river segments make up the 1,012 river kilometers within the Ohio system. There are four scenic river coordinators who are assigned to work with landowners, government agencies, and citizens' organizations to protect these rivers. Ohio's scenic rivers are well-protected, quality streams.

Unprotected Ohio Rivers

The other rivers of the state, in all 70,000 km, receive no protection even though many have also been singled out as quality streams in the National Inventory of Rivers. In fact, state law and local custom have undervalued Ohio rivers. They are often dammed or used as dumps for waste and sewage, and many are stripped of bordering trees and channelized to become drainage ditches.

This frontier attitude is most prevalent in northwest Ohio. When the area was settled in the early 1800s, farmland was created by clearing the dense forest and draining the land, which was known as the Black Swamp, through construction of an extensive network of ditches and drainage tiles. These ditches drain surface water into local streams and rivers, and, ultimately, into Lake Erie. The soils are very fertile and, in recent years, farming operations have become larger and more mechanized. More ditches were constructed because crop yields were thought to depend on getting equipment into the fields early in the season. While this is not necessarily so, especially if conservation tillage methods are used, drainage is essential in this region of flat topography. The size of farm equipment has become larger and the equipment more costly. As a consequence, farmers planted up to the ditch banks to increase their farm income. This practice causes serious erosion and a build-up of sand bars in streams.

Two natural events led to increased degradation of area streams. In the 1950s and '60s the American elms (*Ulmus americana*) that made up a large part of old growth forests died and often littered the flood plains and river borders. Then in 1969, a 100-year flood carried fallen trees and wood debris into the river channels. Many log jams formed. These obstructions were enlarged each year by silt and vegetation from farmed fields until many streams were unable to provide agricultural drainage. An example of these problems is Sugar Creek, a natural stream in an agricultural watershed.

Although Sugar Creek provides drainage for rich, productive farm acreage that is planted in row crops in such specialty foods as tomatoes, pickling cucumbers, cabbage, and sugar beets, it also serves as a corridor for migrating songbirds and waterfowl. Wildlife used its forested floodplain for cover in an area where fencerows have been eliminated to provide for more efficient farming. The creek flows over bedrock limestone and aeration cleanses its waters. It is used for recreation by local landowners and metropolitan area residents. Sugar Creek has been a living laboratory for Ohio State University classes that seined its waters and surveyed its insect and plant populations. Forty-one species of fish are found in its waters.

In 1973 46 massive log jams blocked the drainage capacity of Sugar Creek. Low-lying farmland and a commercial golf course were flooded for as long as 10 days following normal to heavy rains. Roads were covered with water and were sometimes blocked in winter by chunks of broken river ice. Discarded materials created eyesores in many places. A petition to channelize this stream was initiated in May of 1973 by 12 local farmers and the owners of the Sugar Creek Golf Course.

Channelization Under Ohio Laws

Section 6131 of the Ohio Revised Code (ORC) provides a means for watershed landowners to alter watercourses in any way—these modifications are defined as "improvements" by the law. Ohio law states that even a single watershed landowner can introduce a petition to "clean, deepen, widen and straighten" any drainageway, including natural rivers and their tributaries. Drainage projects are designed by the county engineer or by an engineering consultant or company. Even though obstructions are the principal drainage problem, and they occur at specific places along a stream's length, the decision to channelize an entire stream is routine. This is probably because the removal of a tangle of tree parts mixed with soil, rock, and vegetation is considered only a temporary measure whereas "channelization" is interpreted as a more permanent remedy. The design of a trapezoidal channel from the upper to the lower terminus of a project can be quantified in cubic meters of soil removed, graded banks cut to a ratio of height to width, and tons of riprap deposited on outside, eroded curves. In engineering terms, channelization can be drawn and constructed according to plan. Removing obstructions cannot.

In Ohio when a stream is channelized under the ORC, the resulting ditch is placed on "maintenance." Maintenance is similar to perpetual care for cemeteries. Under maintenance, ditches are inspected annually and the banks sprayed for woody vegetation with chemicals that are dangerous to humans and animals. No trees are allowed to grow within a maintenance right-of-way which, by law, can vary from a width of 15–75 ft. There is no possibility that a stream that has been channelized will be allowed to re-naturalize.

Petitioned ditches and their maintenance are paid for solely by the landowners on the watershed. Payment is a part of property taxes, and the assessment for the project appears on the landowner's tax duplicate. Payment is made in advance of the work, and, if there is a discrepancy in the amount assessed, it is made up on the next tax notice. Maintenance charges are reviewed every six years and increased at the request of the county engineer. The decision to construct a ditch is made by vote of the commissioners of the county or counties upstream from the project. The cost of engineering—surveying, aerial photography, design, computing assessments prorated by acreage and distance—are costs of the project. In 1973, if the commissioners did not approve the project, these costs were paid from the general fund of the county. At the present time, an amendment to the law allows the costs to be paid by the landowners in the same ratio as the assessments. A rural county cannot justify paying $100,000 or more to save a natural stream. Engineering costs can be, and usually are, 25% of project costs. In 1973 the total estimated cost for engineering and construction of 16.1 km of channel on Sugar Creek was $739,000. The watershed had approximately 1,500 landowners. Of these, one-third lived on village lots and would have paid the minimum assessment, which at that time was $2.

Action to Restore and Protect Sugar Creek

The Sugar Creek Protection Society (SCPS) was formed by concerned landowners to oppose the channelization of Sugar Creek. The members were joined by two of the petitioners who realized that they had helped to set in motion a process that would be costly and destructive. Before the preliminary hearing that would decide if engineering would begin, members of SCPS started removing instream obstructions. They were helped by church groups and youth groups. The proposed project became very controversial due to statewide publicity about the value of the stream, its beauty, and its wildlife. The Ohio Department of Natural Resources supported the Society. At the August 1 preliminary hearing, petitions containing 1,100 signatures (not legally written or collected, but providing a strong demonstration of public opinion) were presented that opposed the project. Organizations with chapters as far away as Columbus and Cleveland sent members to speak at the hearing. In response to public demand, the Board of Commissioners, composed of nine individuals from three counties, gave the SCPS one year to remove the obstructions from Sugar Creek. In addition, they ordered the farmers who had initiated the original petition to work with the Society to this end.

In the year of grace SCPS members worked through every weekend and all of the summer and came to the hearing able to prove that no obstructions remained in Sugar Creek. The farmers, on the other hand, lost all credibility because none had helped with the stream clearing. As a result, the project was dismissed before any engineering costs had been expended. The SCPS was required by the Board of Commissioners to develop a Drainage Management Program for Sugar Creek. The program was written for a five-year period and was submitted to the counties in November 1974. This plan was subsequently accepted without modification by all commissioners and the Ohio Department of Natural Resources. Thirteen years after, the SCPS is still following the same program. Moreover, a report is provided each year that details the condition of Sugar Creek and the restoration activities of SCPS members.

Restoration Methods

The equipment used to accomplish the initial clearing was provided without charge by SCPS members. Two small, all-purpose tractors with hydraulic buckets were employed to raise loads from the creek. Operators often worked from the higher of the two banks because the ground was firmer. During the first year, felled trees which blocked the creek were cut to manageable lengths and secured in bundles by choker cables, then lifted from the creek by chains. One member brought an A-frame logging truck that could lift and haul whole trees. Chain saws were provided by several. Many of the log jams were solidly packed and cemented with clay. They were very difficult to break up but were successfully removed with brute force and awkwardness. The debris was piled and burned unless the landowner wanted the piles of trees and branches left for firewood or wildlife. The crews that did the work took great care not to destroy crops or cut up farmed fields with their tractor tires. From 10–40 volunteers at a time usually cooperated to clear the creek.

Project Organization and Management

The SCPS is incorporated as a nonprofit organization. Paid membership was 475 landowners and sympathizers in 1973–74. At present, 425 members remain. Dues are $5/yr. for individuals. Many members contribute from $50–$100/yr. Members joined for one or two reasons: to avoid an additional tax and to preserve a valuable natural resource. Under the present documents of incorporation, individual members are not liable for damages. Initially, workers provided landowners with a release form whenever they were on nonmember property. In recent years, this became unnecessary because landowners value the services the SCPS provides. It soon became obvious that Society directors would be most effective if they were chosen to represent specific sections of river. Thus, at least one of the ten directors lives along the stream in each stretch that falls between two roads. They are able to monitor the condition of the creek and report any problems to the work crew. Some former petitioners have now become directors of SCPS.

Annual Restoration Efforts

After the initial work was completed, after the log jams were all removed and the floodplain cleared, caring for the creek became routine. The stream is now inspected by canoe each spring when the water is high. Problem areas are noted. In the fall, when the crops are off the fields, a crew of 3–10 people removes windfalls, leaning trees and branches, and instream debris. A new tractor with a winch is a donated improvement. A snatch block is used in brushy areas where it is difficult to position the tractor above the work. Handheld straw rakes are also valuable tools. The SCPS work crew has had help from creekside landowners and students from Bowling Green State University. More stringent air pollution laws have been enacted, and the workers do not burn debris anymore. All material is now piled in such a way that it will not wash into the creek.

Conclusion: The Mission of SCPS

The Society considers itself a service organization to both the landowners of the area and to other groups that wish to organize to promote alternatives to channelization. The SCPS shared the cost of a stream section relocation at a place where bank erosion threatened to undercut a farm home. The relocation involved 305 m of river channel and helped resolve a problem for a former petitioner. This proved to be an excellent rapport-builder. With the help of students in the local high school, a slide show about the SCPS was developed. The amateur quality has been an advantage to the Society's image as an organization of dedicated volunteers. Members have said that they will go anywhere and do whatever is needed to save the free-flowing streams of Ohio.

Using this presentation, members have spoken at over 100 meetings and rallies. They took their message to Washington, D.C. and presented it to the Agriculture Committees of both the House of Representatives and the Senate. Progress toward the goal of protecting free-flowing rivers has been slow, but the SCPS has, by example, shown that volunteer, no-cost restoration efforts can save threatened streams from channelization.

Justine Magsig is Assistant Director, Center for Environmental Programs, Bowling Green State University, Bowling Green, OH 43403.

Fresh- and Saltwater Wetlands, Seagrass Beds, and Vernal Pools

Hydrologic Techniques for Coastal Wetland Restoration Illustrated by Two Case Studies

Robert Coats and Philip Williams

Abstract: *Successful wetland restoration projects generally include: (a) clearly defined objectives; (b) site-specific analysis of hydrologic, geomorphic, and biological variables; (c) review and feedback by the design team during construction; and (d) post-construction monitoring. Examples of two typical design problems are given. These problems involve: (a) use of a numerical model with field verification, and (b) use of empirical geomorphic relationships.*

Key words: *wetlands restoration, hydrology, mitigation, estuaries, tidal ponds, salt marshes.*

Introduction

Tidal salt and brackish marshes along the California coast have important ecological and hydrologic functions. Over the past century, around 80% of California's coastal wetlands have been diked or filled for agriculture, urban development, and salt production (Dennis and Marcus 1983). As the importance of wetlands has become increasingly understood, attention has been focused on how best to restore, enhance, or manage the remaining tidal wetlands. Some of this interest has its origin in the hope of developers to find "mitigation sites" as replacements for filling in existing wetlands. The purpose of this paper is to describe techniques that we are using to design restoration of tidal wetlands, and to suggest some of the factors in the success or failure of restoration projects.

Rapid urbanization and intensive agriculture adjacent to most of California's coastal wetlands have increased the ecologic and hydrologic value of existing wetlands. Urban runoff and agricultural runoff frequently pass through tidal salt marshes before reaching a bay or estuary, allowing for natural trapping of pollutants, including sediment, heavy metals, toxic organics, and nutrients. Salt marshes often provide natural protection against shoreline erosion. They may provide a valuable base of support for estuarine food chains, as well as habitat for shorebirds and waterfowl. In coastal California, salt marshes are important habitat for at least six rare or endangered animals, including the salt marsh harvest mouse (*Reithrodontomys raviventris*), the black rail (*Laterallus jamaicensis*), California clapper rail (*Rallus longirostris*), light-footed clapper rail (*Rallus longirostris levipes*), the California least tern (*Sterna antillarum browni*), and Beldings savannah sparrow (*Passerculus sandwichensis beldingi*).

Although the value of wetlands has become increasingly recognized, the feasibility of enhancing or restoring wetlands as mitigation is controversial. Race (1985) attempted to evaluate the success of wetland restoration around San Francisco Bay in providing mitigation for disrupted or destroyed wetlands, and concluded that restoration has been substantially unsuccessful. Race's conclusions were disputed by Harvey and Josselyn (1986), who pointed out that she did not provide a clear definition of evaluation procedures or criteria for success or failure, and that she failed to distinguish between experimental planting and restoration. Race also did not count natural recolonization of areas restored to tidal action as successful restoration.

The arguments about feasibility of restoration and mitigation generally have both policy and technological components (Eliot 1985; Josselyn et al. 1989). Among the policy questions currently being debated are: (1) How close to a destroyed wetland must a mitigation site be located? (2) Must the destroyed habitat be replaced in kind, or is it acceptable (for example) to replace a pickleweed (*Salicornia*) marsh with a cordgrass (*Spartina*) marsh? (3) How rapidly must a "restored" marsh become fully vegetated and productive to count as mitigation? (4) Should the enhancement of degraded wetlands be counted as mitigation? This latter question is of special concern to regulatory agencies, since allowing mitigation on degraded wetlands raises their market value as potential mitigation sites above that of healthy and productive wetlands, thus giving landowners an incentive to degrade their wetlands. (5) Can "mitigation banks" provide a feasible mechanism whereby several developers can fund a single major restoration project in exchange for permits?

Once the questions of mitigation and regulatory policy are separated from technical issues of wetland restoration, the problem becomes tractable, albeit complex. The planning and design of projects requires detailed analysis of both biological and hydrologic variables and, in both areas, planning must proceed under uncertainty. Biologists face the uncertainty of climatic events and their effect on biota, the vagaries of propagule dispersal and colonization, and imperfect knowledge about wetland organisms and their life cycles (Zedler et al. 1986). Hydrologists often face the uncertainty of future storm magnitudes, lack of local data on water discharge and sedimentation rates, and an imperfect understanding of the hydraulics of estuaries and wetlands. Nevertheless, the basic steps that must be taken in designing a wetland restoration are straightforward. These steps may be summarized as follows (Harvey, Williams, and Haltiner 1982):

Define the Objectives

The first step in any wetland restoration or enhancement project is to define the objectives. As obvious as this seems, objectives for most of the completed restoration projects in coastal California were never clearly defined.

There are two steps in defining the objectives of a project. First, resource managers must decide what kind of wetland should be created or enhanced. This may involve difficult choices or trade-offs. Creating shallow, seasonally ponded water for ducks and wading birds, for example, may preclude pickleweed habitat for the endangered salt marsh harvest mouse. Public access may encourage intrusion by humans and dogs into wildlife habitat.

Once the biological goals have been defined, they must be translated into hydrologic terms. The hydrologist must know what frequency, duration, and depth of inundation are needed to create the habitat elements desired by the biologists. In some cases, the water quality objectives must also be defined.

Determine the Topography and Tidal Regime at a Site

Designing a wetland requires good topographic information, as differences of only a few centimeters may make a major difference in the duration and frequency of inundation. A minimum requirement is a topographic map at a scale of 1:1200 with one-foot contour intervals and supplementary spot elevations to the nearest 0.1 ft. For some applications, a larger-scale map is desirable. The elevations of existing culverts, if they are to be used in the design, must be known to the nearest 0.1 ft.

Once the topographic information at a site has been defined, the tidal regime must be described. Standard tide tables published by the National Ocean Survey give the mean highs and lows relative to Mean Lower Low Water (MLLW) at many subordinate tide stations along the coast and inside major estuaries. For a given site, these values must be converted to the National Geodetic Vertical Datum (NGVD) in order to provide a reference for construction. The elevations of MLLW relative to NGVD are available from the Office of Oceanographic and Marine Assessment of the National Ocean Service.

Analyze the Opportunities and Constraints Imposed by Local Biological and Physical Conditions of a Site

The basic question that must be answered here is: What kind of ecosystem can be created at the site? This requires a cooperative effort of biologists, hydrologists, and sometimes geomorphologists and soil scientists. Among the important physical parameters that must be considered are local flood protection needs, local runoff, potential shoreline erosion, local sediment transport and deposition rates, historic subsidence, accumulation of floating debris, soil or substrate characteristics, water quality (including salinity and potential pollutants), and presence of buried debris or toxic wastes. Among the biological parameters are the local species of plants and animals that may utilize a site, regional patterns of bird movement, potential vector (mosquito) problems, and control of exotic species.

Develop Design Alternatives for the Site

This may involve determining the dimensions and elevations of channels, culverts, and weirs; determining the elevation and slope for intertidal areas; or determining dredging volumes.

Wetland design must take account of the dynamic physical characteristics of wetlands. Rapid changes can occur in channel morphology, marsh elevation, water quality, and water circulation as a result of changes in wave climate, sea level, watershed inputs, and nearby development projects. These changes may, of course, have major effects on plant and animal populations.

In solving design problems, we use a combination of numerical models and empirical relationships drawn from traditional engineering practice and geomorphology. We have modified and adapted these methods for use in wetlands. The case study discussed below illustrates the use of some of these tools.

Restoration of the HARD Marsh

The Hayward Area Recreation District (HARD) Marsh comprises 11 shallow ponds of various shapes and sizes, covering 30 ha on the eastern shoreline of San Francisco Bay. The ponds were used for salt production from as early as 1853 to the late 1940s. Following abandonment, tidal flow through a failed tide gate provided some circulation, and the levees between ponds developed a dense cover of pickleweed (*Salicornia virginica*) and alkali heath (*Frankenia grandifolia*). In 1986, the adjacent landowner blocked the ditch that provided tidal flow, and the ponds dried out. The biological objectives of this project were to: (1) restore habitat (feeding opportunities and shelter) for shorebirds and migratory waterfowl; (2) maintain healthy pickleweed on levees as habitat for the salt marsh harvest mouse. In hydrologic terms, this required about 0.3 m of tidal range, with water remaining in the ponds at low tide. Making the ponds fully tidal was not acceptable, since sediment from the bay would soon fill the ponds, and access roads behind the bayside levee would be periodically flooded. The first design problem was to select the elevations and dimensions of weirs and culverts to create the desired tidal amplitude and circulation.

The problem was solved with our hydrologic routing model, MPOND (Coats et al. 1987). This model "routes" tidal inflow against storm runoff in a network of up to 20 ponds connected through up to 30 control structures, including culverts, weirs, tide gates, and pumps. Flow rate through a culvert is calculated as a function of the hydraulic head between ponds, and pond elevations are changed in each time step to allow for inflow and outflow.

The data requirements for MPOND include: (1) an elevation-storage capacity curve for each pond; (2) invert elevations, friction coefficients, dimensions and type (box or circular) for culverts, dimensions for weirs, gate settings, and pump capacities; (3) times and heights of high and low tides for one or more points; (4) optional runoff hydrographs; (5) a description of how the ponds and control structures are connected; and (6) a starting water surface elevation for each pond.

The model assumes that the water surface elevation of a pond responds instantly and uniformly to inflow and outflow. This is not quite true; inflow to a pond moves across it as a progressive wave, with the depth varying over space and time. To model unsteady flow across a pond requires a tidal hydrodynamic model linked to flow calculations through control structures. For relatively small ponds, however, the simplifying assumptions of MPOND generally give results that are within 6 cm of actual pond elevations.

To select the dimensions and elevations of water control structures for the HARD Marsh, we tried various options with average and spring tidal days as input, and selected the combination that gave the desired results.

It is useful also to know how water height in the ponds might vary over a period longer than a day, since the ponds respond very slowly to changes in tide height in the bay, and daily tidal cycles vary considerably over a year. To answer this question, we used a mean tidal month for San Francisco generated by the Tidal Analysis Branch of the National Ocean Service from the short-term tidal constituents. We corrected for the time and height dif-

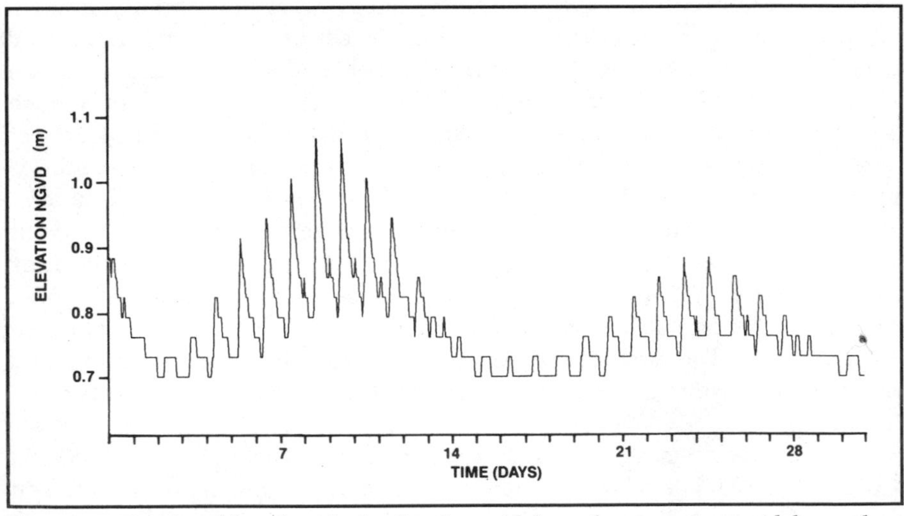

FIGURE 1. *Simulated hourly tide heights in a tidal pond over a mean tidal month at Hayward, CA.*

ference between San Francisco and Hayward, and ran MPOND for the mean tidal month. Figure 1 shows a typical plot of simulated hourly values in one of the ponds; Figure 2 shows a set of height-duration curves for three ponds, as well as for the Bay.

The plot of hourly values (Figure 1) reveals an interesting problem: there is a zone between 0.73 m and 0.79 m NGVD that is inundated continuously for six days, and then exposed continuously for six days. In the largest pond, this amounts to about 0.8 ha. The zone must be a harsh environment for most benthic organisms, but given the objectives of maintaining shallow ponded water and preventing flooding of roads and levees, the harshness is unavoidable.

Following construction of the chosen design and reopening of the marsh to tidal flow, we tested and "calibrated" the model. We set up a portable recording tide gage in the bay, and took readings at tide gages in the marsh over a 22-hr period. We then ran the model with the real tide height as input and improved the "fit" by adjusting the friction coefficients (Manning's "n" values) in the model. Since the model neglects friction across the ponds, it was necessary to assign rather high "n" values to openings between ponds (up to 0.07). Figure 3 shows the real and simulated tide heights, before and after calibration, in the largest pond.

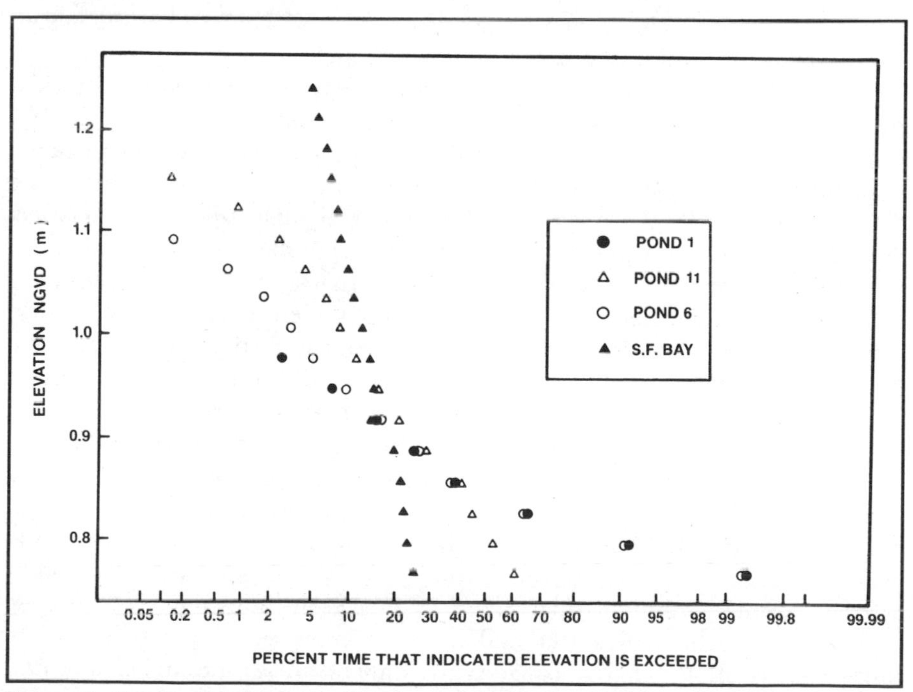

FIGURE 2. *Height-duration curves for three interconnected tidal ponds at Hayward, CA.*

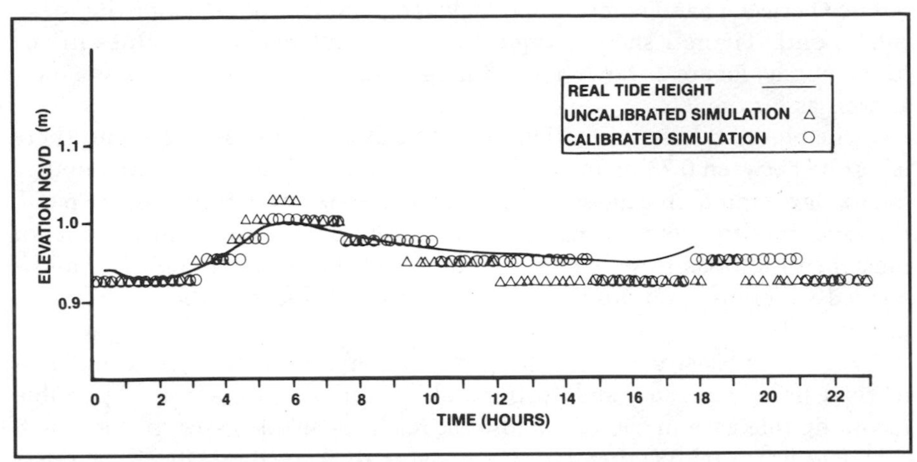

FIGURE 3. *Real and simulated tide height in a tidal pond at Hayward, CA.*

The calibrated model was then used to compare alternative settings for adjustable tide gates and weirs, and to identify two constrictions where internal circulation could be improved.

In restoring the HARD Marsh, it was necessary not only to install new culverts and weirs, but also to excavate a new slough channel between the marsh and the bay. The second design problem was to determine the design depth and cross-sectional area of the new channel.

In a tidal marsh, there is an equilibrium relationship between the tidal prism (volume of water exchanged between high and low tide) and the hydraulic geometry of the channel (Myrick and Leopold 1963). We have gathered data from several tidal marshes around San Francisco Bay, and plotted the relationships between mean diurnal tidal prism, channel cross-sectional area, and channel depth (Haltiner and Williams 1987a). If the upstream tidal prism of a slough channel is reduced (by diking of the marsh, for example), sediment will accumulate until bed shear stress is just sufficient to resuspend recently deposited unconsolidated sediment. If, however, the tidal prism is increased, the channel geometry will approach a new equilibrium somewhat more slowly, since consolidated cohesive sediment is resistant to erosion.

Figures 4 and 5 show the relationships between mean diurnal tidal prism, depth below mean higher high water, and cross-sectional area, based on data from eight tidal salt marshes. In order to estimate the mean diurnal tidal prism at the HARD Marsh, we ran MPOND with the selected culvert and weir configuration, and added the diurnal prism volumes for all of the ponds. For the estimated diurnal tidal prism of 11,100 m^3, a channel 1.2 m deep (below mean higher high water) and 8.4 m^2 in cross-sectional area should be sufficient. To allow a margin of safety, we chose a cross-sectional area of 12.3 m depth of 1.8 m below MHHW, and a top width of 10.4 m. If a channel is slightly oversized, it will soon adjust through deposition and bank slumping.

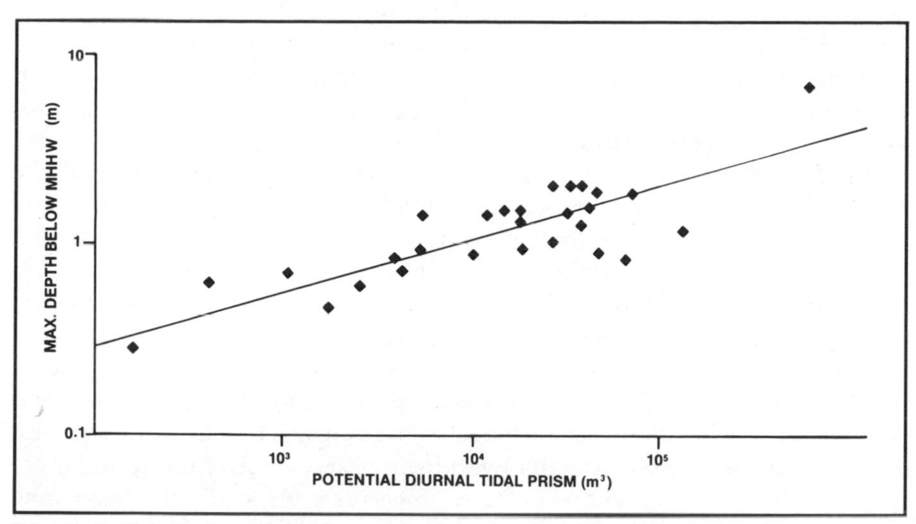

FIGURE 4. *Slough channel depth versus tidal prism in eight tidal salt marshes.*

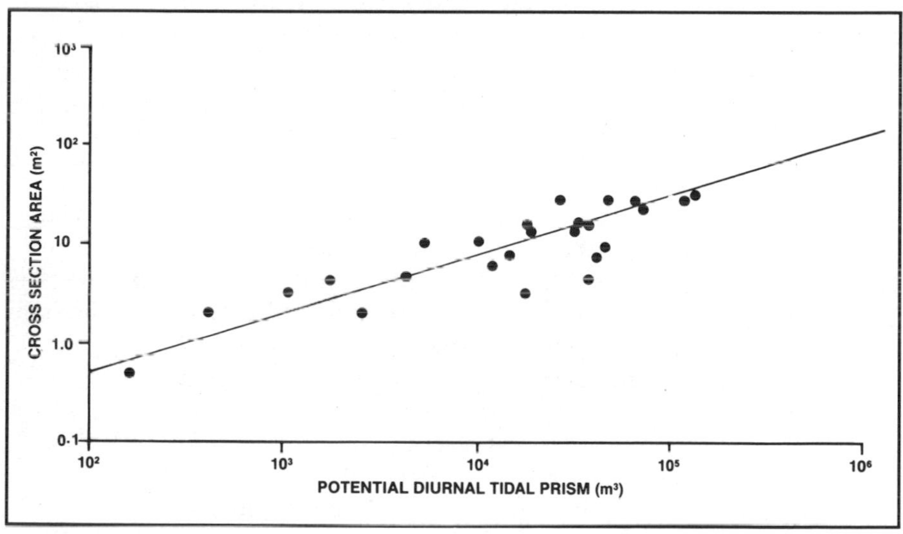

FIGURE 5. *Channel cross-section area versus tidal prism in eight tidal salt marshes.*

Haltiner and Williams (1987b) have extended the hydraulic geometry relationships to include marsh area-tidal prism and tidal prism-channel top width and tidal prism-width:depth ratio. With these relationships, a good first approximation of channel dimensions can be obtained from the marsh area alone.

Implementation and Monitoring

The success or failure of a wetland restoration project rests not only on the adequacy of the design, but also on the faithful execution of the design. We recently participated in a critical review of 18 wetland mitigation projects for the San Francisco Bay Conservation and Development Commission. Of the projects examined, six were successful in meeting defined objectives; seven did not meet all stated objectives, but did produce productive wetlands. Two were judged complete failures, and three were in progress or too recently completed to judged. Many lacked clearly defined biological goals. Of the failures, the most common problems seemed to be poor implementation and lack of agency follow-through. Wetland restoration seems to present a special problem with respect to project implementation. If a flood control project is poorly designed or constructed, the problem may eventually be noticed, and can be corrected. But if a wetland project doesn't work properly, the problem may go unnoticed. Ducks can't talk, and they don't have standing to sue.

Because of the unique problems with project implementation, wetland restoration projects should have a built-in post-construction monitoring phase that includes both biological and hydrologic observations. Monitoring must go beyond a "final inspection" of the engineering trade. Up to five years may be required to assess wildlife usage, vegetation viability, water quality, and water circulation. A well-designed monitoring program can not only suggest minor improvements (such as screw gate settings and weir elevations) in an individual project, but it can also help provide a base of information for use in planning and designing future projects.

Summary

The increasing value of tidal wetlands in coastal California has led to a surge of interest in wetland restoration and enhancement. The question of mitigating destruction of some wetlands by restoring others remains controversial, but once the policy questions are separated from technical questions, the steps in a successful project are generally straightforward. On the basis of our experience with design and review of over 60 projects, we concluded that successful projects include the following elements:

1. Clear definition of biological objectives.
2. Translation of biological to hydrologic alternatives.
3. Good definition of site topography and tidal regime.
4. Analysis of physical constraints and opportunities, including consideration of flood control, local runoff, erosion problems, sediment transport and deposition, subsidence, debris problems, soil characteristics, water quality, and presence of toxic wastes.
5. Analysis of biological constraints and opportunities on the site.
6. Development of design alternatives using numerical computer models and empirical geomorphic relationships along with calibration and verification of models with field data.
7. Review of design drawings and specifications by the design team.

8. Inspection by members of the design team during construction, with approval required for any design changes.
9. Post-construction monitoring of both biological and hydrologic parameters.

Conclusions

Wetland restoration is no longer in its infancy, but neither is it a mature technology. Early projects were planned and designed largely by biologists, with little help from hydrologists. Future projects will no doubt involve use of increasingly sophisticated hydraulic and hydrologic models for runoff, estuarine circulation, water quality, and sedimentation. They will also involve closer collaboration among biologists, geomorphologists, and engineers.

References

Coats, R., C. Farrington, and P. Williams. 1987. Enhancing diked wetlands in coastal California. In *Proceedings of the fifth symposium on coastal and ocean management,* 3688–3700. Seattle, WA: American Society of Civil Engineers.

Dennis, N. B., and M. L. Marcus. 1983. Status and trends of California wetlands. Prepared for the California Assembly Resources Subcommittee on Status and Trends. T. Goggin, Chairman. Novato, CA: ESA/Madrone.

Eliot, W. 1985. Implementing mitigation policies in San Francisco Bay: A critique. Oakland, CA: California State Coastal Conservancy.

Haltiner, J. and P. Williams. 1987a. Slough channel design in salt marsh restoration. In *Proceedings of the eighth annual meeting of the Society of Wetland Scientists: Wetland and riparian ecosystems of the American West,* eds. K. M. Mutz and L. C. Lee, 125–130. Wilmington, NC: Society of Wetland Scientists.

———. 1987b. Hydraulic design in salt marsh restoration. In *Proceedings national symposium: Wetland hydrology,* eds. J. A. Kusler and G. Brooks, 293–299. Berne, NY: Association of State Wetland Managers.

Harvey, H. T., and M. N. Josselyn. 1986. Wetlands restoration and mitigation policies. *Environmental Management* 10:(5):567–569.

Harvey, H. T., P. Williams, and J. Haltiner. 1982. *Guidelines for enhancement and restoration of diked historic baylands.* San Francisco, CA: San Francisco Bay Conservation and Development Commission.

Josselyn, M., J. Zedler, and T. Griswold. 1989. Wetland mitigation along the Pacific coast of the United States. In *Wetland creation and restoration: The status of the science, 1: Regional Reviews,* eds. J. A. Kusler and M. E. Kentula. Corvallis, OR: U.S. Environmental Protection Agency, Environmental Research Laboratory.

Myrick, R. M. and L. B. Leopold. 1963. Hydraulic geometry of a small tidal estuary. U.S. Geological Survey Professional Paper 422-B. Washington, D.C.: Government Printing Office.

Race, M. S. 1985. Critique of present wetlands mitigation policies in the United States based on an analysis of past restoration projects in San Francisco Bay. *Environmental Management* 9:71–82.

Zedler, J. B. 1984. *Salt marsh restoration: A guidebook for southern California.* California Sea Grant College Program.

Zedler, J. B., J. Covin, C. Nordby, P. Williams, and J. Bolland. 1986. Catastrophic events reveal the dynamic nature of salt marsh vegetation in southern California. *Estuaries* 9(1):75–80.

Robert Coats and Philip Williams are principals of Philip Williams and Associates, Pier 33 North, The Embarcadero, San Francisco, CA 94111.

SEASONAL FRESHWATER WETLAND DEVELOPMENT IN SOUTH SAN FRANCISCO BAY

GARY S. SILVERMAN AND EMY CHAN MEIORIN

ABSTRACT: *A 20-ha fallow agricultural field in Fremont, California, was converted into a seasonal freshwater wetland, creating habitat historically characteristic of South San Francisco Bay but largely missing today. This project was unusual in using urban stormwater runoff for its water supply. Primary design considerations were water quality enhancement and long term fate of pollutants. The success of this system would indicate a potential for developing similar systems around the Bay. Local agencies may find wetlands to be effective tools to meet nonpoint source water quality control requirements now being developed at the federal level.*

KEY WORDS: *freshwater wetlands, urban runoff, stormwater, Coyote Hills.*

INTRODUCTION

LOSS OF WETLANDS habitat is now well understood as a serious threat to environmental quality. In the San Francisco Bay Area, the largest estuarine complex on the west coast of North America south of Alaska, there has been massive loss of both open water and wetlands. Before 1850, the Bay and associated wetlands covered approximately 1,850 km². Today, the Bay has an area of about 1,100 km² with approximately 325 km² of adjacent tidal marshes and 210 km² of seasonal wetlands behind dikes (Wakeman 1982). Very little of this surviving wetlands is freshwater, most being brackish or salt. Furthermore, most wetlands creation or renovation activities focus on brackish and saline systems, largely ignoring freshwater marshes. Yet the San Francisco Bay Conservation and Development Commission (BCDC), the state agency serving as a principal advocate of Bay wetlands protection, has recognized the critical need for freshwater wetlands: "Wetlands of all kinds around the Bay have been diminished by development . . . the freshwater marshes are possibly the most important because they are so few in number and support such a wide variety of wildlife species" (BCDC 1982). The topic of this paper is the development of a seasonal freshwater wetlands in South San Francisco Bay. This project not only created some of this critical habitat, but also provided an opportunity to examine the potential of freshwater wetlands to use and treat urban stormwater runoff.

The project grew out of an interest in exploring mechanisms for protecting the Bay's water quality from pollutants introduced in urban stormwater runoff. Pollutants from nonpoint sources are a substantial threat to water quality throughout the nation. For example, six out of ten U.S. Environmental Protection Agency (EPA) regional offices reported that nonpoint source pollution was the principal remaining cause of water quality degradation in their regions (EPA 1984). Similarly, the Association of State and Interstate Water Pollution Control Administrators (ASIWPCA 1985) reported that 12% of rivers and streams, and 24% of lakes and reservoirs are threatened by nonpoint source pollution.

A number of researchers have documented the substantial pollutant contribution from urban nonpoint sources. Field and Turkeltaub (1981) reported that suspended solids, 5-day biological oxygen demand (BOD_5), and coliform levels from nonpoint sources may equal or exceed levels characteristic of secondary effluents. Connel (1982) reported that urban runoff is the predominant source of oil and grease loading to the Hudson Estuary. Whipple and Hunter (1979) reported that urban runoff is anticipated to carry about 40% of the total oil and grease load into the Delaware Estuary, and a number of other studies confirm that substantial oil and grease loading typifies urban sources (Hermann 1981; Hoffman et al. 1984, among others). Stenstrom et al. (1984) reported that concentrations of oil and grease in urban runoff entering San Francisco Bay often exceed that permitted in point source discharges. Furthermore, Silverman et al. (1985) showed that future growth will result in a dramatic increase in oil and grease loading to the bay.

Given this threat, an opportunity was seen to intercept runoff with a wetlands and allow natural mechanisms to remove pollutants. This idea also appeared desirable as a means to create or renovate freshwater wetlands using a cost-free water supply. Since water is very expensive in the West, it would be difficult to develop a wetlands system dependent upon potable water sources (perhaps accounting for most activity focussing on brackish and saline systems). However, by using a source without commercial value, we hoped to see a dual benefit: removal of a pollutant source and creation of additional freshwater wetlands habitat.

Wetlands Design

The project is situated on a site of approximately 20 ha in Fremont, California, on an undeveloped section of Coyote Hills Regional park (Figure 1). Prior to

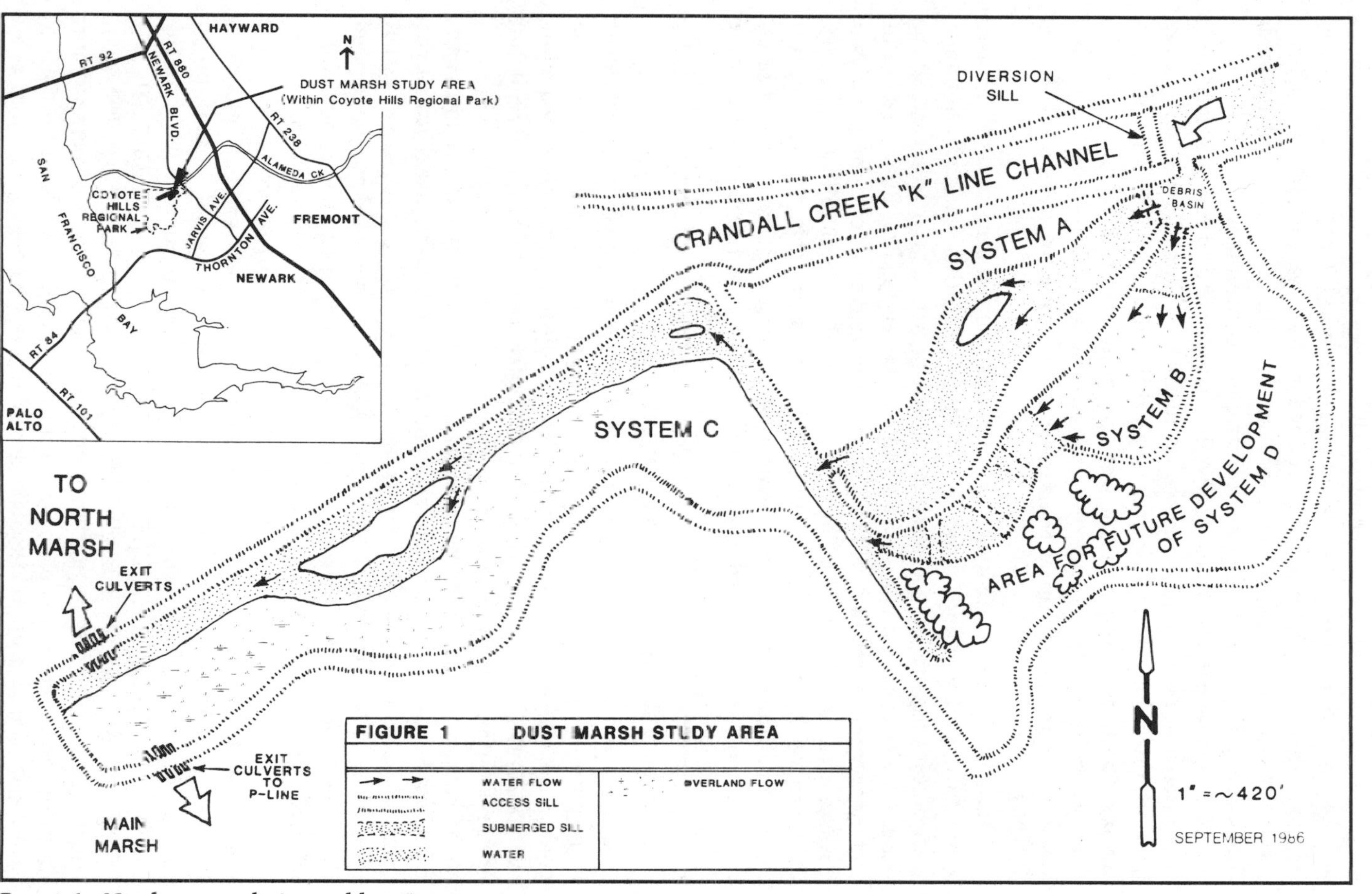

FIGURE 1. *Marsh system design and location.*

development, the site consisted of an abandoned agricultural field, a dense willow grove, a stand of pickleweed (*Salicornia virginica*), and a meandering slough with no surface outlet that drained a small agricultural area. Historically, the site was within the floodplain of Crandall Creek and Alameda Creek at the Bay margin. Diking converted it to an upland.

A design objective was for the wetlands to serve as a research facility to study wetlands creation for stormwater treatment. Thus, the wetlands design includes several configurations that offer different pollutant-removal mechanisms. The designs reflect data on wetlands pollutant-removal characteristics chiefly obtained from municipal wastewater treatment systems, since data are almost nonexistent on wetlands systems treating urban runoff (Chan et al. 1982).

Both the chemical and hydrological nature of input into municipal wastewater treatment wetlands differ substantially from this wetland. Municipal wastewater tends to have a fairly consistent flow and quality. Urban stormwater flow varies drastically with rapid changes in quality. Biota in a stormwater treatment wetlands will be present in large part due to their ability to withstand extremely variable conditions, rather than their ability to optimize use of energy sources supplied in the influent stream. Thus, it is critical to examine pollutant removal mechanisms and measure performance of prototype systems before widespread implementation of urban stormwater wetlands enhancement programs.

Physical Pollutant-Removal Mechanisms

Sedimentation, adsorption, and filtration are physical processes which result in capture of pollutants in wetlands. Subsequently, pollutants may be degraded, isolated, reintroduced into the water column, or concentrated. However, prolonged retention in the system generally is a prerequisite for reduction of the pollutant hazard potential.

Sedimentation occurs as spreading water loses velocity and energy to carry particles. Evaluation of particle size and flow characteristics revealed that insufficient area was available to adequately remove particulates by sedimentation, indicating that the system could not be designed as a large pond. However, large ponds were seen as offering a potential to provide removal of relatively large particles and to treat base flow.

Physical adsorption and filtration of pollutants will occur on vegetative surfaces and in the sediments and should be important removal processes. However, adsorption and filtration in the sediments depend, in part, on porosity. The dense clays found at this site present an almost impermeable barrier to water movement, limiting removal by these mechanisms. The project design incorporates a number of different depths and flow designs to encourage a variety of vegetation types with the objective of being able to monitor the effectiveness of these habitats in capturing pollutants.

Evaporation and photochemical degradation processes differ from those processes already discussed in that they result in pollutant removal. In particular, hydrocarbons will lose their lower molecular weight fraction to the atmosphere relatively quickly (MacKenzie and Hunter 1979). Photochemical reactions involving the absorption of light energy may also be a significant removal mechanism. In particular, polynuclear aromatic hydrocarbons and unsaturated chlorinated alkanes have been found susceptible to photochemical degradation reactions in water (Larson et al. 1977; Lockhart and Blakeley 1975).

Chemical Pollutant-Removal Mechanisms

Important chemical pollutant-removal processes in a wetland include precipitation, adsorption, and decomposition. Precipitation may occur in response to changes in water quality. For example, pollutants in stormwater are usually in an oxidized state, since runoff is usually aerobic. If reducing conditions are present in a wetlands, metals such as cadmium, copper, lead, mercury, silver, and zinc can form insoluble sulfides (Framson and Leckie 1978; Gambrell et al. 1980). While this will probably not be a major process for pollutant removal, it may have some significance if the local drainage area is contaminated.

Chemical adsorption differs from physical adsorption (discussed above) by the bond strength. Chemical adsorption occurs if the bonds between the adsorbate and adsorbent are strong, and the reaction is seldom reversible. Physical bonding predominates in a wetland, so re-release of pollutants typically can occur following a water quality change.

Retention of pollutants in a wetland also will allow opportunity for chemical decomposition. Processes such as chemical oxidation and reduction should account for some pollutant removal, although decomposition products may remain hazardous.

Biological Pollutant-Removal Mechanisms

Biological processes in wetlands are known to play important roles in degradation of organic compounds, pathogen destruction, and nutrient recycling. However, some synthetic organic compounds, including many chlorinated hydrocarbons, degrade slowly and can persist or accumulate in wetlands. The rate of petroleum hydrocarbon degradation in wetlands is largely undetermined, although many hydrocarbons are known to be effectively decomposed through land treatment (Kincannon 1972). Therefore, a key consideration in assessing the utility of wetlands will be to determine the rate of hydrocarbon decomposition.

Most human pathogenic organisms entering a wetland will die, since the wetland will be an unfavorable environment. While pathogens released directly to San Francisco Bay would also die, the rate of introduction in stormwater may be sufficient to prevent use of valuable offshore shellfish beds.

Nutrient cycling in wetlands is very complex. For example, nitrogen may be removed by ammonia volatilization, depending to a large extent upon temperature and re-dox potential (Rheinheimer 1974). Ammonia is the most prevalent nitrogen species in many wetlands, due to the anaerobic conditions of most wetlands soils which prevent the presence of nitrifying bacteria (Lance 1972). However, if nitrate is found in the system, denitrification at the sediment-water interface may be a significant removal mechanism (Kaplan et al. 1979). Nitrogen-fixing bacteria may be present, although probably not favored in a system receiving adequate nitrogen supply from urban runoff. Algae and higher plants further complicate nutrient cycling, as they will use nutrients during periods of growth, and re-release them during periods of dieoff.

Anticipated Wetlands Performance

Review of wetlands performance in treating wastewater reveals great variability in system effectiveness. Shown on Table 1 are water quality treatment efficiencies for four natural wetlands which receive urban runoff. Removal

TABLE 1
URBAN RUNOFF TREATMENT EFFICIENCY OF FOUR WETLANDS SYSTEMS
(HICKOK ET AL. 1977; LYNARD ET AL. 1980; MORRIS ET AL. 1980)

Wetlands	Constituent	Removal efficiency (%)
Montgomery County, MD	Ammonia nitrogen	99
Wetlands detention basin	Total phosphorus	99
	Orthophosphate	93
	BOD_5*	97
	Cadmium	98
	Iron	96
	Lead	96
	Zinc	99
University of Central Florida	Total nitrogen	95
Cypress dome	Total phosphorus	97
	Suspended solids	99
	BOD_5*	89
Lake Tahoe, CA	Ammonia nitrogen	≤67
High altitude meadows	Nitrate nitrogen	≤96
	Total phosphorus	≤93
	Suspended slids	≤99
Wayzata, MN	Ammonia nitrogen	Net increase
Peat bog	Total phosphorus	78
	Suspended solids	94
	Cadmium	25–80
	Copper	73–83
	Lead	90–97
	Zinc	78–86

*Five-day biological oxygen demand.

varies considerably among systems and during different times within systems. Thus in the design of the Coyote Hills wetlands, we sought to maximize pollutant removal while anticipating that removal effectiveness would be quite variable. Particular attention should be focussed on wetlands performance as vegetation matures for several years following construction.

Coyote Hills Wetlands Design

The main supply of water to the site is from a diversion of surface runoff from Crandall Creek into a 0.16 ha pond (debris basin) at the northeast corner of the system (Figure 1). This basin captures large debris and distributes flow into systems A and B over two 18-m long concrete sills. Concrete was used because design flow velocities would have resulted in severe erosion to earthen berms. The system was designed to accommodate runoff from the 10-year, 6-hour storm, with maximum flow of 13.8 m^3/s. Greater flows will cause the diversion structure to fail, allowing excess water to be diverted into the flood control basin to the north.

System A is a 2-ha lagoon with a central island. Most of the pond margins have a 1:4 slope, as steep a slope as practical for long-term stability, while providing minimal habitat at needed depths for cattails. The lower margin has a 1:10 slope to provide a "shelf" to encourage wetlands vegetation. Thus, this system has a large basin to promote sedimentation, with thick stands of emergent and submerged vegetation to promote filtration, absorption, and biological degradation and transformation.

System B receives discharge initially into a small pond which distributes flow across a 1.6-ha overland flow area. Water levels in the overland flow area are shallow, with most of the ground exposed during the dry season. Vegetation which can tolerate wet-dry cycles will predominate, such as alkali bulrush (*Scirpus robustus*). Following the overland flow area, water enters a 0.7-ha pond transected by three submerged sills designed to support narrow stands of cattails.

Flow from Systems A and B join in System C. System C has a main channel of 1.6-2 m depth covering approximately 4.6 ha. About 3.8 ha adjoining the channel is a broad flat overflow area thickly vegetated with cattails and alkali bulrush. During the peak flow periods, the water level rises about 0.33 m, inundating most of this area. This system provides a large water-soil-plant interface with substantial storage capacity.

Water discharges from the west end of System C through four 1.5-m metal culverts under the northern boundary levee into the Coyote Hills Park North Marsh, and also through a series of three 0.9-m metal pipes under the southern boundary levee into Coyote Hills Park Main Marsh (Figure 1).

Wetlands Performance Monitoring

The Coyote Hills Wetlands offers an opportunity for monitoring the performance of wetlands designed specifically to treat urban stormwater runoff. Programs could be (and have been) designed to monitor such wetlands attributes as recreational opportunity, wildlife use, etc. However, since our concern was to develop a system that could use urban stormwater as its water supply and produce a high quality effluent, with minimal adverse environmental effects, this discussion will be limited to considerations of water quality and environmental fate.

Metals will be of particular interest in a wetlands monitoring program because of the potential for substantial metal depositions into the watershed from upstream activities, and the ability of many metals to accumulate to toxic levels. Data are not available to adequately evaluate the potential for accumulation of metals in wetlands receiving stormwater runoff. However, since most treated wastewaters applied to land should not result in significant metal accumulation (EPA 1977), and since urban stormwater usually does not contain significantly higher levels of heavy metals than wastewater, detrimental effects due to accumulation would not be anticipated. Upstream control of point sources containing high concentrations of heavy metals may be critical to the long-term success of the system.

Solids typically will be removed as water passes through a wetlands. However, some potential exists for re-release of particles during extreme events, or as vegetative material dies. Thus, a substantial effluent solid load may be expected periodically.

Wetlands appear to offer a high potential for biodegradation of organics. The fate of petroleum hydrocarbons in wetlands is of particular interest, both because of its toxicity and because of the frequent occurrence of relatively high concentrations in urban runoff (Fam et al. 1987). Hydrocarbon degradation in wetlands would permanently remove its toxicity threat; however, the anaerobic condition expected in the sediments may result in hydrocarbon accumulation with potential for re-release.

Monitoring also needs to focus on the primary nutrients responsible for cultural eutrophication: nitrogen and phosphorus. Nutrient removal in a wetlands may be very high; Prenti et al. (1978) report removal of 330 kgN/ha and 467 kgP/ha by a cattail marsh in Wisconsin. However, nutrient re-release may also be anticipated as plant populations fluctuate. Of particular interest should be the timing of nutrient release to receiving waters. Also of interest will be accumulation of nutrients in harvestable vegetation, offering the opportunity of removing nutrients from the system and perhaps using this plant growth for food or fuel.

Conclusions

Most wetlands preservation and development activities have been justified on the basis of habitat and recreational values. Largely untapped is the potential for wetlands development to treat urban stormwater runoff. The project at Coyote Hills serves as a mechanism to evaluate this potential.

Development of freshwater wetlands in the West is hindered by the scarcity and cost of water, particularly in urban areas. Systems designed to import and treat potable water do so at too great a cost to consider its use to maintain a freshwater wetlands. However, stormwater runoff generally has no commercial value. The ecological value of runoff is degraded by the pollutant load obtained by urban area drainage; treatment of this supply in a wetlands should increase this value.

Proposals to develop freshwater wetlands using urban runoff will face difficulty due to insufficient data to ensure that these systems will not lead to toxic accumulation of materials. Data on the Coyote Hills wetlands are not available to provide empirical evidence of the potential for toxic materials to accumulate, or design and maintenance procedures which can prevent such occurrences. However, this system offers the potential for use as a demonstration project to study these issues, perhaps resolving concerns for other wetlands development in the region.

Clearly, the need for controlling nonpoint sources of pollution is growing. For example, the Clean Water Act Reauthorization of 1987 (CWA) requires states to develop a management program for waters not meeting CWA goals due to nonpoint source runoff. Wetlands have an intrinsic capability to treat much of the pollutant load characteristic of runoff, but have not been designed specifically to fulfill this function. While federal programs to implement monitoring and quality requirements on nonpoint source pollution in urban areas have been deferred, their eventual adoption is probably inevitable and will require the use of innovative mitigation techniques.

However, development of wetlands can probably best proceed not on the basis of providing a particular value, but on their ability to fulfill multiple objectives. These values include wildlife habitat, flood protection, high primary productivity, groundwater recharge, recreational opportunities and aesthetics, as well as water quality control. It may be in the best interest of those individuals promoting a system for a particular value to consider its multiple-use potential. In particular, as regulatory systems require reduction and control of pollutants from nonpoint sources in urban areas, wetlands advocates may consider this an opportunity for enhancing the seasonal freshwater wetlands resource.

References

ASIWPCA (Association of State and Interstate Water Pollution Control Administrators). 1985. America's clean water. The states' nonpoint source assessment 1985. Washington, DC: ASIWPCA..

BCDC (San Francisco Bay Conservation and Development Commission). 1982. Diked historic bayland of San Francisco Bay. Finding 3, San Francisco Bay Conservation and Development Commission. San Francisco, CA: San Francisco Bay Conservation and Development Commission.

Chan, E., T. Bursztynsky, N. Hantzsche, and Y. J. Litwin. 1982. The use of wetlands for water pollution control. Completion report. Municipal Environmental Research Laboratory. Cincinnati, OH: U.S. Environmental Protection Agency. EPA-600/S2-82-086.

Connel, D. W. 1982. An approximate petroleum hydrocarbon budget for the Hudson Raritan Estuary-New York. *Marine Pollution Bulletin* 12(3):89–93.

EPA (U.S. Environmental Protection Agency). 1977. Process design manual for land treatment of municipal wastewater effluents. EPA-6225/1-77-008. Cincinnati, OH: U.S. Environmental Protection Agency.

EPA (U.S. Environmental Protection Agency). 1984. Nonpoint source pollution in the U.S.: Report to Congress. Office of Water, Criteria and Standards Division. Washington, DC: U.S. Environmental Protection Agency.

Fam, S., M. K. Stenstrom, and G. S. Silverman. 1987. Hydrocarbons in urban runoff. *Journal of Environmental Engineering* 113(5):1032.

Field, R. and T. Turkeltaub. 1981. Urban runoff receiving water impacts: program overview. *Journal of the Environmental Engineering Division.* American Society of Civil Engineers, 107(EE1).

Framson, P. E. and J. A. Leckie. 1978. Limits of coprecipitation of cadmium and ferrous sulfide. *Environmental Science and Technology* 12(4):464.

Gambrell, R. P., R. A. Khalid, and W. H. Patrick, Jr. 1980. Chemical availability of mercury, lead and zinc in mobile bay sediment suspensions as affected by pH and oxidation reduction conditions. *Environmental Science and Technology* 14(4):431.

Hermann, R. 1981. Transport of polycyclic aromatic hydrocarbons through a partly urbanized river basin. *Water, Air and Soil Pollution* 25:349–364.

Hickok, E. E., M C. Hannaman, and N. C. Wenck. 1977. *Urban runoff treatment methods: I—nonstructural wetlands treatment.* EPA-600/2-77-217. Cincinnati, OH: U.S. Environmental Protection Agency.

Hoffman, E. J., J. S. Latimer, G. L. Mill, and J. G. Quinn. 1984. Urban runoff as a source of polycyclic aromatic hydrocarbons to coastal water. *Environmental Science and Technology* 18(8):580–587.

Kaplan, W., I. Valiela, and J. M. Teal. 1979. Denitrification in a salt marsh ecosystem. *Limnology and Oceanography* 24(4):726.

Kincannon, C. B. 1972. Oily waste disposal by soil cultivation processes. EPA-R2-72-110. Cincinnati, OH: U.S. Environmental Protection Agency.

Lance, J. C. 1972. Nitrogen removal by soil mechanisms. *Journal of the Water Pollution Control Federation* 44(7):1352.

Larson, R. A., L. L. Hunt, and D. W. Blankenship. 1977. Formation of toxic products from a #2 fuel oil by photo oxidation. *Environmental Science and Technology* 11(5):492.

Lockhart, H. B., Jr., and R. V. Blakeley. 1975. Aerobic photodegradatio of Fe(III)—(Ethylenedinitrilo) tetraacetate (Ferric EDTA). *Environmental Science and Technology* 9(12):1034.

Lynard, W. G., E. J. Finnemore, J. A. Loop, and R. M. Finn. 1980. Urban stormwater management and technology: Case histories. EPA-600/8-80-035. Cincinnati, OH: U.S. Environmental Protection Agency.

Mackenzie, M. J. and J. V. Hunter. 1979. Sources and fates of aromatic compounds in urban stormwater runoff. *Environmental Science and Technology* 13(2):179–183.

Morris, F. A., M. K. Morris, T. S. Michaud and L. R. Williams. 1980. Meadowland natural treatment processes in the Lake Tahoe Basin—a field investigation. Las Vegas, NV: U.S. Environmental Protection Agency, Environmental Monitoring Systems Laboratory.

Prenti, R. T., T. D. Gustafson and M. S. Adams. 1978. Nutrient movements in lakeshore marshes. In *Freshwater wetlands: Ecological processes and management potential,* eds. R. E. Good, D. F. Whigham, and R. L. Simpson. New York: Academic Press.

Rheinheimer, G. 1974. *Aquatic Microbiology.* London: John Wiley and Sons.

Silverman, G. S., M. K. Stenstrom, and S. Fam. 1985. Evaluation of hydrocarbons in runoff to San Francisco Bay. Completion report to U.S. Environmental Protection Agency. Oakland, CA: Association of Bay Area Governments.

Stenstrom, M.K., G. S. Silverman, and T. A. Bursztynsky. 1984. Oil and grease in urban stormwaters. *Journal of the Environmental Engineering Division,* American Society of Civil Engineers 10(1):58–72.

Wakeman, N. 1982. Development of regional wetlands restoration goals: San Francisco Bay. In *Wetlands Restoration and Enhancement in California,* ed. M. Josselyn. Proceedings of a workshop held at California State University, Hayward, CA. February, 1982. La Jolla, CA: California Sea Grant College Program.

Whipple, W., Jr., and J. V. Hunter. 1979. Petroleum hydrocarbons in urban runoff. *Water Resources Bulletin* 15(4):1096–1105.

Gary S. Silverman is Director and Associate Professor, Environmental Health Program, Bowling Green State University, Bowling Green, OH 43403. Emy Chan Meiorin is Senior Environmental Engineer, Association of Bay Area Governments, Oakland, CA 94607.

Construction of Cattail Wetlands Along the East Slope of the Front Range in Colorado

David L. Buckner and Robert L. Wheeler

Abstract: *A 5-hectare cattail wetland was constructed in eastern Boulder County, Colorado, in November 1986. Two of the 4.9 hectares were planted using "live topsoil" removed directly from a marsh doomed by highway construction several miles away. The remaining 3 hectares were seeded in March 1987 with cattail seed collected locally in November 1986. Plant materials in "topsoiled" areas germinated slowly but steadily during May and June 1987, resulting in a cover by hydrophytes of 48% (with 30% open water) by September 1987. Seeded areas experienced germination beginning in mid-May and progressed to a 77% cover by hydrophytes (with 8% open water) by September.*

Key words: *cattail/bulrush wetland, wetland construction, live topsoil, cattail seeding.*

Introduction

Boulder Reservoir is a water storage facility serving both the City of Boulder municipal water supply system and the agricultural water supply system operated by the Northern Colorado Water Conservancy District. The reservoir was designed to store 15.7 million m^3 of water but had previously been operated at a maximum capacity of 10.9 million m^3. Extensive wetlands around the western margins had been identified by local naturalists as important habitat for certain birds of prey that had become increasingly rare as urbanization of the region had increased. Thus, plans for raising reservoir water levels generated concern for the future of these wetlands. A study of the ecology and history of these wetlands showed that their presence along the margin of the reservoir was coincidental, that they had little direct support from reservoir waters, and that they represented a small portion of what was once a much larger expanse of wetlands present in the valley before the reservoir was built (CDM 1986). Analysis of past wetland response to variations in water levels showed that a loss of 5 ha of cattail wetlands could be expected as a result of the increased water levels. The desirability of establishing new cattail wetlands to serve the habitat needs of wildlife species of concern was identified as a result of this study. Potential areas for this mitigation were also evaluated and ranked.

Following the study, during the fall of 1986, the City of Boulder Public Works Department (BPWD) identified a set of circumstances that would allow mitigation to proceed on two of the three potential reconstruction sites. The Colorado State Department of Highways (CDH) was in the advanced preconstruction stages of extending a four-lane highway in the northeastern part of Boulder. An approximately 0.6 ha cattail wetland in the path of the highway had been identified and its mitigation of proposed disturbance addressed in an application to the U.S. Army Corps of Engineers under section 404 of the federal Clean Water Act (33 U.S. C. 1344). As part of a complex, mutually beneficial arrangement, the City of Boulder was given the marsh soil and plant materials for use in Boulder Reservoir wetland mitigation in exchange for allowing the CDH to construct its mitigation wetland on other city-owned property. This paper describes the objectives, methods, and results of construction of 5 ha of cattail wetland sponsored by the City of Boulder in response to citizen concerns about the loss of wildlife habitat as a result of higher reservoir water levels.

Design Objectives

Criteria used in the design of the new wetlands included (1) emphasis of tall species, especially broadleaf cattail (*Typha latifolia*) and hardstem bulrush (*Scirpus acutus*) to provide potential nesting and hunting habitat for northern harriers (*Circus cyaneus*) and appropriate winter hunting cover for short-eared owls (*Asio flammeus*); (2) location of created wetlands as near to those to be lost as possible, to ease transfer of use by wildlife; and (3) minimization of construction and upkeep costs.

Construction Design

The site selected to receive created wetlands, Little Dry Creek, was the area ranked highest in the previous study of mitigation alternatives at Boulder Reservoir (CDM 1986). At the Little Dry Creek site, topography suitable to sustain wetlands extended over about 5 ha. This site rated highest because (1) it had, until the 1940s, been the site of an extensive wetland that was drained at that time, and (2) the soils of the site were clay-textured. The latter was an advantage because soil with minimum infiltration would allow a given amount of water to spread over the widest possible area after the drainage ditch was filled. Little Dry Creek, the water source for any wetlands to be constructed at this site, is small, flowing at about 12 liters per second or less for much of the time, so little of the available water could be lost to infiltration if 5 ha were to be adequately wetted.

Of the 5 ha to be transformed to wetlands, slightly less than 2 ha were to be treated using soil and plant material ("live topsoil") removed from the highway project marsh. The remaining approximately three ha were to be seeded with cattails.

To encourage even spreading of water, three 46 cm high spreader berms (contour dams to back up shallow water) were envisioned, spaced evenly through the area. In order to achieve the required conditions with the least effort and expense, the spreader berms were to be simple earthen dikes without reinforced drop structures (spillways) of any type; since it was assumed that coarse wetland vegetation cover would eventually function in spreading water flow across the area, the long-term integrity of the spreader berms was not seen as critical.

Overall, to avoid problems with construction equipment, the most practical strategy seemed to be that all earthwork and planting in the area should be completed before the drainage ditch was filled and water began to spread through the area.

Live Topsoil Areas

The term "live topsoil" has arisen in the field of coal surface mine reclamation and refers to the removal and immediate use of topsoil without storage. This procedure allows the most beneficial use of the substantial living component of soil, in the form of seeds, roots, rhizomes, tubers, and bulbs as well as the microflora and fauna. Areas to receive "live topsoil" from the highway project marsh were located immediately upslope from the two lower spreader berms. This area would expose them to the deepest water the site could offer and would discourage development of undesirable weeds that had infested parts of the donor marsh, especially teasel (*Dipsacus sylvestris*). The latter is a biennial introduced species that has, over the past 20 years, expanded rapidly in Boulder County. Although very aggressive, it will not germinate in standing water and, thus, it was deemed important that imported live topsoil materials should be located in standing water to prevent establishment of this undesirable weed in the Boulder Reservoir area. The presence of teasel in the donor marsh may have reflected a diminished availability of water as development had crowded around it in recent years. To make sure that the land surface of the reconstructed wetland area would be well supplied with water, at least 15 cm of soil were removed from the treated areas so that when 15 cm of marsh topsoil was added, the land surface would be no higher than it was originally.

Seeded Areas

In the slightly more than 3 ha where cattail was to be seeded (Figure 1), no adjustments to surface elevations were to be necessary, but to allow young

FIGURE 1. *Wetlands resulting from direct seeding of cattails, five months after germination; Boulder Reservoir, Boulder County, Colorado.*

cattail seedlings a chance to establish, it was deemed desirable to reduce the competition with pre-existing vegetation by removing at least some of it to open up space for development of cattail seedlings.

FIELD CONSTRUCTION

Construction activities in the field ensued in early November 1986. The first step was establishment of the location of spreader berms exactly along equal elevation contours. This was accomplished by City of Boulder surveying personnel who also marked the top of the berm 46 cm above the ground on each location stake. Earthmoving was carried out by a private contractor. The first earthmoving work was establishment of the spreader berms using a motor grader that bladed material from upslope into an 46 cm high berm and then compacted it carefully with the wheels of the machine. The grader operator frequently used a 1.5 m level to check berm-top elevation against grade stakes.

Next, removal of soil from areas destined to receive live marsh topsoil was begun. Although an elevating scraper was tried at first, it was not effective because the clay soil, even when slightly moist, offered too little traction. A D-8 bulldozer proved the most practical machine. Fifteen to twenty cm of soil were pushed from the areas to receive live topsoil to the edge of the channel draining the valley bottom.

During this same time, the removal of live topsoil from the marsh at the highway project was begun. It was fortuitous that this operation could occur during a season when the wetland plants were dormant and mortal damage as a result of excavation would be minimized. The marsh had been dewatered by pumping several weeks previously. The boundaries of the desirable marsh vegetation had been marked by flagging to guide heavy equipment operators during the topsoil removal. Actual removal of the marsh soils was accomplished using a large track-mounted backhoe. Soils were removed to a depth of about 46 cm, the limit to which the dark organic muck and useful plant parts extended. The backhoe piled the material on an adjacent upland location, whereupon a rubber-tired front-loader placed the material in trucks for transport to the Boulder Reservoir site. The trucks used were semi tractor-trailer machines of two types: belly-dump and end-dump. For this type of application, under somewhat slippery, muddy conditions, the two types have different strengths and weaknesses. Belly dump trucks, upon arriving at the Boulder Reservoir site and releasing their load, typically ended up immobilized with masses of slimy marsh soil in front of their trailer wheels. Belly dump trucks usually have a "stinger" protruding from the rear of the trailer, for the purpose of accepting a push by a bulldozer, which process, while time consuming, at least gets them out of the way of the next truck waiting to dump. End dump trucks release their load to the rear of the trailer wheels and are thus not nearly as likely to cause themselves to get stuck; however, if they do become mired, they are much more difficult to extricate since they cannot be pushed from behind. Over a period of six hours, approximately 3,060 m^3 cubic yards of marsh soil were delivered to the Boulder Reservoir site. The D-8 bulldozer along with a sheeps-foot dozer (a wheeled dozer with large, blunt spikes covering the wheels) were used to spread and smooth the marsh soils material. Because of the fine-textured and very wet nature of the marsh soil, it was impossible to spread perfectly. The sheeps-foot dozer seemed preferable in this regard, as its broad footprint accomplished more desirable spreading than the blade on either machine.

In the 3 ha not treated with live topsoil, the original plan had been to disc plow the pre-existing upland vegetation to reduce late competition with seeded cattails as much as possible. However, because it had snowed twice by late November, it was simply too wet to operate a disc plow or any other farm implement in the clay soils of the site. As a compromise, the rippers on the rear of the D-8 were used to achieve at least some destruction of pre-existing vegetation and expose bare soil for cattail seeds to contact when they were sown the following spring.

The final step in field construction was to fill the drainage ditch with soil and extend the spreader berms across this area. This operation was difficult because, in the soft conditions, it was very easy to mire even a D-8 bulldozer.

The spreader berm sections across the drainage ditch fill were necessarily built by the dozer and were crude by comparison with the berm built and compacted by the motor grader.

Cattail Collection and Seeding

Approximately 32 kg of broadleaf cattail seed was collected from Boulder Reservoir wetlands in November and December 1986. By this time, the cattail spikes were on the verge of shattering, a condition taken to indicate that seed was ripe. The spikes were placed in paper or plastic mesh bags and stored in unheated buildings until March 1987. At this time, three methods of seeding were employed: (1) shattering cattail spikes in a bucket, then mixing with wet masonry sand which then acted as a "carrier" as the mixture was hand broadcast; (2) poking unshattered spikes into the soil, anticipating slow dispersal of seed onto the soil surfaces around the implanted spike; and (3) release of cattail seed into the wind from locations just upwind from the project area.

First Year Results

Vegetation

Beginning in approximately mid-May, cattail seeds started to germinate; germination percentage seemed to be quite high judging from the carpets of seedlings found surrounding some of the implanted spikes (see planting method 2, above). By this time, only a very few shoots could be found emerging from the live topsoil areas. Even by mid-June, the vegetation response in the live topsoil areas was slight compared to the seeded areas where the cattails were already as tall as 60 cm. During the mid- to late-summer, however, vegetation in the live topsoil areas developed rapidly. Heavy stands of cattail and hardstem bulrush (*Scirpus acutus*) were present along with certain of the aquatic broadleaf flowering forbs, especially arrowhead (*Sagittaria latifolia*) and water plantain (*Alisma plantago-aquatica*). Other graminoids present included barnyard grass (*Echinochloa crus-galli*), common spikerush (*Eleocharis macrostachya*), Baltic rush (*Juncus arcticus ssp. balticus*), and Torrey rush (*Juncus torreyi*). Cattails and bulrushes were mostly in the range of 1.2 to 1.5 m in height. Point-intercept samplings in the live topsoil areas showed a cover of 48% by hydrophytes, with 30% open water. In the seeded areas, cover by hydrophytes was 77% and open water comprised 8% cover. Cattails in the seeded areas were fully as tall as those in the live topsoil areas. No teasel germination was observed except where chunks of marsh topsoil extended above water level.

Wildlife

Use of the area by birds was remarkable even during the development phase of the new wetlands. Species observed during the first summer include kildeer

(*Charadrius vociferus*), Canada goose (*Branta canadensis*, possibly including a nesting pair), mallard duck (*Anas platyrhyncos*), widgeon (*Mareca americana*), blue wing teal (*Anas discors*), cinnamon teal (*Anas cyanoptera*), great blue heron (*Ardea herodius*), snowy egret (*Leucophoyx thula*), and avocet (*Recurvirostra americana*), as well as the red-wing and yellowheaded blackbirds (*Xanthocephalus xanthocephalus*), as well as red-winged blackbirds (*Agelaius phoeniceus*), typical of cattail stands of almost any sort.

Conclusions

Construction of cattail wetlands at Boulder Reservoir was very successful because (1) municipal, state, and federal government agencies as well as private industry cooperated closely to the advantage of all parties and (2) construction efforts addressed the establishment of wetland biota and hydrologic conditions with methods as simple and direct as possible.

Acknowledgments

This project was conducted with the financial and technical support of the City of Boulder, Colorado Public Works Department. Special cooperation was also given by the City of Boulder Parks Department, and the Real Estate and Open Space Department; the Colorado Department of Highways; and Flatirons Construction Company.

References

Camp Dresser & McKee (CDM). 1986. *Boulder Reservoir Environmental Study Final Report.* Prepared for City of Boulder Public Works Department, Parks and Recreation Department, Real Estate and Open Space Department, and the Northern Colorado Water Conservancy District. Prepared by Camp Dresser & McKee, 2300 15th Street, Suite 400, Denver, CO 80202.

David L. Buckner is Plant Ecologist with ESCO Associates, Inc., P.O. Box 13098, Boulder, CO 80308. Robert L. Wheeler is Assistant Director of Public Works/ Utilities with the City of Boulder, 1739 Broadway, Boulder, CO 80302.

Restoration of Mangroves and Seagrasses — Economic Benefits for Fisheries and Mariculture

Anitra Thorhaug

Abstract: *This paper reviews mangrove and seagrass restoration in terms of its ecological and economic benefit to fisheries and aquaculture. In the tropics and subtropics, mangroves are a critical habitat in the intertidal zone and on the upper shoreline. Seagrasses are a critical habitat from the intertidal zone seaward to the coral reef. Mangroves and seagrasses serve parallel functions as nursery grounds and critical habitats for fish, as a direct food source for fish, and as surfaces for growth of epizonts, which serve as food for fish.*

This paper contains a review of coastal restoration efforts, including an assessment of development impacts on mangroves and seagrasses. Management solutions to nearshore fisheries problems are also discussed, and general recommendations for using seagrass and mangrove restoration to sustain fisheries are made.

Key words: *seagrass restoration, mangrove restoration, fisheries.*

Objective

The objective of this paper is to review mangrove and seagrass restoration in terms of its ecological and economic benefit to fisheries and aquacultures. In the tropics and subtropics, mangroves are a critical habitat in the intertidal zone and on the upper shoreline. Seagrasses are a critical habitat from the intertidal zone seaward to the coral reef. Mangroves and seagrasses serve parallel functions as nursery grounds and critical habitats for fish, as a direct food source for fish and as surfaces for growth of epizonts, which serve as food for fish.

For those unfamiliar with coastal restoration techniques for seagrasses and mangroves and their results, reviews of global coastal restoration efforts will be made. Economic and ecological benefits of seagrass and mangroves are reviewed. Development impacts on mangroves and seagrasses are assessed. Management of nearshore fisheries in a sustainable manner are discussed. General recommendations for using restoration of seagrasses and mangroves to sustain fisheries and fisheries nurseries are made.

Impacts on Mangroves and Seagrasses in the Tropics

Human impacts which we wish to manage and control fall into several major categories: infrastructure, industry, urban expansion, agricultural effects, and direct removal of mangroves (for firewood, timber, and mariculture ponds).

Infrastructure development is generally planned by the government, although it may be carried out by the private sector. These developments include roads, causeways, ports, marinas, dredged channels, coastal airports, power plants, sewage and water facilities, and waste dumps. Many of these can be planned with avoidance of impact or minimal impact, but usually are not.

Industry frequently not only locates factories or industrial parks along estuaries or coastlines (to take advantage of "free" water facilities), but cuts down mangroves and dredges and fills seagrasses to build sites. Industry releases effluents into water and air, some of which are highly toxic.

Urban expansion of coastal cities is occurring rapidly in most nations as populations grow and migrate from inland to coastal cities. Urban wastes dumped on seagrasses and mangroves include storm-water runoff from streets, including oil, fertilizers, and chemicals of all nature. During urban expansion people frequently chop down mangroves and dredge and fill mangrove swamps.

Agricultural effects include soil erosion, which increases turbidity, thereby killing seagrasses, which no longer receive enough light in deeper waters. Fertilizers, pesticides, and other substances can also adversely affect estuaries.

The causes of direct removal of mangroves for human purposes have been documented by Kapetsky (1985) and in a series of publications by World Resource Institute (WRI). Natural disasters can also decimate mangroves and seagrasses, but we can restore them rapidly to production. Typhoons, tropical cyclones and hurricanes are the most frequent natural impacts. Tidal wave and storm surges, and drought—with incumbent high salinity and hypersalinity—can effect mangroves and seagrasses. Biological stressors can also decimate populations. Microbial disease has been recorded to afflict mangrove species. Grazing, for example, by goats, cattle, and camels, has reduced mangroves. Manatees and dugongs also graze on seagrasses as do fish and invertebrates, including echinoderms.

Economic and Ecological Benefits of Mangroves and Seagrasses

Mangroves provide a series of direct economic benefits from the uses of their wood (Christensen 1984; Kapetsky 1985). Seagrasses provide more indirect economic benefits through their function as fishery nurseries, by providing food and habitat for adult fisheries species, and through erosion control. Mangroves also function in many of these above capacities. (Frequently the fisheries data has not separated mangrove functions from seagrass functions and has attributed a fisheries nursery to mangroves, when a large seagrass bed was present.) Several studies have specifically separated the two (Roessler et al. 1974; Lugo and Cintron 1975).

Once mangroves and seagrasses are removed, none of these economic values are available. Once mangroves and seagrasses are restored, all of these economic values return. The United Nations Environment Programme has placed a $215,000/ha value on these resources. (See Table 1 for values of Mangroves.) It costs about $1,236/ha to restore them in developing nations. Therefore, it is of economic benefit for government, private sector, and citizens who use mangroves and seagrasses to restore them. The decision-making criterion is how much the development project is worth in benefit to the people, and how much multi-resource benefit could be derived from some environmental modification of the project.

The cost/benefit analysis of restoration would include an evaluation of the cost to plant mangroves or seagrasses versus the benefits of multiple use or single use. A cost/benefit analysis for mangrove restoration can be done for the direct uses of mangroves or for the indirect cost/benefits from the derived fisheries or water clarity. Since there is almost no direct use of seagrasses except for sediment stabilization, the costs and benefits of seagrasses derive primarily from the nursery and adult fisheries in seagrass beds. Water clarity can be thought of as a secondary product.

Many studies of the fisheries associated with mangroves and some with seagrasses have been carried out. Unfortunately for Asian countries, most are carried out in the Greater Caribbean region. The nursery function of a dense, large, intertidal mangrove swamp is far greater than a thin patchy layer of shoreline mangroves. The fisheries product of a large, dense estuarine seagrass meadow is quite different from a thin ridge of shoreline mangroves along a coral rock or a developed coast. Most studies of fisheries in mangroves or seagrasses do *not* preclude the presence of the other habitat, although they center on the primary habitat. Many situations have predominantly mangroves (very turbid river mouths with large tidal incursions) or predominantly seagrass (high energy or open ocean coastlines, rocky intertidal and upper tidal areas with sand or sediment sublittoral, large shallow clear estuaries with no vegetation or other vegetation on shoreline).

The economic benefit in the form of catch fisheries associated with either mangroves or seagrasses is frequently multiplied several orders of magnitude by the nursery function of the habitat. This has not been specifically quantified.

As a protective nursery habitat, seagrasses and mangroves are places in which marine fauna lay eggs and in which early larvae of hundreds of species are sheltered. Both seagrasses and mangroves act as a place of attachment for adult attached forms, ranging from foraminifera to sponges, and commercial molluscs, such as scallops and oysters. Adult fishes caught in seagrasses

TABLE 1
SOME GROSS ECONOMIC VALUES OF MANGROVES FOR FISHERIES BASED ON VALUES OF LANDINGS (TAKEN FROM KAPETSKY 1985)

Country/Region	Value in U.S. Dollars	Basis for Value Measurement	Year of Estimate	Principal Species	References
Panama/Gulf of Panama	26,350 (/km)	Value per kilometer of mangrove shoreline.	1978	*Penaeus, Trachypenaeus*	D'Croz and Kwiecinski, 1980
	65,164 (/km)	Value per kilometer of mangrove shoreline.	1978	*Cetengraulis, Mysticatus*	
	3,114 (/km)	Value per kilometer of mangrove shoreline.	1978	*Micropogon, Tutlanus, Cantropomus*	
Brazil/Cururuca Estuary	76,886 (/km²)	Includes only finfishes. Estimate based only on areal extent of open water.	1981/82	*Mugil, Canyactemus, Macrodon, Sagra, Macropengenisa*	Universidad Federal de Maranhoe, 1983
Malaysia/Sabah	133 (/km²)	Value based on areal extent of mangroves.	1977	*Scylla serrata*	Foo and Wong, 1980
Malaysia/Peninsula	277,235 (/km²)	Areal extent of mangroves plus estuaries and lagoons.	1979	*Penaeus, Scolopherus, Pampus, Polynemus, Lucjanus*	Gedney, Kapetsky and Kuhnhold, 1982
Thailand/Khiung District	3,000 (/km²)	Value of fishery products captured inside the mangrove system.	1977	*Lita, Eleutheroneus, Arias, Ophichchus, Lates*	FAO, 1982
	10,000 (/km²)	Value of mangrove-associated species caught elsewhere.	1977	*Penaeus*	
Bangladesh/ Sundarabans	2,076 (/km²)	Value based on mangrove area plus open water area.	1982/83	*Hilsa, Penaeus*	Rabanal, 1984
Papua New Guinea Gulf Province	426 (km²)	Value of shrimp caught outside the mangroves and of subsistence fishing and crabbing inside.	1977	*Penaeus, Metapenaeus, Scylla serraca,* ambassids, gobies, gudgeons, catfishes	Liem and Haines, 1977

include snapper, grouper, bonefish, barracuda, sea trout, tarpon, sea bass. Invertebrates include shrimp, crabs, lobster, scallops, sponges, molluscs. Adult fishes caught in mangroves include carangids, clupeids, pomadadasids, serranids, scianids, mulids, siganids, and lutjanids. Invertebrates include clams, crabs, and lobsters.

Aquaculture and mariculture economic costs and benefits from mangroves have been detailed by several authors including Kapetsky (1985) and Christensen (1984). There appears a direct trade-off between the land near coastline needed for mariculture and mangroves. Evidently a cost-beneficial, mixed system of mangroves co-existing with mariculture facilities has not yet been worked out so that the most productive part of the forest and its water flow can remain in proximity to limited mariculture facilities.

Seagrass and Mangrove Restoration Efforts

General Results

Mangrove and seagrass restoration are much younger technologies than upland reforestation, especially in temperate zones, and have received far less attention, funding, and research. Mangroves have been planted seriously for two decades; seagrasses for one. The conditions under which restoration takes place are difficult and usually polluted. The sometimes high currents, wave action, and periodic heavy storm events (typhoons, hurricanes, etc.) of disturbed areas usually makes the process more complex than upland reforestation.

Yet, success has occurred in projects of many sizes. Most work has occurred in the American Caribbean, although some substantial work has recently occurred in Southeast Asia so that all the species of mangrove and (as of 1988) seven of the most important species of seagrass are under cultivation.

In terms of planning for restoration, programs which can incorporate the final restoration of mangroves or seagrasses into the plan (when they must necessarily be removed) are far ahead in cost-effectiveness and ability to rapidly regain the fisheries function than those projects which do not plan well in advance. An example is a coastal road bordering a mangrove swamp. Many such roads have killed many square kilometers of mangroves and seagrasses with concomitant loss of fisheries throughout the developed and developing world. But if the appropriate culverts under the road are placed to retain tidal flushing, and the slope of side of the road is made appropriate for replanting mangroves (intertidal) and seagrass (sublittoral), development can occur while resources are wisely managed.

Technology of mangrove and seagrass planting can be transmitted and taught to laborers in nations throughout the world. Fishermen have frequently been the planters (Thorhaug et al. 1985, 1986). The planting techniques themselves, once established, are not difficult. The design, choice or preparation of planting site, choice of species, planting method, and anchoring system are still only well executed by experts in restoration of mangroves or seagrasses. Work is underway to design foolproof site criteria from physical-chemical data so that design can be done by others.

Seagrass Restoration

Methods

1. Plugs

In this historically first successful technique (Kelly et al. 1971), plugs of combined blades, roots, rhizomes, and sediment are extracted from a natural bed and transported to a polluted site where a second hole is excavated to receive the plug. Usually a round cylinder between 10 and 20 cm, such as a posthole digger, soil auger, or a PVC pipe with handles, is used. A floating machine with a cutting arm extracting a 1 x 2 meter piece which was then cut into 20 cm pieces has been recently employed (Thorhaug 1986). Various methods of anchoring the plug in turbulent areas have been devised (cement plug collars, low barriers to energy-walls [Cooper 1979], chicken wire [Carangelo et al. 1979], various plug placement design [Goforth and Peeling 1979]). See footnote for research done by other workers using plugs.[1]

2. Seeds

The first seagrass transplant was done with *Zostera* seeds by Addy (1947a and b) who hand-planted a small number of seeds, which died. Thorhaug in 1973 had the first successful seed transplant (Thorhaug 1974). Fruit was hand-gathered from a Caribbean seagrass bed by scuba, dehisced (separated from the fruit pod), and seeds transported back to Miami where they were planted; they had a high survival rate. Since then, other workers have succeeded with other seagrasses.[2]

3. Turfs, Sods, and Grids

Turfs differ from plugs in that no attempt is made to get the entire root rhizome system by cutting deeply into the sediment, but rather the seagrass is gathered by trowel or shovel by a worker standing in shallow water. At the recipient site, a second hole is sometimes made to receive the turf. For researchers using the turf technique see Backman (1973), Carangelo et al. (1979), Larkum (1976), and Ranwell et al. (1974).

[1]Addy 1947b; Backman 1973; Carangelo et al. 1979; Churchill et al. 1978; Continental Shelf Associates 1983; Cooper 1974, 1976, 1977, 1978, 1979; Goforth and Peeling 1979; Larkum 1976; Phillips 1974, 1978; Rogers and Bisterfeld 1975; Thorhaug 1986; Thorhaug et al. 1985, 1986; Van Breedveld 1975; Winter 1986.

[2]Churchill 1983; Cooper 1974, 1976, 1977, 1978, 1979; Derrenbacker and Lewis 1983; Fonseca et al. 1985; Orth and Moore 1983; Phillips 1974; Thorhaug 1974, 1980, 1982, 1985; Thorhaug and Hixon 1975; Thorhaug et al. 1983, 1985, 1986.

4. Shoots (Turions or Sprigs)

Since 1971, individual rhizomes containing multiple blade bundles and usually (although not always) the apical meristem have been planted at sites by a series of workers.[3]

Results

General Results

This review of seagrass restoration shows that 21 groups of workers have made over 165 attempts (including subplots) at restoration in sites from Europe and the Americas to Australia and the Philippines. There were 75 successful attempts. Many of the failures were initial attempts by workers followed by technical improvements by the same workers which eventually resulted in final success. Most efforts were made in the United States Coastal Zone (southeast Florida, the Gulf of Mexico, the Carolinas, Oregon, California and Washington) with more in the southeast United States than any other region in the world. Here the United States has enormous stretches of seagrass resources.

When surveying the total results one sees that a great deal more plugging (71 attempts with 37 successes) has been attempted than seeding, shoots, or turfs. However, those who have succeeded with seeds have found this method very cost-effective. Projects using shoots and turfs can also be cost-effective and report a higher ratio of success to failure than projects using plugs.

Lack of knowledge about the general physiology of seagrass has impeded restoration. More detailed knowledge about physical environmental factors affecting restoration is needed in order to predict and accomplish future successful restoration. These factors are critical for site selection. Much less is known about chemical requirements of the species such as nutrients, trace metals, etc.

Plugs

In general, plugs have been reported to be successful 37 out of 71 times in restoring seagrass. They are especially useful for those seagrasses in shallow water that have meristematic material throughout (Cooper 1979; Goforth and Peeling 1979; Thorhaug 1985, 1986; Thorhaug et al. 1983; Van Breedveld 1975). Positive aspects of plugging consist of extraction allowing the use of unskilled workers and the ubiquitousness of technique for many species and types of sites. Plugging has been successful in the Mediterranean and Caribbean, in Australia, the Philippines, the Gulf of Mexico, and on the east and west coasts of the United States.

[3]Continental Shelf Associates 1983; Derrenbacker and Lewis 1983; Fonseca et al. 1979, 1984, 1985; Kelly et al. 1971; Kenworthy and Fonseca 1977; Ogden 1980; Phillips 1974; Thorhaug 1980, 1983, 1986; Thorhaug et al. 1985, 1986.

Seeds and Seedlings

There have been 25 seeding attempts of which 14 have been successful. In the tropics, low turbulence areas are more easily seeded. Seeds survived in as many locations as sprigs in a comparative study (Thorhaug et al. 1985). The seeding method is usually cost-efficient. Severe storms can be destructive for seeds, as well as turions and plugs (Thorhaug 1980, 1985). Seeds can grow either in barren sediment, in invading or successional stages of alpha-colonizing seagrasses, or in benthic algae. Full restoration can occur within four years when using seeding technique.

Turfs (Sods)

Investigations of restoration with turfs have had mixed success. Of 16 attempts, half have succeeded.

The greatest concentration of turfing (sodding) work has been done by Carangelo et al. (1979) in Texas estuaries and coastlines. One of the goals was to explore seagrass restoration on dredged material. To accomplish this, different types of seagrasses were used, such as *Ruppia maritima, Thalassia testudinum,* and *Halodule wrightii,* to ascertain which were the most appropriate. *R. maritima* was only partially successful but *H. wrightii* was more so.

An extensive effort was made by Ranwell's group (Ranwell et al. 1974) in Norfolk, England, wherein 2,000 turfs of *Zostera noltii* and *Zostera marina* were spaded by prisoners and students into an impacted area. This work appeared highly successful after only one year and in many cases coalesced after two.

Sprigs or Shoots (Turions)

Fifty-one attempts to plant shoots have been made (many more if each test plot were counted), of which nine were successful and twenty are pending final results.

Workers in North Carolina (Fonseca et al. 1984) have attempted to use *H. wrightii* and *Z. marina* in a mechanized shoot method. However, the success rate is uncertain. Turions (rhizomes up to 30 cm with apical meristem and several packets of blades) of three species have also been planted in a series of comparative test plots in the subtropics and tropics (Thorhaug 1985; Thorhaug et al. 1985). The results showed species differences in survival by sprig. However, the success rate is uncertain. *Thalassia* was the most widely viable in both test areas, primarily due to better survival than *Halodule* in exposed areas. *Syringodium* was the least successful in both studies. High growth and coalescence of successful *Halodule* sites were excellent. *Thalassia* grew at a slower rate due to much subsediment tissue production.

Animal Community Recovery in Restored Seagrass

The rate of return of members of the animal community to restored seagrass beds versus areas that have not been restored is a critical question in seagrass restoration as fisheries nursery rehabilitation is a primary justification for efforts. Nevertheless, there has only been one quantitative study of this phenomenon (McLaughlin et al. 1983). Results were extremely encouraging; at least four years after seeding (intervening years were not studied) animal communities within the restored *Thalassia* were not statistically different than communities in unimpacted *Thalassia* beds next to the restored area. Thorhaug and McLaughlin (1979) and Eleuterius (1975) found about 300 animal species in a restored area, while only three or four were noted in the barren control. All researchers have noted substantial increase in animals within restored seagrasses versus nonrestored areas.

Cost

The cost of planting seagrasses depends on a series of factors (Thorhaug and Austin 1976) including: type of labor used, depth of planting, experience with planting technique, equipment needed, accessibility to planting site, proximity of dissemules and species of seagrass planted. The costs as reported by workers range from $1,500–20,000/ha. This is in contrast to the value of approximately $150,000/ha for estuaries (unit value calculated by the United Nations Environment Programme in 1980). The cost of more large-scale commercial efforts must be better documented, including the differences in cost due to geographic factors (islands or mainlands, temperate or tropical climates) and levels of technological development. Prices should include cost per unit and take into account the scale of operation. The analysis should separate labor, materials, transportation and other costs (insurance, overhead, payroll, taxes, etc.) per unit planted. Unfortunately, most seagrass planting to date has been extremely expensive as it has been experimental, using new techniques, few specimens, and small, highly trained crews of biologists. Repeated efforts with a successful set of techniques, however, should bring labor costs down.

Potential for Large-Scale Restoration

This kind of project may be effective over hundreds of thousands of hectares throughout the world's coastal zones where human perturbations have created extensive underwater wastelands. Installations are presently labor-intensive due in part to the heterogeneity of most sublittoral habitats; however, in developed countries machines could be employed to lower the costs. In developing countries fishermen can supply labor needs, which will also lower costs.

Mangroves

Techniques

Seeds or Propagules

Propagules have been harvested directly from the tree (only mature propagules), from litter under trees, from rack on beaches, and from naturally planted propagules. Seeds have been collected similarly. Sorting for nondiseased ones should occur. Many investigators have nurseried propagules or seeds, others have planted many more to compensate for early mortality. Davis (1970), Savage (1972), Rabinowitz (1975), Teas et al. (1975), Lewis (1980), Schroeder and Thorhaug (1980), Thorhaug et al. (1983) and Thorhaug and Colin (in preparation) were among those who had some success with propagules or seeds. Some seeds have been nurseried for up to a year and then planted. Higher success rates were found for these.

Small Trees

Many investigators have grown mangroves to a young tree stage (0.3–1.2m). This solves the mortality-from-disease stage, and provides more rapid growth and substrate stabilization. It is more costly to use this method, but survival, especially under difficult conditions, is greater (Kinch 1975; Hannan 1975; Schroeder and Thorhaug 1980; Teas 1977; Pulver 1976; Hoffman and Rodgers 1981; Gibbs 1981). Frequently, in higher energy areas, anchors or stakes are used to ensure stability of trees.

Large Trees

Large trees are the obvious finally desired end product of projects and are frequently available at the site to be impacted by development. Ideally, one mangrove forest could be moved to a previously impacted site prior to a development. Trees are top-pruned (one quarter to one third) and frequently root pruned prior to removal from donor site (see Pulver [1976]). Cost is the factor presently preventing relocations.

More research should be done on transplanting larger trees, especially in Southeast Asia. Costs should be estimated.

In eroding or high energy areas black mangroves or species with extensive sediment holding devices generally survive far better. In low energy areas a variety of species can survive.

Results

There have been several hundred mangrove restoration efforts around the world which have been highly confined to First World and Caribbean nations. Due to the pressing need, as well as the availability of many more trained mangrove scientists, work throughout Southeast Asia has taken place on mangrove restoration. Therefore the major species are under cultivation and many technical problems have been solved.

Highest successes have been in low energy areas, especially where a one-time impact (dredging, filling, hurricane, etc.) occurred. There are some cases where, despite continuting pollution, mangroves survived and grew. Higher energy areas have had better successes with either larger specimens (more extensive root systems) or with barriers to the high energy (riprap in front, planters, floating tires). Stakes have been successful in coral rock (Schroeder and Thorhaug 1980; Goforth and Thomas 1980) but less successful in sand or mud.

All authors have noted proper elevation within the intertidal zone is necessary for the seaward edge of mangroves. Measuring where each species occurs in relationship to the intertidal zone in a particular area for the purposes of duplication is a rule of thumb.

Cost

Person-hour estimates range from 450 to 1,800 person-hours/ha for both collection and installation of propagules (seeds). Small trees require approximately twice as much time (2,550 to 3,100 person-hours/ha). The larger trees would be about two orders of magnitude higher than propagules (Teas 1977).

Costs for red, black, and white mangroves are estimated by Teas (1977) and Lewis (1982), the latter on Table 2.

Animal Recolonization into Restored Mangroves

Most observers have noted an immediate recolonization by attached invertebrates and by fishes and larvae on higher tides. Large animal populations

TABLE 2
ESTIMATED COST ($/HA) FOR PLANTING MANGROVES BY USING VARIOUS TECHNIQUES

Species and technique	Spacing (m)			Reference
	0.30	0.61	0.91	
Rhizophora mangle				
Propagules (collected)	10,175	2,470	1,140	Teas 1977
			12,500*	Lewis 1988
			6,250	Lewis 1988
	26,000	13,000	6,545*	Lewis 1981
Propagules (purchased)	11,251	2,742	1,261	Teas 1977
R. mangle, Avicennia germinaus, Laguncularia racemosa				
6-month-old seedlings	22,400	5,400	2,510	Teas 1977
(purchased)	107,593	27.2	12,103	Lewis 1981
3-year-old trees (purchased)			216,130	Teas 1977
			40,755	Teas 1977
			70,000	Teas 1977
R. mangle				
3-year-old trees (transplanted)			45,386	Goforth and Thomas 1980
A. germinaus, L. racemosa (transplanted)			11,459	Hoffman and Rodgers 1981

*Actual cost of commercial project
Modified from Lewis 1981.

have been noted in mature restored mangrove areas. To date no one has noted differences between restored and natural mangrove population in a location and of the same size and density. Due to the difficulty of sampling and lack of funds for restoration research, no careful quantitative study has been undertaken, although one is planned presently in Florida (comparing seagrass and marsh recolonization by animals).

Acknowledgments

The author most gratefully acknowledges the support of the United Nations Food and Agriculture Organization for the Seagrass Restoration Project in the Philippines which has supported many ideas of this paper. In particular, Dr. Kapetsky, Messrs. Henderson and Cortez and, at Philippine FAO, Drs. Grieb and Barfoed. The author also gratefully acknowledges the former grant #DAN-5542-6-SS-2101-00 (Office of Science Advisor) which supported Caribbean seagrass and mangrove restoration efforts.

References

Addy, C. E. 1947a. Eelgrass planting guide. *Maryland Conservation* 24:16–17.

Addy, C. E. 1947b. Germination of eelgrass seeds. *Journal Wildlife Management* 11:279.

Backman, T. W. 1973. Transplantation of plugs of eelgrass into San Diego Bay. Personal communication.

Carangelo, P. D., D. H. Oppenheimer, and P. E. Picarazzi. 1979. Biological applications for the stabilization of dredged materials, Corpus Christi, Texas: submergent plantings. In *Proceedings of the 6th annual conference on restoring coastal vegetation in Florida,* ed. D. P. Cole, 79–80. Tampa, FL: Hillsborough Community College.

Christensen, B. 1984. Los manglares para que sirvin? *Unasylva* 35(179):2–5.

Churchill, A. C. 1983. Field studies on seed germination and seedling development in *Zostera marina. Journal Aquatic Botany* 16:21–29.

Churchill, A. C., A. E. Cok, and M. I. Riner. 1978. Stabilization of subtidal sediments by the transplantation of the seagrass *Zostera marina L.* Report No. NYSSGP-RS-78-15. Garden City, NY: New York Sea Grant.

Continental Shelf Associates. 1983. Final report-seagrass revegetation studies in Monroe County. Tallahassee, FL: Florida Department of Transportation. FL-ER-20-82.

Cooper, G. 1974. Les Posidonies . . . la pollution. Importance écologique et physique. *Bulletin observatoire de la mer.* Hyères, France.

______. 1976. La Posidonie, plante étonnante. La pêche ou la mariculture? . . . *Jardinier de la mer.* Hyères, France: L'Assoc.-Fond. G. Cooper. Cahier no. 1.

______. 1977. Rôle physique de *Posidonia oceanica,* dans la formation et le maintion du double tombolo de Giens. *Jardinier de la mer.* Hyères, France: L'Assoc.-Fond. G. Cooper. Cahier no. 2.

______. 1978. *Posidonia* et la desertificaiton sous la mer. *Peuples Mediterranées* 4:43–76.

______. 1979. *Posidonia oceanica* = un arbe. Jardinier de la mer. Hyères, France: L'Assoc.-Fond. G. Cooper. Cahier no. 3.

Davis, J. H. Jr. 1970. The ecology and geologic role of mangroves in Florida. *Carnegie Institute Washington Publication* 517.

D'Croz, L., and B. Kwiecinski. 1980. Contribución de los manglares a las pesquerias do la Kahia de Panama. *Review Biology of the Tropics* 28(1):13–29.

Derrenbacker, J. Jr., and R. R. Lewis, III. 1983. Seagrass habitat restoration, Lake Surprise, Florida Keys. In *Proceedings of the 9th annual conference on wetlands restoration and creation,* ed. D. P. Cole, 28–32. Tampa, FL: Hillsborough Community College.

Eleuterius, L. N. 1975. Submergent vegetation for bottom stabilization. *Estuarine Research,* vol. II, 439–456. New York: Academic Press.

FAO (United Nations Food and Agriculture Organization). 1982. Management and utilization of mangrove in Asia. FAO Environment Paper 3. Rome: United Nations Food and Agriculture Organization.

Fonseca, M. S., W. J. Kenworthy, J. Homziak, and G. W. Thayer. 1979. Transplanting of eelgrass and shoalgrass as a potential means of mitigating a recent loss of habitat. In *Proceedings of the 6th annual conference on wetlands restoration and creation,* ed. D. P. Cole. Tampa, FL: Hillsborough Community College.

Fonseca, M. S., W. J. Kenworthy, K. M. Cheap, C. A. Currin, and G. W. Thayer. 1984. A low-cost transplanting technique for shoalgrass (*Halodule wrightii*) and manatee grass (*Syringodium filiforme*). Instruction Report EL 84-1. Vicksburg, MS: U.S. Army Engineer Waterways Experiment Station/National Marine Fisheries Service.

Fonseca, M. S., G. W. Thayer, and W. J. Kenworthy. 1985. The use of ecological data in the implementation and management of seagrass restoration. Gainesville, FL: American Institute Biological Sciences Symposium. *American Journal Botany* 73(S):11.

Foo, H. T., and Wong, J. T. S. 1980. Mangrove swamp and fisheries in Sabah. *Tropical Ecological Development* 10:1157–61.

Gedney, R. H., Kapetsky, J. M., and Kuhnhold, W. W. 1982. Training on assessment of coastal aquaculture potential. South China Sea Fisheries Development and Coordinating Programme. SCS/GEN/82:3562. Rome: United Nations Food and Agriculture Organization.

Gibbs, P. 1981. New South Wales State Fisheries, Sydney, Australia. Personal communication.

Goforth, H. W., and T. J. Peeling. 1979. Intertidal and subtidal eelgrass (Zostera marina L.) transplant studies in San Diego Bay, California. In *Proceedings of the 6th annual conference on the restoration and creation of wetlands,* ed. D. P. Cole, 324–356. Tampa, FL: Hillsborough Community College.

Goforth, H. W., and Thomas, J. R. 1980. Plantings of red mangroves (*Rhizophora mangic L.*) for stabilization of marl shorelines in the Florida Keys. In *Proceedings of the 6th annual conference on wetlands restoration and creation,* ed. D. P. Cole, 207. Tampa, FL: Hillsborough Community College.

Hannan, J. 1975. Aspects of red mangrove reforestation in Florida. In *Proceedings of second annual conference on restoration of coastal vegetation in Florida,* ed. R. R. Lewis, 112–121. Tampa, FL: Hillsborough Community College.

Hoffman, W. E., and J. A. Rodgers. 1981. Cost benefit aspects of coastal vegetation establishment in Tampa Bay, Florida. *Environmental Conservation* 8(1):39.

Kapetsky, I. M. 1985. Mangroves, fisheries and aquaculture. In Report of the 11th session of the advisory committee on marine resources research, Supplement. FAO *Fish. Rep.* (338): Suppl.

Kelly, J. A., Jr., C. M. Fuss, Fr., and J. R. Hall. 1971. The transplanting and survival of turlegrass, *Thalassia testudinum,* in Boca Ciega Bay, Florida. *Fisheries Bulletin* 69(2):273–280.

Kenworthy, W. J., and M. S. Fonseca. 1977. Reciprocol transplant of the seagrass *Zostera marina L.* Effect of substrate on growth. *Aquaculture* 12:197–213.

Kinch, J. C. 1975. Efforts in marine revegetation in artificial habitats. In *Proceedings of second annual conference restoration of coastal vegetation in Florida,* ed. D. P. Cole, 100–111. Tampa, FL: Hillsborough Community College.

Larkum, A. W. D. 1976. Ecology of Botany Bay, vol. I. Growth of *Posidonia australis* (bBrown) Hoak. f. in Botany Bay and other bays of the Sydney Basin. *Australian Journal Marine Freshwater Research* 27:117.

Lewis, R. R. 1980. Techniques for restoring mangroves. In *Proceedings of Wetland Restoration Conference,* ed. R. R. Lewis, Hillsborough Jr. College, Tampa, FL.

Lewis, R. R. 1981. Economics and feasibility of mangrove restoration. In *Proceedings U.S. Fish and Wildlife Service Workshop Coastal Ecosystems Southeastern U.S.,* ed. P. S. Markouts, 88–89. Washington, DC: U.S. Fish and Wildlife Service.

Lewis, R. R. 1982. Mangrove forests. In *Creation and restoration of coastal plant communities,* ed. R. R. Lewis, 153–171. Boca Raton, FL: CRC Press.

Lewis, R. R. 1988. Economics and feasability of mangrove restoration. In *Proceedings of U.S. Fish and Wildlife Service workshop on coastal ecosystems of southwestern U.S.,* ed. P. S. Markowitz, ed. Big Pine Key, FL.

Liem, D. S., and A. K. Haines. 1977. The ecological significance and the economic importance of the mangrove and estuarine communities of the Gulf Province. *Environmental Studies* 3:1–35.

Lugo, A. E., and G. Cintron. 1975. The mangrove forests of Puerto Rico and their management. In *Proceedings international symposium biological management Mangroves,* ed. G. Walsh, S. Snedaker, and H. Teas, 825. Gainesville: Food and Agricultural Sciences, University of Florida.

McLaughlin, P. A., S. A. Treat, A. Thorhaug, Rafael Lematril. 1983. Restored seagrass and its animal communities. *Environmental Conservation* 10(3):247–252.

Ogden, J. C. 1980. Faunal relationships in Caribbean seagrass beds. In *Handbook of seagrass biology, an ecosystem perspective,* eds. R. C. Phillips and C. P. McRoy, 173–198. New York: Garland STPM Press.

Orth, R. J., and K. A. Moore. 1983. Seed germination and seedling growth of *Zostera marina* L. (eelgrass) in the Chesapeake Bay. *Aquatic Botany* 15:117–131.

Phillips, R. C. 1974. Transplantation of seagrasses, with special emphasis on eelgrass *Zostera marina* L. *Aquaculture* 4:161–176.

Phillips, R. C. 1978. Seagrass bed development on dredged spoil at Port St. Joe, Florida. In *Proceedings of the 4th annual conference on restoration of coastal vegetation in Florida,* ed. R. R. Lewis and D. P. Cole, 1–11. Tampa, FL: Hillsborough Community College.

Pulver, T. R. 1976. Transplant techniques for sapling mangrove trees, *Rhizophora mangle, Laguncularia racemosa,* and *Avicennia germinans,* in Florida. *Florida Marine Resources Publication* 22. Tallahasse, FL: Florida Department of Natural Resources.

Rabanal, H. R. 1984. Fisheries integrated development in the sundarbans. Bangladesh: Sundarban Forest Development Planning Mission. Rome: United Nations Food and Agriculture Organization. FD:TCP/PGO/2309.

Rabinowitz, D. 1975. Planting experiments in mangrove swamps of Panama. In *Proceedings international symposium biology mangroves,* eds. G. Walsh, S. Snedaker, and H. Teas, 385. Gainesville, FL: Institute of Food and Agricultural Sciences, University of Florida.

Ranwell, D. S., D. W. Wyer, L. A. Boorman, J. M. Pizzey, and R. J. Waters. 1974. Zostera transplants in Norfolk and Suffolk, Great Britain. *Aquaculture* 4(3):185–198.

Roessler, M. A., D. C. Tabb, R. Rehrer, and J. Garcia. 1974. Studies of effects of thermal pollution in Biscayne Bay, Florida. EPA Ecological Resources Services. U.S. EPA 660/3-74-014. Miami, FL: University of Miami.

Rogers, R. G., and F. T. Bisterfeld. 1975. Seagrass vegetation attempts in Escambia Bay, Florida, during 1974. In *Proceedings 2nd annual conference restoration of coastal vegetation in Florida,* ed. R. R. Lewis, 26–28. Tampa, FL: Hillsborough Community College.

Savage, T. 1972. Florida mangroves as shoreline stabilizers. Florida Department of Natural Resources, Professional Papers Series no. 19, 1–25. Tallahassee, FL: Florida Department of Natural Resources.

Schroeder, P. B., and A. Thorhaug. 1980. Trace-metal cycling in Thalassia community. *American Journal Botany* 67:1075–1076.

Teas, H. J. 1977. Ecology and restoration of mangrove shoreline in Florida. Environmental Conservation 4:51.

Teas, H. J., W. Jurgens, and M. C. Kimball. 1975. Plantings of red mangroves (*Rhizophora mangle* L.) in Charlotte and St. Lucie Counties, Florida. In *Proceedings 2nd annual conference restoration of coastal vegetation in Florida,* ed. R. R. Lewis, 26–28. Tampa, FL: Hillsborough Community College.

Thorhaug, A. 1974. Transplantation of the seagrass *Thalassia testudinum Konig. Aquaculture* 4:177–183.

———. 1980a. Biological effects of thermal effluents in the marine environment: Tropics and subtropics with a guideline. FAO Report to UNESCO, 1–250. Rome: United Nations Food and Agriculture Organization.

———. 1980b. Environmental management of a highly impacted, urbanized tropical estuary: Rehabilitation and restoration. *Helgolander Meersuntersuchen,* 33:614–623.

———. 1982. Primary productivity of seagrasses. *Handbook of biosolar resources, II.* Boca Raton, FL: CRC Press.

———. 1983. Habitat restoration after pipeline construction in a tropical estuary: Seagrasses. *Marine Pollution Bulletin* 14(11):422–425.

———. 1985. Large-scale seagrass restoration in a damaged estuary. *Marine Pollution Bulletin* 16(2):55–62.

———. 1986. Report on Port of Miami seagrass mitigation phase 3. Dade County Seaport Department. Dade County, FL: Dade County Seaport Department.

———, and C. B. Austin. 1976. Restoration of seagrassses with economic analysis. *Environmental Conservation* 3(4):259–268.

———, A. R. Cruz, and M. Fortes. 1986. Seagrass restoration in the Philippines. Report to FAO, Rome, 1–55. Rome: United Nations Food and Agriculture Organization.

———, and R. Hixon. 1975. Revegetation of *Thalassia testudinum* in a multiple-stressed estuary, North Biscayne Bay, Florida. In *Proceedings 2nd annual conference on restoration of coastal vegetation in Florida,* ed. R. R. Lewis, 12–27. Tampa, FL: Hillsborough Community College.

———, and P. McLaughlin. 1979. Restoration of *Thalassia testudinum* animal community in a maturing four-year-old site—preliminary results. In *Proceedings 5th annual conference on restoration of coastal vegetation in Florida,* ed. D. P. Cole, 149–161. Tampa, FL: Hillsborough Community College.

———, B. Miller, and B. Jupp. 1983. Progress reports, 1–3. Seagrass restoration in Caribbean nearshore areas. U.S. Agency for International Development Grant. DAN-5542-G-SS-2101-00. Miami, FL: Florida International University.

———, B. Miller, B. Jupp, and F. Booker. 1985. Effects of a variety of impacts on seagrass restoration in Jamaica. *Marine Pollution Bulletin* 167(9):355–360.

Universidade Federal de Maranhão. 1983. Relatorio final. Caracterizacão ambientale prospecção pesqueirra do estuario do Rio Cururuca-Maranhão. Belém, Brazil: Universidade Federal de Maranhão.

Van Breedveld, J. F. 1975. Transplanting of seagrasses with emphasis on the importance of substrate. *Florida Marine Resources Publication* 17(1):1–25.

Winter, P. A. 1976. Unpublished. Culture studies of the seagrass *Ruppia*. Report to Sea Grant.

Anitra Thorhaug is Research Professor of Biological Sciences, Florida International University, 555 NE 15th St., #PH-H, Miami, FL 33132.

Creation and Monitoring of Vernal Pools at Santa Barbara, California

David A. Pritchett

ABSTRACT: *Vernal pools are endangered wetland ecosystems that flood annually during the winter and support a unique biota. In 1986, six pools were created by excavating shallow depressions into a clay soil, and three of the six were inoculated with a seed bank obtained from local natural vernal pools. Results from the first year of monitoring after creation of the pools showed that: (1) the duration of flooding was longer and fluctuated more often in the created pools than in local natural pools serving as a comparison, (2) more native plants occurred in the inoculated created pools than in the uninoculated pools, and (3) two annual plants endemic to vernal pools were more abundant in the inoculated pools than in the natural pools.*

KEY WORDS: *restoration, ecology, wetlands, rare plants.*

Introduction

THE CREATION OF artificial habitat is one technique to compensate for the loss of natural habitat. This approach can be employed as a direct mitigation for development that degrades or destroys habitats for rare or endangered species (Rieger 1988). Furthermore, whole ecosystems also can be created as a means to compensate for the loss of rare or endangered types (Zedler 1982).

Vernal pools are one type of endangered ecosystem that could be created. They are endemic to California, but similar types of wetlands occur in other areas of the world that have a Mediterranean climate, characterized by mild temperatures, moderate winter rainfall, and a summer drought. Vernal pools form in shallow depressions underlain by claypan or hardpan. They retain water from winter rains, yet desiccate completely by spring or summer. The ultimate indicator of a vernal pool, however, is the presence of certain plants that are restricted locally or entirely to this type of habitat. Many plant species are endemic to vernal pools, but other vernal pool plants occur in various types of wetlands.

Naturally rare in California, vernal pools have become endangered because of agricultural and urban expansion. As a result of this endangerment, various federal, state, and local agencies recently have required some kind of mitigation when vernal pools have been degraded by development (Bartel and Knudsen 1984; Rieger 1984). Vernal pools have been created serendipitously or intentionally throughout California. Many scientifically oriented projects have occurred in San Diego in recent years (Scheidlinger et al. 1984; Zedler 1986) and in Santa Barbara (Ferren and Pritchett 1988). Ferren and Pritchett (1988) give a review of vernal pools restoration and creation projects and provide a comprehensive analysis of the research reported in this paper.

The vernal pools considered in this report were created by the staff of the Herbarium at University of California, Santa Barbara (UCSB) in the summer of 1986 as part of a larger effort that included enhancement, restoration, and creation of vernal pools at Del Sol Reserve (Ferren and Pritchett 1988; Pritchett 1988). The Reserve is public property, 5 ha in area, and is located in the coastal community of Isla Vista, near Santa Barbara, California. Several natural vernal pools occur at the Reserve, and about 20 exist within a distance of 2 km. The pools were not created to provide habitat for a particular species, but rather to compensate for the historic loss of vernal pool ecosystems in the Santa Barbara area.

The purpose of this study was to design and construct the created pools and to compare physical and biological conditions of three types of pools: (1) created pools inoculated with a seed bank, (2) created pools not inoculated, and (3) natural vernal pools. Physical conditions that were monitored include the size of pools and the extent and duration of flooding. Biological conditions include the abundance of five native plant species and the floras of the pools. For the criteria monitored, the goals were to demonstrate: (1) the degree to which the created pools resembled the natural pools, and (2) the effect of inoculating the created pools with seed bank obtained from the natural pools.

Methods

Creation and Inoculation of Pools

To create the pools, a Skiploader tractor excavated a cluster of six circular depressions into an upland habitat comprised largely of introduced grasses. The excavations at Del Sol Reserve were limited to six depressions because of finite space available. The protected status of the Reserve should afford permanent preservation of the pools. The depressions were a depth comparable with local natural pools (ca. 30 cm), and did not penetrate the claypan in the soil, which occurred 50 cm deep, measured about 1.5 m thick, and contained 40% clay (Jones 1987).

Shortly after excavation of the six pools, three randomly selected pools were inoculated with seed bank obtained from two natural vernal pools located 2 km from the site. These donor pools were selected because they appear to support a higher diversity of native plant species than natural pools at Del Sol Reserve. To acquire the inoculum, up to 1 cm of material was scraped with a hoe from scattered areas of the natural pools. A volume of 150 liters of material was collected and subsequently dispersed evenly among the three created pools.

Monitoring of Pools

Hydrologic conditions in the six created and two natural pools were assessed by using permanent markers (rebar segments imbedded in the ground) that form a transect bisecting the lowest point in each pool. The extent and duration of flooding in a pool were determined by periodically noting the points along the transect that were inundated by water. To note the highest and lowest water levels, the pools were monitored the days before and after a rainfall event. These data allow the duration of flooding to be determined for any point along the transect. However, for use with the data on plant abundances, the hydrologic data were combined into three classes of flooding: (1) areas flooded for one or more days (flooded), (2) areas within 1 m of flooding (low), and (3) areas further than 1 m from flooding (high).

The abundance of plant species in the pools were assessed with the same permanent transects utilized for the hydrologic conditions. The vegetation along the transects was sampled by using 0.5 m^2 quadrats placed at 2 m intervals. A transect spanned an entire pool, and included the upland above the level of highest water. Cover of each species was estimated for the entire quadrat, and plant density was counted in twenty 10x10 cm cells that were selected randomly within each quadrat. An Index of Abundance, which is the sum of cover and density, was calculated for each species in each quadrat.

To increase the sample size used in the analysis, data were combined from the quadrats of the six created pools and two natural pools to form three treatments: (1) inoculated created vernal pools, (2) uninoculated created pools, and (3) natural pools. Thus, for each species, an Index of Abundance was determined for three flooding classes in each of three pool treatments. The sample sizes (n) were as follows: natural pools high, 4; natural pools low, 1; natural pools flooded, 17; inoculated pools high, 18; inoculated pools low, 3; inoculated pools flooded, 12; uninoculated pools high, 10; uninoculated pools low, 5; and uninoculated pools flooded, 10.

The five native plant species reported herein for the quantified analysis of abundances include: wooly heads (*Psilocarphus brevissimus*) (*Asteraceae*), and popcorn flower (*Plagiobothrys undulatus*) (*Boraginaceae*), two annual species endemic to vernal pools in California; coyote-thistle (*Eryngium vaseyi*) (*Apiaceae*), a perennial species endemic to vernal pools in California; and loosestrife (*Lythrum hyssopifolia*) (*Lythraceae*) and needle spike-rush (*Eleo-*

charis acicularis) (*Cyperaceae*), two perennial species with cosmopolitan distributions.

The floras of the pools were surveyed four times during spring and summer 1987 by noting the presence of all vascular plant species occurring in the flooded areas of the pools and within 2 m of flooded areas. Voucher specimens are deposited at the UCSB Herbarium. The floristic data are combined for each of the three treatments.

RESULTS

Physical Conditions

Results from creating the six pools, labelled A to F on Figure 1, show that they were excavated below their original elevation by the following amounts: A, 58 cm; B, 46 cm; C, 43 cm; D, 37 cm; E, 34 cm; and F, 37 cm. The pools are approximately circular, and range in diameter from 7–10 m.

The hydrologic conditions of the six created pools and the two natural pools are shown in Figure 3. The area within a polygon on each graph indicates the extent (vertical axis) and duration (horizontal axis) of flooding in a pool. Pool A, for example, first held standing water after 16 mm of rainfall on the sixth day of the year 1987. On day ten, standing water existed from 7 m to 13 m on the transect bisecting the pool. By day 24, the pool had dried, but it filled again after 19 mm of rainfall on day 40. The deepest part of the pool occurred at 10.4 m on the transect, and the extent of the pool was between 6 m and 13.6 m on the transect.

The water level fluctuated more often in the created pools than in the

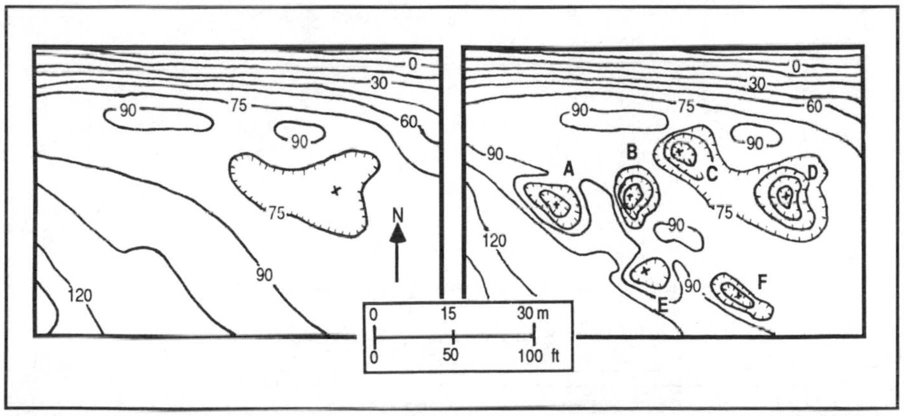

FIGURE 1. *Topography of the six created pools. The left map shows the site (entirely upland habitat) before excavation. The right map shows the site (upland and wetland habitat) after excavation. Contour interval is 15 cm. Reference datum (0 cm contour) is 9.8 m elevation.*

natural pools. The created pools also required less rainfall to initiate flooding. For example, the four rains occurring before day 65 produced little or no flooding in Pools H or G (the two natural pools), but Pools A to F (the six created pools) did support standing water after these rains. On day 65, the peak of flooding for all pools, the extent of flooding was greater in either of the two natural pools that in any of the six created pools. The overall duration of flooding, however, was greater in the created pools, but upper elevations in the created pools had a total duration of flooding comparable with the natural pools.

Biological Conditions

In Figure 2, an Index of Abundance is presented for each of the five native plants studied. Nine possible histograms represent each species according to the three classes of flooding designated for the three treatments of pools. Histograms that do not appear on the graphs have an abundance rating of

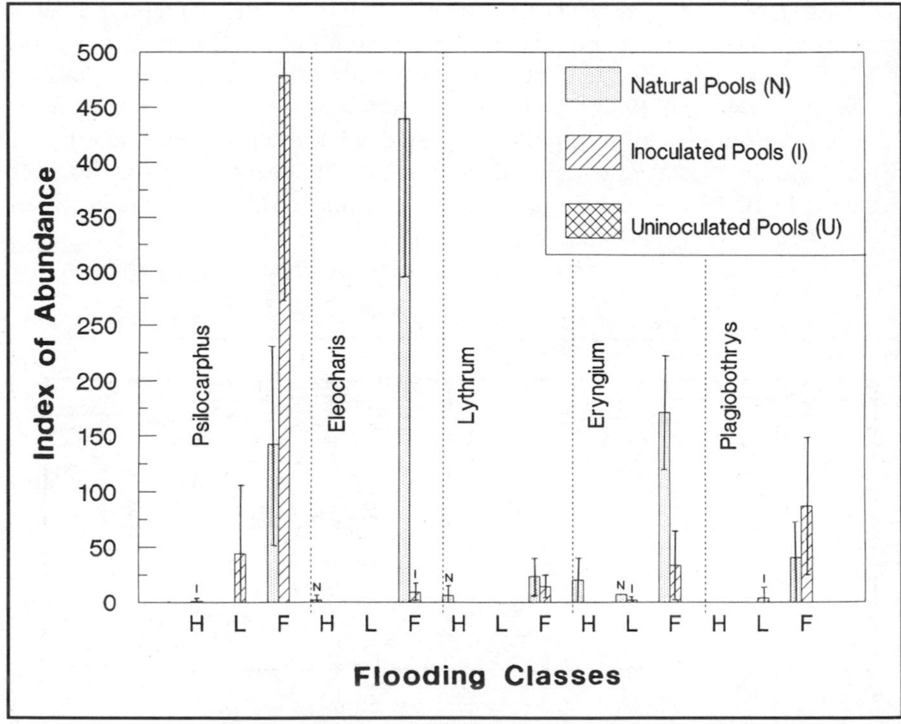

FIGURE 2. *Plant abundances at the three flooding classes. An Index of Abundance based on density and cover provides a comparison of five native plants occurring in the three treatments of pools at the three flooding classes: areas flooded for one or more days (F), areas within 1 m of flooding (L), and areas further than 1 m from flooding (H). Error bars are two standard errors of the mean.*

zero, which occurred for each of the five plants in the uninoculated created pools.

Each hydrologic class of the inoculated created pools had a higher abundance of *Psilocarphus* and *Plagiobothrys* than in the same hydrologic class of the uninoculated pools or the natural pools (Figure 2). *Eleocharis, Lythrum,* and *Eryngium,* however, were more abundant in each hydrologic class of the natural pools than in the same class of either type of created pool.

The floras from the three treatments of pools (Table 1) indicate that the natural pools had 14 native species. Ten of those 14 occurred in the inoculated created pools (eight of the ten were native hydrophytes), and five of those 14 occurred in the uninoculated created pools (four of the five were native hydrophytes).

Discussion

Physical Conditions

The extent of flooding in the six created pools was not as great as in the two natural pools (Figure 3) because the depressions excavated to form each of the created pools were not as large as the depressions that form each of the two natural pools. The depth of the created pools (Figure 1), however, approximates measured depths of natural pools elsewhere in southern California (Zedler 1987).

The overall longer duration of flooding at the deepest parts of the created pools (Figure 3) could have been a result of greater pool depth, larger watershed area in relation to pool size, and higher soil impermeability. Rainfall and evaporation, two other significant variables affecting the flooding in pools (Zedler 1987), were presumably the same at the created and natural pools because of their close proximity (2 km).

Biological Conditions

The higher similarity of native flora (Table 1) between the natural pools and the inoculated pools rather than between the natural pools and the uninoculated created pools probably was a result of the introduced seed bank obtained from the natural pools. Although *Eleocharis acicularis, Eryngium vaseyi,* and *Lythrum hyssopifolia* had a calculated abundance of zero in the uninoculated pools (Figure 2), the presence of these three species in the uninoculated pools (Table 1) suggest that wind, birds, or other animals (e.g., ecologists) could have dispersed the seeds of these three plants from the inoculated pools to the uninoculated pools (Zedler 1987).

The higher abundance of the native plants *Eleocharis, Lythrum,* and *Eryngium* (Figure 2) in the inoculated pools than in the uninoculated pools suggests that the introduction of seed bank was an effective method to establish these four plants in the created pools. As the populations and individual

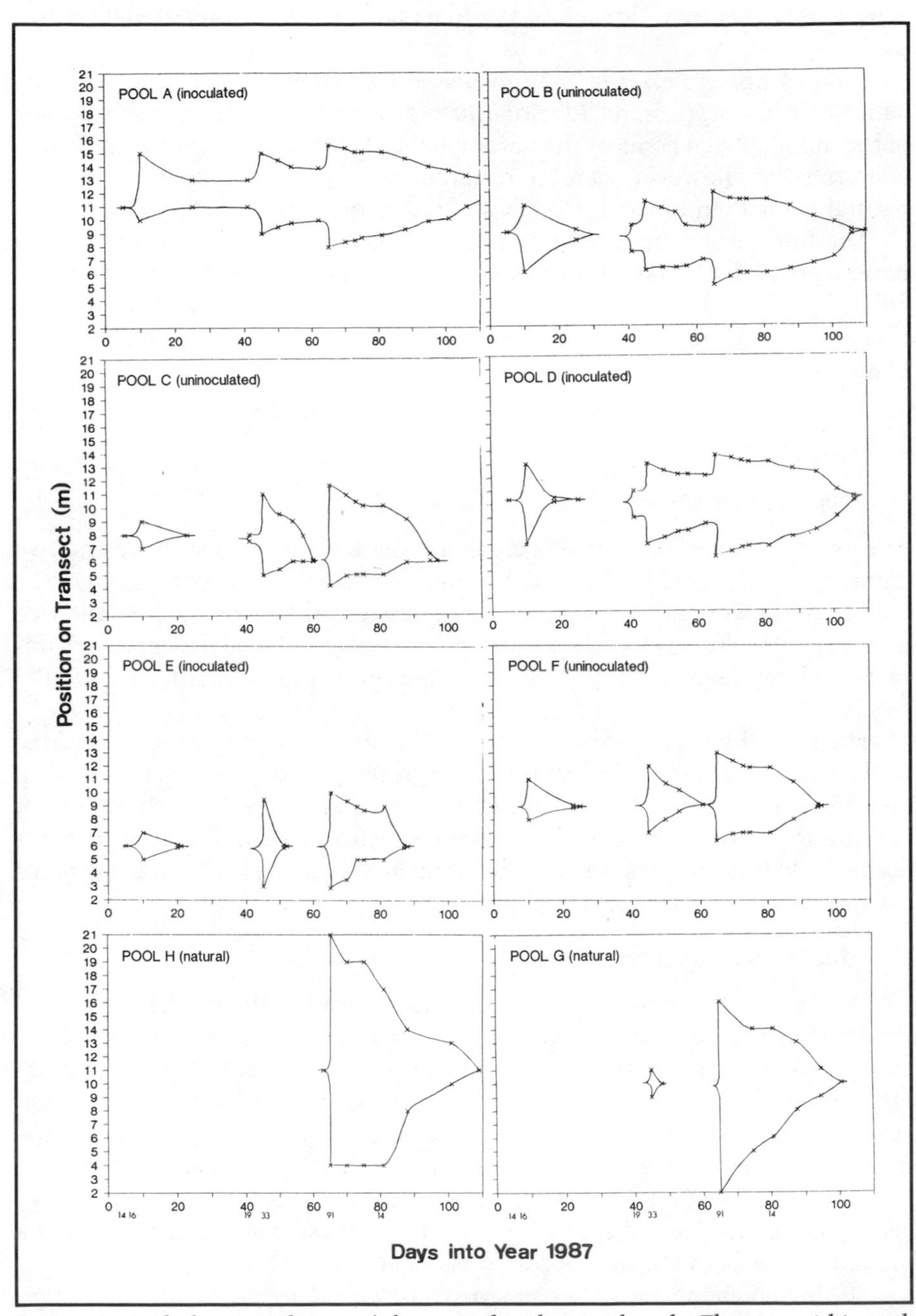

FIGURE 3. *Hydrologic conditions of the created and natural pools. The area within each polygon indicates the extent and duration of flooding in a pool. Days on which rain occurred are noted on the bottom axes by mm of precipitation.*

TABLE 1
FLORAS OF NATURAL AND CREATED POOLS

Species	Natural vernal pools	Inoculated created pools	Uninoculated created pools
Ambrosia psilostachya DC var. *californica* *(Rydb.) Blake		x	x
Anagallis arvensis L.		x	x
Avena fatua L.	x	x	x
Boisduvalia glabella *(Nutt.) Walp.	x		
Bromus diandrus Roth.		x	x
B. hordeaceus L.	x	x	
Calandrinia maritima *Nutt.		x	x
Callitriche marginata *Torrey	x	x	
Convolvulus arvensis L.		x	x
Cotula coronopifolia L.	x	x	
Crassula aquatica *(L.) Schoenl.	x	x	
Cynodon dactylon (L.) (Pers.)	x		
Cyperus eragrostis Lam.	x		
Distichlis spicata *(L.) Greene var. *spicata*		x	x
Elatine brachysperma *Gray		x	x
Eleocharis acicularis *(L.) R. & S.	x	x	x
E. palustris *(L.) R. & S.	x		
Eremocarpus setigerus *(Hook.) Benth.	x	x	x
Erodium botrys (Cav.) Bertol.	x	x	x
E. cicutarium (L.) L'Her.	x	x	x
Eryngium vaseyi *Coulter & Rose	x	x	x
Geranium dissectum L.	x		
Hemizonia australis *(Keck) Keck	x	x	
H. fasciculata *(DC.) T. & G.	x		
Hordeum brachyantherum *Nevski	x		
H. geniculatum Allioni	x	x	
H. leporinum Link.		x	
Hypochoeris glabra L.	x		
Juncus bufonius *L.	x	x	x
Lactuca serriola L.			x
Lolium multiflorum Lam.	x	x	x
Lythrum hyssopifolia *L.	x	x	x
Medicago polymorpha L.	x		
Phalaris lemmonii *Vasey	x		
Plagiobothrys undulatus *(Piper) Jtn.	x	x	
Plantago lanceolata L.	x	x	x
Poa annua L.	x	x	
Polygonum arenastrum Bor.	x		
Polypogon interruptus HBK.	x		
P. monspeliensis (L.) Desf.	x	x	
Psilocarphus brevissimus *Nutt.	x	x	
Rumex crispus L.	x		
Sonchus asper (L.) Hill			x
S. oleraceus L.			x

TABLE 1, CONTINUED
FLORAS OF NATURAL AND CREATED POOLS

Species	Natural vernal pools	Inoculated created pools	Uninoculated created pools
Spergula arvensis L.	x	x	x
Spergularia bocconnii (Scheele) Foucaud	x		
S. villosa (Pers.) Camb.	x		
Vicia benghalensis L.	x	x	x
V. sativa L.	x		
Vulpia bromoides (L.) S. F. Gray		x	x

*Native species.

sizes of these perennial species grow, the created pools eventually might support abundances comparable with the natural pools, as has occurred in San Diego, where, four years ago after excavation and inoculation, a created pool with hydrologic conditions akin to a natural pool supported vegetation statistically indistinguishable from a natural vernal pool (Scheidlinger et al. 1984).

The absence of *Psilocarphus* and *Plagiobothrys* from the uninoculated pools, and the higher abundances of these two plants in the inoculated pools (Table 1 and Figure 2), suggests that the introduced seed bank also was an effective method to establish these native plants in the created pools. Furthermore, the higher abundances of *Psilocarphus* and *Plagiobothrys* in the inoculated created pools compared with the natural pools (Figure 2) indicates that the created pools, for the first year of their existence, were superior habitat for these two plants. The open substrate of the created pools probably allowed many seeds of these two annual plants to germinate and grow without inhibitions from other plants, a condition that undoubtedly would occur in natural vernal pools, which could support denser vegetation dominated by more perennials than were present in the created pools.

CONCLUSIONS

The methods used to design and construct the created pools resulted in six depressions that held water and supported native plants indicative of vernal pool ecosystems of the Santa Barbara area. Most portions of the six created pools and two natural pools underwent the same duration of flooding. Two annual plants endemic to vernal pools were more abundant in the inoculated created pools compared with the natural pools, but three perennial plants were less abundant. Introduced seed bank had a significant effect in the created pools, as indicated by the presence of more native species and the higher abundance of five native species in the inoculated created pools compared with the uninoculated created pools. Long term monitoring of the created pools at Del Sol Reserve will provide data about the degree to which the created pools resemble local natural vernal pools, and will provide a means to evaluate further the methods used to create the pools.

Acknowledgments

I thank Wayne R. Ferren, Jr., curator of the UCSB Herbararium, for supervising the creation and monitoring of the pools. Holly Forbes surveyed the floras, and Charles Hutchins, Brad Penkala, and Carolyn Stange assisted with field-work. Funding was provided by grant no. 85-073-86-020 from the California State Coastal Conservancy to the Isla Vista Recreation and Park District for the Del Sol Vernal Pools Enhancement Plan.

References

Bartell, J. A., and Knudsen, M. D. 1984. Federal laws and vernal pools. In *Vernal pools and intermittent streams,* eds. S. K. Jain and P. Moyle, 263–268. Davis, CA: UCD Institute of Ecology Publication No. 28.

Ferren, W. R. Jr., and D. A. Pritchett. 1988. Enhancement, restoration, and creation of vernal pools at Del Sol Open Space and Vernal Pool Reserve, Santa Barbara County, California. Santa Barbara, CA: UCSB Herbarium Environmental Report no. 13.

Jones, S. A. 1987. Vernal pools project soils analysis: particle size distribution. Report to UCSB Herbarium, 16 April 1987.

Pritchett, D. A. 1988. Enhancement, restoration and creation of vernal pools (California). *Restoration and Management Notes* 5:93.

Rieger, J. P. 1984. Vernal pool mitigation—the CALTRANS experience. In *Vernal pools and intermittent streams,* eds. S. K. Jain and P. Moyle, 269–272. Davis, CA: UCD Institute of Ecology Publication no. 28.

Rieger, J. P. 1988. Three riparian restoration projects in San Diego County. In *Proceedings, second native plant revegetation symposium,* eds. J. P. Rieger and B. Williams, 141–152. Madison, WI: Society for Ecological Restoration and Management.

Scheidlinger, C. R., C. C. Patterson, and P. H. Zedler. 1984. Recovery of vernal pools and their associated plant communities following surface disturbance. Report to U.S. Environmental Protection Agency and U.S. Fish and Wildlife Service. San Diego, CA: Department of Biology, San Diego State University.

Zedler, J. B. 1982. The ecology of southern California coastal salt marshes: A community profile. U.S. Fish and Wildlife Service Biological Report no. FWS/OBS 81/54. Washington, DC: U.S. Department of the Interior, Fish and Wildlife Service.

Zedler, P. H. 1986. Problem analysis and proposal: creation of vernal pool habitat on Del Mar Mesa, San Diego County. Report to California Department of

Transportation, Grant no. 4584-145. San Diego, CA: Department of Biology, San Diego State University.

Zedler, P. H. 1987. The ecology of southern California vernal pools: a community profile. U.S. Fish and Wildlife Service Biological Report no. 85(7.11). Washington, DC: U.S. Department of the Interior, Fish and Wildlife Service.

David A. Pritchett is an Herbarium Associate, UCSB Herbarium, Department of Biological Sciences, University of California, Santa Barbara, CA 93106.

PART III

Strategic Planning and Land Acquisition for Restoration

RESTORATION OF RIPARIAN WETLANDS ALONG A CHANNELIZED RIVER: OXBOW LAKES AND THE MIDDLE MISSOURI

JOHN H. SOWL

ABSTRACT: *Oxbow lakes on the Middle Missouri River were examined as unique ecological features in the fluvial landscape whose continued existence is threatened by the channelization of the river. Several existing plans designed to preserve some of these lakes and to mitigate for fish and wildlife losses resulting from channelization were examined using matrix decision analysis based on four criteria. Plans by the U.S. Fish and Wildlife Service and the U.S. Army Corps of Engineers did not preserve and enhance these oxbow lakes as completely as could a Compromise Plan, developed as a result of this research to provide more comprehensive benefits.*

KEY WORDS: *channelization, Missouri River, oxbow lakes, rivers, wetland restoration.*

INTRODUCTION

OXBOW LAKES ON the Middle Missouri River in central North America were examined as unique ecological features in the fluvial landscape. Oxbows are arcuate or horseshoe-shaped bodies of water which occupy the channel of an old river meander or bend. They are commonly formed when river meanders develop sufficiently to the point where the entire loop may be "cut off" or isolated by silting, leaving a mostly shallow, crescentic lake (Hutchinson 1957). While the phenomenon of oxbow lakes themselves is not uncommon, it is becoming increasingly rare to find them associated with large rivers and streams in the United States. This is primarily due to river channelization for flood control and navigation, which restricts the natural meandering pattern of a river and thus its ability to create oxbows (as is the case with the Middle Missouri), and to human encroachment as a result of draining and infilling of these wetland areas (Frederickson 1979).

These oxbow lakes and the study area corridor in which they reside along the Missouri River represent one of the last bastions for the native plants and animals that inhabit what remains of the once extensive floodplain ecosystem in this region (Hallberg et al. 1979). Since the river was channelized by the U.S. Army Corps of Engineers to promote navigation and to provide flood protection, there has been a steady loss of native flora and fauna, as well as wetland, riparian, and forest habitat (Funk and Robinson 1974; Fredrickson 1979; Hallberg et al. 1979). This is primarily due to such factors as accelerated riverbed degradation rates and detrimental land use patterns resulting from the channelization process (Bragg and Tatschl 1977; SIMPCO 1978), and it is now impossible, under present conditions, for new oxbows to form to replace the vanishing old ones (Hallberg et al. 1979). Since the study area was the last reach of the Missouri to undergo channelization, it also holds the greatest potential for successful rehabilitation and restoration.

Purpose and Methodology

Purpose

The purpose of this research was to examine ways to preserve, enhance and manage a maximum number of the threatened remaining oxbow lakes and their associated native floodplain ecosystem on the Middle Missouri River. The research study area is located along the middle section of the Missouri River for 148 km between Omaha, Nebraska, on the south and Sioux City, Iowa, on the north (see Figure 1). Specifically, the study area forms a corridor consisting of 296 km situated between River Mile 633 as the southern limit, River Mile 725 as the northern limit, and extending for 1.6 km on either side of the river (for River Mile destinations see U.S. Army Corps of Engineers 1983). This corridor accommodates 12 man-made oxbow lakes which were created when the river was channelized, one naturally formed oxbow lake, and one naturally formed chute, which together represent the focal points of this research.

Methodology

The oxbow lakes in the study area were examined in a physical, biological, and cultural context over time to determine what the study area was like more than a century ago (i.e., approximately presettlement conditions), how this area became the way it is today, and the physical, biological, and cultural consequences of these actions. This examination was guided by the principles and theories of landscape ecology.

Landscape ecology can be defined as a study of the structure, function, and change in a heterogeneous land area composed of interacting ecosystems (Forman and Godron 1986). It deals with an ecological mosaic of patches with continuously varying degrees of connectedness and recognizes the importance

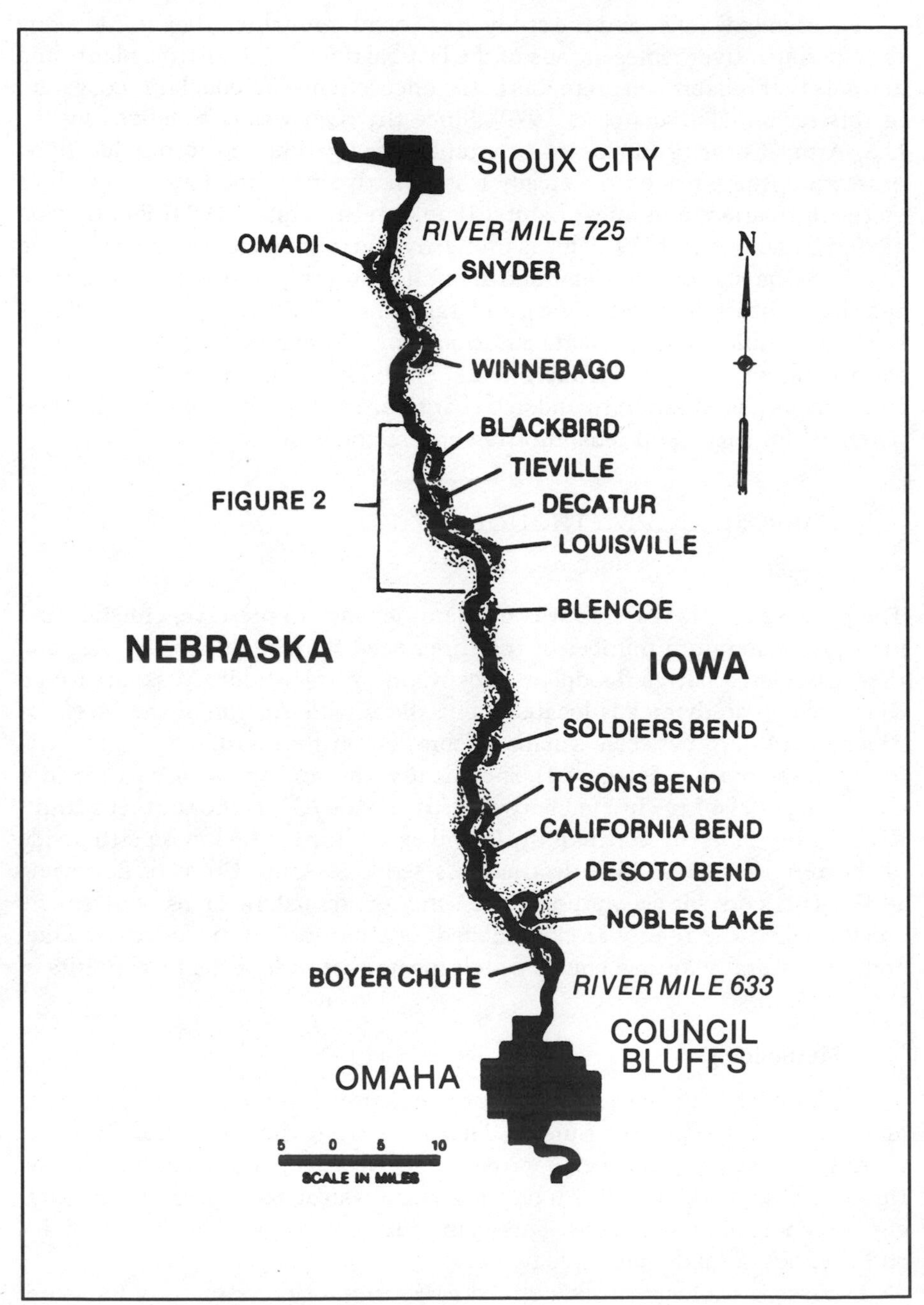

FIGURE 1. *Oxbow lakes and corridor design.*

of matrix and corridors to terrestrial habitat island dynamics (Noss 1983). One of the primary thrusts of landscape ecology is the preservation of native ecosystems and their components. For any landscape, the model natural ecosystem complex is the presettlement vegetation and the associated biotic and abiotic elements. Preservation activities would ideally maintain high quality examples of their former abundance in the region. This does not mean trying to hold nature static. Rather, preservation should imply perpetuating the dynamic processes of presettlement landscapes (Forman 1981). Thus the ideal plan for preserving, enhancing, and managing the remaining oxbow lakes and remnants of floodplain ecosystem on the Middle Missouri River would work to rehabilitate the biotic and abiotic elements and processes of this riparian corridor to approximate presettlement conditions to the greatest extent possible.

Based on these principles, four criteria were developed which would aid in the determination and design of an acceptable, adequate, and appropriate plan for the ongoing preservation, enhancement, and management of the oxbow lake corridor. These criteria and the assumptions from which they evolved are as follows:

1. The management plan should preserve a maximum ratio of aquatic acreage (i.e., oxbow lakes) to terrestrial acreage.
   Assumption: Oxbow lake preservation is the primary focus of this research.
2. The management plan should contain the greatest expanse of aquatic and terrestrial acreage as possible.
   Assumption: The larger the preserve size, the better for protecting and maintaining the ecological diversity and integrity of aquatic and terrestrial ecosystems (Soulé and Wilcox 1980).
3. The management plan should have the greatest political feasibility.
   Assumption: While several management plans have heretofore been proposed, none have been accepted. Past research has found that the primary reason for this lack of acceptability seems to be the reluctance of private landowners (e.g., farmers) to give up or convert their land into a preserve status, or their inability to manage existing wetlands and forests for multipurpose values (Missouri Basin States Association 1983). Therefore, it is assumed that the less a management plan requires the status quo to be altered, the more acceptable (i.e., politically feasible) it is.
4. The management plan should utilize the largest amount of available public lands in its total acreage.
   Assumption: Public lands can more easily be converted (with less cost and time involved) into a governmental preserve than can private lands (Missouri Basin States Association 1983).

Next, five management plans were analyzed: four were existing official proposals by the U.S. Fish and Wildlife Service and the U.S. Army Corps of

Engineers for the mitigation of fish and wildlife losses on the river due to channelization. The fifth was an alternative "compromise" proposal which was developed as a result of this research to rehabilitate fish and wildlife habitat based on a series of restored oxbow lakes. These management plans were then subjected to a matrix format decision analysis. The management plans were evaluated against the four established criteria and against each other using quantitative data and replicable methods.

Data and Analysis

The U.S. Fish and Wildlife Service (USFWS) and the U.S. Army Corps of Engineers (USACE) plans were designed to treat over 1,182.3 km of the Missouri. In contrast, the Compromise Plan treats over 148 km. This difference in scope exists because the primary purpose of the Compromise Plan is not simply mitigation for fish and wildlife losses but rather to preserve, enhance and manage the Missouri's greatest remaining concentration of oxbow lakes and their associated native floodplain ecosystem as unique ecological features in the fluvial landscape. Once this is accomplished, this corridor will form a solid ecological foundation on which these biotic and abiotic natural resources can be restored to the region in approximate proportion to their former abundance and managed on a sustainable basis.

Both the USFWS plan and the three-level USACE plan were compiled into a report by the USACE's Missouri River Division which was then forwarded to the Secretary of the Army for review in May 1984. The chief of engineers has essentially recommended the USACE's Level B plan, which was the plan recommended by the Missouri River Division. The project was then organized by the Water Resources Development Act of 1986 (P.L 99-662). To date, no funding has been appropriated by Congress to carry out this project, however. There is also much disagreement on the recommended plan's comprehensiveness and effective scope between the agencies involved, as well as by opposing technical and political factions. Therefore, the Compromise Plan represents a "middle ground" proposal to break the existing deadlock and begin to rehabilitate the corridor area and its natural resources before they are lost.

Data

Table 1 compares the major components of the USFWS plan, the three-level USACE plan (U.S. Army Corps of Engineers 1981), and the Compromise Plan.

Analysis

Tables 2–5 analyze the components of the five plans to establish how well each plan meets the requirements of the four criteria previously described. A five-point system was created where five points equals the greatest potential for meeting the criterion and one point the least potential. Point values for

TABLE 1
MAJOR PLAN COMPONENTS

	USFWS	USACE A	USACE B	USACE C	Compromise
Terrestrial Hectares					
Restored and Preserved	60,649	22,819	18,166	6,837	12,931
Aquatic Hectares					
Restored and Preserved	3,155	1,942	1,011	404	1,497
Constructed	13,195	404	283	242	–
Total Plan Hectares	76,999	25,165	19,460	7,483	14,428

TABLE 2
CRITERION I
GREATEST AQUATIC TO TERRESTRIAL AREA RATIO

Management Plan	Hectares % Aquatic	Hectares % Terrestrial	Ratio	Point Value
USFWS	21.2	78.8	.269	5
Corps "A"	9.3	90.7	.103	3
Corps "B"	6.6	93.3	.071	1
Corps "C"	8.6	91.4	.094	2
Compromise	10.4	89.6	.116	4

Derived by: $\frac{\text{\% Aquatic Hectares}}{\text{\% Terrestrial Hectares}} = \text{Ratio}$

Point value: 5 = highest, 1 = lowest.

TABLE 3
CRITERION I
GREATEST TOTAL PLAN AREA

Management Plan	Total Plan Hectares (Aquatic and Terrestrial)	Point Value
USFWS	76,999	5
Corps "A"	25,165	4
Corps "B"	19,460	3
Corps "C"	7,483	1
Compromise	14,428	2

Derived by ranking plans according to size of total plan hectares (aquatic and terrestrial).
Point value: 5 = highest, 1 = lowest.

TABLE 4
CRITERION III
GREATEST POLITICAL FEASIBILITY

Management Plan	Nonpublic Hectares	% Nonpublic Land to be Acquired	Point Value
USFWS	60,816	79.0	1
Corps "A"	17,681	70.3	2
Corps "B"	11,288	62.2	3
Corps "C"	0	0.0	5
Compromise	7,994	44.6	4

Derived by: $\frac{\text{Nonpublic Hectares x 100}}{\text{Total Hectares}}$ = % Nonpublic Lands to be Acquired

Point value: 5 = highest, 1 = lowest.

TABLE 5
CRITERION IV
GREATEST PUBLIC LAND INCORPORATION

Management Plan	# Public Hectares	% Public Lands in place	Point Value
USFWS	16,184	21.0	1
Corps "A"	7,485	29.7	2
Corps "B"	7,363	37.8	3
Corps "C"	7,483	100.0	5
Compromise	6,433	55.4	4

Derived by: $\frac{\text{\# Public Hectares x 100}}{\text{Total Plan Hectares}}$ = % Public Lands in Place

Point value: 5 = highest, 1 = lowest.

each plan are totaled and the plan with the greatest overall point value (Table 6) is deemed to be the most appropriate restoration and management plan for the oxbow lake study area.

Table 7 evaluates the number of ha of habitat per km that each plan restores, preserves and/or constructs. These new point values are then incorporated in Table 8 with the findings from Table 6 to make a more refined determination of the most appropriate oxbow lake corridor management plan.

Discussion

The Matrix Decision Analysis in Table 6 totaled the scores for each management plan with regard to each criterion and determined which plan would be the most appropriate management alternative for the oxbow lake study area based on its total score.

TABLE 6
MATRIX DECISION ANALYSIS TABLE

Management Plan	Management and Plan Selection Criteria and Point Values 1	2	3	4	Point Total	Rank
USFWS	5	5	1	1	12	C
Corps "A"	3	4	2	2	11	D
Corps "B"	1	3	3	3	10	E
Corps "C"	2	1	5	5	13	B
Compromise	4	2	4	4	14	A

Rank: "A" = Most appropriate plan

TABLE 7
PLAN EQUIVALENCY

Management Plan	River Miles in Plan	Total Plan Hectares	Total Number Restored, Preserved, Constrcted Hectares/ Kilometer of Plan	Point Value
USFWS	1,182.5	76,999	104 ha/km	4
Corps "A"	1,182.5	25,165	34 ha/km	3
Corps "B"	1,182.5	19,460	26 ha/km	2
Corps "C"	1,182.5	7,483	10 ha/km	1
Compromise	148	14,428	156 ha/km	5

Derived by:

$$\frac{\text{\# Hectares in Plan (Aquatic + Terrestrial)}}{\text{\# Kilometers in Plan}} = \text{\# Restored, Preserved, Constructed Ha/Km}$$

Point Value: 5 = Highest, 1 = Lowest

In spite of the USFWS's twice being considered to be the plan of choice based on its high scores using matrix decision analysis, as was the USACE's Plan "C," the newly developed Compromise Plan, which was based on this research, was determined to be the most appropriate management alternative overall, according to these analyses.

The USACE's Plan "C" was determined to be the second most appropriate management plan under matrix decision analysis. While it had poor scores for preserving the greatest ratio of aquatic area to terrestrial area, as well as in preserving the greatest total habitat area, it did prove to have the greatest political feasibility and, therefore, potential implementation efficiency as defined by that criterion.

The USFWS's management alternative was determined to be the third most appropriate plan. It offered the greatest ratio of preserved aquatic area

TABLE 8
FINAL PLAN SELECTION
MATRIX DECISION ANALYSIS TABLE INCLUDING RESTORED HA/KM VALUES

Management Plan	Management and Plan Selection Criteria and Point Values 1	2	3	4	Restored ha/km Hectares/km Points	Point Total	Rank
USFWS	5	5	1	1	4	16	B
Corps "A"	3	4	2	2	3	14	D
Corps "B"	1	3	3	3	2	12	E
Corps "C"	2	1	5	5	1	14	C
Compromise	4	2	4	4	5	19	A

to terrestrial area, and provided the greatest total habitat area to be found in any of the plans. At the same time, however, it was determined that this plan offered the least political feasibility and, therefore, the lowest potential implementation efficiency as defined by that criterion.

The USFWS's plan and the USACE's Plan "C" are essentially opposites, each plan being strong in two categories but weak in the remaining two. However, the Compromise Plan possessed the most consistently high across-the-board scores of all the proposed management alternatives, thus giving it the broad-based strength and point total to make it the most appropriate management plan alternative for the oxbow lake study corridor.

An additional analysis as described previously was employed in Table 7 to aid in the decision process. When the score for each management plan under this last analysis is added in Table 8 to the matrix decision analysis presented earlier, the Compromise Plan has an even stronger showing as the plan of choice for the oxbow lake corridor. It provides almost twice as much potential habitat area per km as the USFWS plan and approximately 16 times the potential habitat area per km as the Corps' Plan "C." This is consistent with the analysis presented in Table 6.

While the management plans proposed by the USFWS and the USACE appear, at face value, to restore and preserve more habitat area than the Compromise Plan, this analysis has proven that the habitat area under these plans is in reality stretched thinly over 1,182.5 km, resulting in a minimal and rather less-than-ideal treatment per km. These plans would probably not be as effective in their design and results as would the Compromise Plan, with its more extensive and thorough habitat restoration and preservation treatment per kilometer over a smaller area. The Compromise Plan also appears to be the only management alternative that would significantly benefit the stated purpose of this research; namely, the preservation, enhancement, and management of the last remaining oxbow lakes and their associated native floodplain ecosystem, which, for the most part, occur only in the designated 148 km oxbow lake study area corridor.

Hence, these analyses have determined that the Compromise Plan is the management plan of choice for the oxbow lake study area corridor.

Overview of the Compromise Plan

In addition to the plan's physical dimensions as described in Table 1, the Compromise Plan (see Figures 1 and 2) recommends that the USACE should purchase all non-public lands necessary for inclusion in the oxbow lake study area corridor. Iowa public lands already situated within the study area would be incorporated into the corridor design. The Corps would also purchase any privately owned lands situated in Iowa as are deemed necessary to maintain continuity in the corridor design. It would also be the Corps' responsibility to purchase any lands necessary to establish a 100-meter wide buffer strip of restored tallgrass prairie habitat to border the entire project. If the lands necessary for inclusion in the corridor cannot be purchased from willing sellers or acquired through conservation easements and similar agreements, federal powers of condemnation should be invoked with fair market value compensation paid to the displaced owners. If condemnation cannot secure these lands, then tax credits should be employed as the best means of controlling the necessary lands and making them available for the oxbow lake corridor. After oxbow lake corridor rehabilitation work is completed, the acquired lands in the corridor will then be turned over to their respective states of Nebraska and Iowa for management by the appropriate state-level agencies.

The Corps would also be responsible for the construction and/or alteration of all necessary river control structures to allow for the preservation, enhancement, and management of the oxbow lakes and their associated diverse habitats. This responsibility includes any construction, dredging, or other appropriate techniques that are deemed necessary to rehabilitate the oxbow lakes and wetlands. The remaining lands in the corridor would be rehabilitated and managed by the respective states through a self-perpetuating trust fund established by the Corps to facilitate timely and judicious use of rehabilitation resources to preserve, enhance and manage the oxbow lakes and riverine habitats, as well as to sidestep undue bureaucratic "red tape" and delays in achieving the desired goals. The daily administration of the corridor would be supervised by a joint commission comprised of representatives of the Corps and the Nebraska and Iowa agencies charged with carrying out their plan's provisions. All Native American lands will be exempt from inclusion in this project because of legal complications.

Conclusion

The implementation of the Compromise Plan for oxbow lake preservation and enhancement would make significant advances toward the formation of

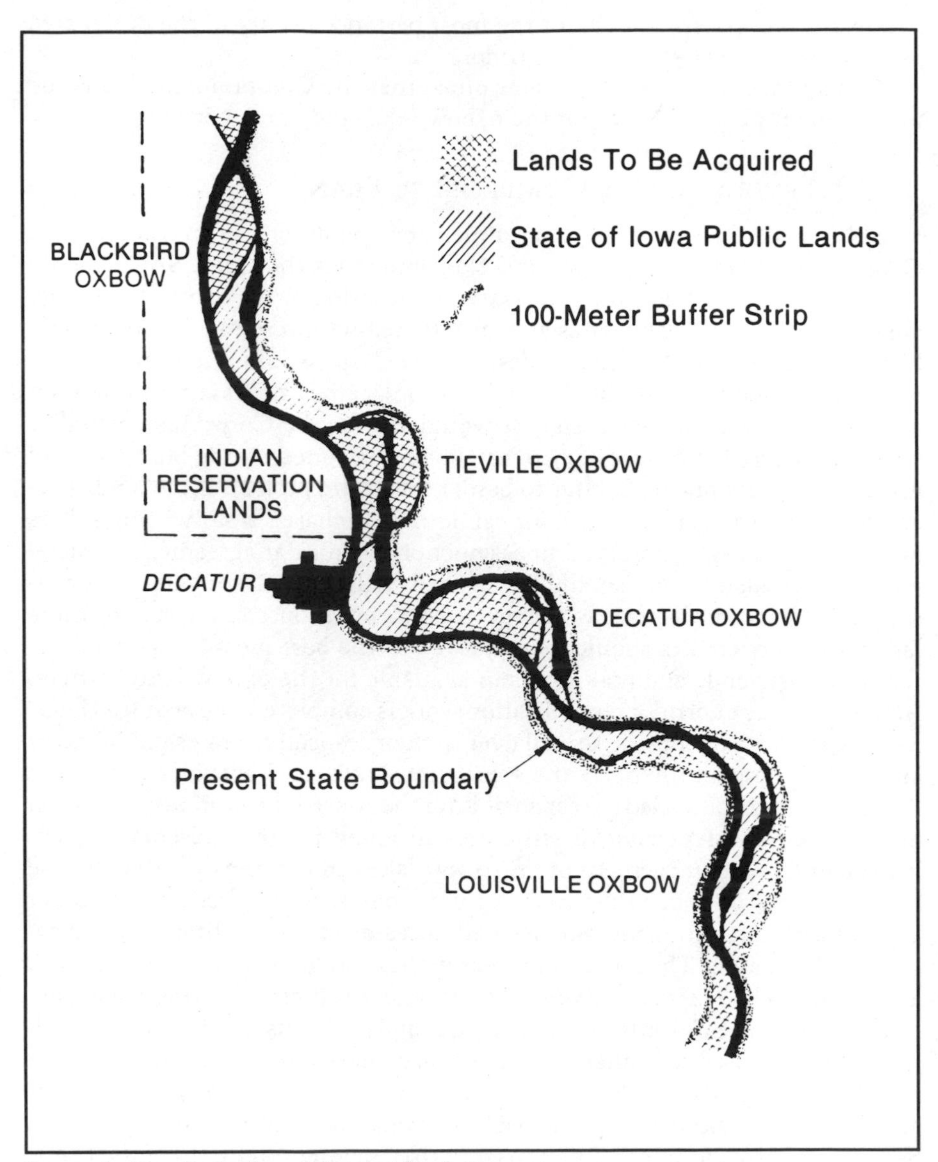

FIGURE 2. *Detail of oxbow lake corridor design.*

a solid ecological foundation on which the biotic and abiotic natural resources of the Middle Missouri can be restored and managed on a sustainable basis. Once established, there would be a myriad of opportunities for cultural, recreational, and economic development and enhancement for the communities located along the oxbow lake corridor based on these resources, as well as entrepreneurial spin-offs and future linkages throughout the region.

References

Bragg, T. B. and A. K. Tatschl. 1977. Changes in flood-plain vegetation and land use along the Missouri River from 1826 to 1972. *Environmental Management* I(4):343–348.

Forman, R. T. 1981. Interactions among landscape elements: a core of landscape ecology. In *Perspectives in landscape ecology: Proceedings of the International Congress of Landscape Ecology*, 35–48. Veldhoven, Netherlands: Pudoc Publishing Co.

Forman, R. T., and M. Godron. 1986. *Landscape ecology.* New York: John Wiley & Sons.

Fredrickson, L. H. 1979. *Floral and faunal changes in lowland hardwood forests in Missouri resulting from channelization, drainage and impoundment.* Washington, DC: Eastern Energy and Land Use Team, U.S. Fish and Wildlife Service.

Funk, J. L., and J. W. Robinson . 1974. *Changes in the channel of the lower Missouri River and effects on fish and wildlife.* Jefferson City, MO: Missouri Department of Conservation.

Hallberg, G. R., J. M. Harbaugh, and P. M. Witinok. 1979. *Changes in the channel area of the Missouri River in Iowa, 1879–1976.* Special Report Series no. 1. Iowa City, IA: Iowa Geological Survey.

Hutchinson, G. E.. 1957. *A treatise on limnology, I.* New York: John Wiley & Sons.

Missouri Basin States Association. 1983. *Final report: Missouri River flood plain study.* Omaha, NB: Missouri Basin States Association.

Noss, R. F. 1983. A regional landscape approach to maintain diversity. *Bioscience* 33(11):700–706.

SIMPCO (Siouxland Interstate Metropolitan Planning Council). 1978. *Missouri River woodlands and wetlands study.* Sioux City, IA: Siouxland Interstate Metropolitan Planning Council.

Soulé, M. E., and B. A. Wilcox. 1980. *Conservation biology: An evolutionary-ecological perspective.* Sunderland, MA: Sinauer Associates, Inc.

U.S. Army Corps of Engineers. 1981. *Missouri River bank stabilization and navigation project—fish and wildlife mitigation report, plus appendices.* Omaha, NB, and Kansas City, MO: U.S. Army Corps of Engineers—Omaha and Kansas City Districts.

U.S. Army Corps of Engineers. 1983. *Missouri River navigation charts—Sioux City, Iowa to Kansas City, Missouri.* Omaha, NB: U.S. Army Corps of Engineers, Omaha District.

John H. Sowl is a landscape architect and ecologist with the Midwest Region, U.S. National Park Service, 1709 Jackson St., Omaha, NE 68102, and was formerly with the U.S. Army Corps of Engineers—Omaha District.

Land Acquisition for Restoration and Protection

Barbara C. Brumback and
Richard A. Brumback

Abstract: *Ways in which public agencies can acquire lands for restoration are discussed as is the use of land acquisition to assure that the natural functions and values which have been restored are permanently protected. A range of potential acquisition techniques and funding sources are explored.*

Key words: *acquisition, fee simple acquisition, easement, purchase and lease-back, purchase and resale.*

Introduction

> Getting and spending, we lay waste our powers:
> Little we see in Nature that is ours . . .
> –Wordsworth

Many land acquisition techniques and potential revenue sources can enable public agencies to get and spend for natural resource restoration. Land acquisition is a necessary component in a restoration strategy. In many cases, acquisition is necessary to transfer land from full private ownership to public or nonprofit ownership so that restoration efforts can begin. In other cases, acquisition is necessary to assure that restored natural areas are protected from later actions that can degrade natural resource values.

Getting Land for Restoration

Land can be acquired through purchase, donation, or "exaction" through the land use planning and regulatory process. When land is acquired as an exaction, the landowner, wishing to modify his land in some way and requiring the government's permission to do so, can be required to provide land or interest in land before being permitted to make the change. The many legal and economic limitations on the exactions process are beyond the scope of this paper. The use of exactions has been increasing since the 1970s, but some kinds of exactions have been called into question by decisions of the U.S. Supreme Court (Bauman 1987).

Among the many different types of land acquisition techniques, some may be used to acquire land to restore it, and some may be used to protect areas once restoration has been completed. Three major categories of acquisition techniques include fee-simple acquisition, acquisition of interest, and post-acquisition strategies (Brumback 1987). Each has its own benefits and drawbacks.

Fee-Simple Acquisition

When most people think of acquiring property, they are thinking of a fee-simple acquisition. Property ownership comes with a bundle of rights, such as the right to develop the land, mine its minerals, or cut its timber. In fee-simple acquisitions, the entire bundle of property rights is transferred from one owner to another (Whyte 1968). With absolute ownership of property comes control over the land's use and the responsibility for managing the land. Of all the options, fee-simple acquisition is the most expensive, both initially, in acquiring all the property rights, and over the long term, in retaining those property rights and managing the land (Florida Atlantic University/Florida International University Joint Center for Environmental and Urban Problems 1986).

Acquisition of Interest

Instead of acquiring the land in fee-simple, in the acquisition-of-interest approach only the property rights that could be used to harm the resource are acquired. The landowner keeps the rest of the property rights as well as the land itself. Easements are a key acquisition-of-interest device.

An easement is a specially tailored document that either grants rights to others (positive easements) or restricts actions available to the landowner (negative easements) (Stover 1985). Easements can be granted for a specific term or in perpetuity. Positive easements can provide access (the most frequent use of this technique), hunting or fishing rights, or hiking trails. Negative easements can prevent actions, like filling wetlands, cutting trees, or developing property.

Since all the property rights are not acquired, easements can be a cost-effective acquisition approach. Long-term costs can also be reduced. The landowner can be made responsible for managing the land. This can reduce long-term cost for the easement holder since land management can prove costly over time, however, the easement holder is still responsible for enforcing the easement's terms. A true assessment of an easement's cost includes the costs of acquisition and the costs of enforcement.

Conservation easements (negative easements placed on property to protect environmental or other values) can be a useful option to assure that the effects of restoration are maintained. In effect, conservation easements allow the easement holder to control property rights that the landowner could otherwise use to degrade the property's resource values. Public agencies and private nonprofit organizations in the U.S. have used conservation easements for the past 50 years to protect 728,000 ha. Most conservation easements protect roadside scenery or the open space character of a community; but this technique has also been used to protect habitat, freshwater resource areas, and resource-based land uses, such as agriculture or timber (Land Trusts' Exchange 1985). In south Florida, over 40,470 ha are in conservation easements in Everglades areas to protect the natural sheet flow of water over the land (Florida Atlantic University/Florida International University Joint Center for Environmental and Urban Problems 1986).

Post-acquisition Strategies

Using post-acquisition strategies, land is acquired fee simple, but some or all of those property rights are disposed of, either permanently, through selective resales, or temporarily, through lease-backs. Typically, selective resales are accompanied by restrictions that limit the land's later use. Post-acquisition strategies can recoup a portion of the acquisition costs and reduce the costs of management.

Using purchase and lease-back arrangements, the agency remains the landowner, but leases the land for another's use under conditions or limitations that are compatible with the agency's needs (Osgood & Koontz 1984). For example, at Land-Between-the-Lakes, a national recreation area on the Kentucky Reservoir, the Tennessee Valley Authority leases some of its land to farmers to grow row crops with the stipulation that some of the crop remain in the field to provide forage for wildlife (Florida Atlantic University/Florida International University Joint Center for Environmental and Urban Problems and Florida State University Institute for Science & Public Affairs 1987).

Using the purchase and resale option, land can be acquired, its future use limited by easement or deed restriction (a recorded legal limitation noted in the deed), and resold. Colorado Open Lands, a private nonprofit land conservation organization, used purchase and resale to protect Evans Ranch, a 1,313 ha ranch about 64 km outside of Denver. Existing zoning allowed the ranch to be divided into 1,600 lots. Colorado Open Lands received a low-interest loan, secured by the land, and developed a master plan for the ranch that divided it into five smaller ranches, each with a 16 ha homesite. As of 1986, three ranches had been sold, and those sales generated enough money to repay the loan (Land Trusts' Exchange 1986).

Notes of Caution

All of the land-acquisition techniques discussed have potential for assuring that the resource values gained by restoration can be retained. However, no acquisition technique is without limitations. The principal limitation of fee-simple acquisition is the expense of acquiring all the property rights and managing the land.

When an alternative to fee-simple acquisition is used, the transaction can be challenging and time consuming. Landowners may be confused about the ramifications of giving up some of their property rights. Assigning value to the rights that are relinquished can be a source of controversy. Both Colorado Open Lands and California's State Coastal Conservancy, an agency with considerable experience in alternative acquisition techniques, have found that, in resales, the value of restricted land is often much higher than anticipated. With acquisition of interest techniques, however, the cost rises with the restrictions on the property's use. If protecting the land's resource values requires acquiring most of the property rights, the cost of acquisition of interest approaches the cost of fee-simple acquisition (Florida Atlantic University/ Florida International University Joint Center for Environmental and Urban Problems 1986).

Finally, while techniques like easements, lease-backs, and resales typically eliminate the need for actual management of the land, monitoring and enforcement are needed to assure that the restrictions to protect natural values are being followed.

There is no one best way to acquire land for natural resource protection. Problems with land-acquisition methods typically arise not from the methods themselves, but from the way those methods are used. This is a lesson that was learned in the 1930s through the National Park Service's experiences along the Blue Ridge Parkway, and relearned in the 1960s through Wisconsin's easement protection program. In the latter case, easements were chosen as the low-cost alternative and used as the one-and-only way to protect resources on a large scale. In both cases, the easement acquisitions fell short of their goals (Hoose 1981). The techniques outlined above are only tools, and the next section suggests potential means of paying for those tools.

Spending for Acquisition

Funding land acquisition can be a major stumbling block, since acquisition costs can be high, especially in locales desirable for development, such as coastal areas. Nonetheless, there are many acquisition programs at the federal, state, and local government levels as well as in private, nonprofit resource conservation groups.

In recent years at the federal level, the largest funding source for the acquisition of natural resource lands has been the Land and Water Conservation Fund, established in 1965. This fund receives revenues from the sale of surplus federal real estate, taxes on motorboat fuels, and receipts from both recreation fees and outer continental shelf leases (Owens 1983).

Several state and local governments have created substantial acquisition efforts. Florida, for example, has developed a natural resource land acquisition program that will generate an estimated $1 billion over the next decade. Revenue sources for this effort have included bonds, general revenues (largely from sales taxes), documentary stamp taxes on real estate transfers, severance taxes on minerals, and an optional local tourist tax on temporary accommodations (Florida Atlantic University/Florida International University Joint Center for Environmental and Urban Problems 1986).

While these examples do not represent the entire range of possible options, they do provide a general framework for choosing sources of money for acquisition. There are two logical and general kinds of funding sources for land acquisition and restoration: (1) those that are somehow linked to the activity that causes the need to acquire and restore land, such as severance taxes on oil, gas, and minerals, or documentary stamp taxes on real estate transactions; and (2) those that are somehow linked to the people who will benefit from the acquisition and restoration. The second funding source includes tourist taxes on temporary accommodations and increased hunting and fishing license fees.

Using the benefit approach, if it is assumed that the general population will benefit, logical funding sources include general revenue sources and bonds. Other possible sources include a variety of user fee approaches, such as recreation fees, hunting and fishing licenses, or recreational vehicle registration fees. A creative look at potential funding techniques and the use of the appropriate acquisition technique can help assure that land necessary for restoration can be acquired and that the restored natural functions and values are protected.

References

Bauman, G. 1987. Overview of recent decisions. ALI-ABA *Course of study: Land use institute planning, regulation, litigation, eminent domain, and compensation.* Philadelphia, PA: American Law Institute, American Bar Association.

Brumback, B. C. 1987. Protecting places: A new look at land acquisition. *California Waterfront Age* 3:4.

Florida Atlantic University/Florida International University Joint Center for Environmental and Urban Problems. 1986. *Alternatives to fee-simple land acquisition: Approaches and opportunities.* Fort Lauderdale, FL: Florida Atlantic University/Florida International University Joint Center for Environmental Problems.

______ and Florida State University Institute for Science and Public Affairs. 1987. *Land management issues: A guide for Florida's water management districts.* Tallahassee, FL: Florida State University Institute for Science and Public Affairs

Hoose, P. 1981. *Building an ark: Tools for the preservation of natural diversity through land protection.* Covelo, CA: Island Press.

Land Trusts' Exchange. 1985. Innovative loan package. *Land Trusts' Exchange* 3:4.

______. 1985. Special issue: Report on 1985 national survey of government and non-profit easement programs. *Land Trusts' Exchange* 4:3.

______. 1986. Case study: Colorado open lands, Denver, Colorado. *Land Trusts' Exchange* 4:4.

Osgood, R. and R. Koontz, 1984. A summary of forms and tax consequences of land acquisition by a charity. In *Land Saving Action,* eds. R. Brenneman and S. Bates. Covelo, CA: Island Press.

Owens, D. 1983. Land acquisition and coastal resource management: A pragmatic perspective. *William and Mary Law Review* 24:625.

Stover, E., ed. 1975. *Protecting nature's estate: Techniques for saving land.* Washington, DC: U.S. Department of the Interior, Bureau of Outdoor Recreation.

Whyte, W. 1968. *The last landscape.* New York: Doubleday & Co.

Barbara C. Brumback is senior research associate, Florida Atlantic University/ Florida International University for Environmental and Urban Problems, Fort Lauderdale, FL 33301. Richard A. Brumback is assistant professor, School of Public Administration, Florida Atlantic University, Boca Raton, FL 33301.

Watershed Planning and Restoration: Achieving Holism Through Interjurisdictional Solutions

Kittie E. Ford, Karen A. Glatzel, and Rocky E. Piro

ABSTRACT: *This paper examines watershed planning and management in an interjurisdictional context. Conflicts arising from a fragmented regulatory framework impede progress toward watershed restoration. Watershed planning considers the watershed boundary as defining an elemental planning unit. The focus on nonpoint sources as they affect water quality requires that activities in an entire watershed be considered in planning efforts. Two regional approaches to watershed planning are examined and evaluated as applied by: (1) the Marine Resources Council of East Florida for the Indian River Lagoon, and (2) the Puget Sound Water Quality Authority in Washington State. Interjurisdictional solutions are a prime mechanism for implementing watershed planning and restoration efforts and are often the only realistic approach in light of a fragmented political and governmental structure.*

KEY WORDS: *watershed planning, interjurisdictional planning, Indian River Lagoon, Puget Sound, nonpoint source pollution, estuaries, coastal lagoons, restoration.*

Introduction

HOLISM IN THE context of environmental planning implies planning for an entire system and all of the interrelated components. In watershed planning, holism is predicated on the idea that runoff from the watershed impacts various receiving bodies, such as coastal estuaries. Hence, estuarine management implies watershed management.

Problems such as restoring degraded estuarine resources typically transcend existing political boundaries. Local and state governments usually are not able to act comprehensively toward resolving restoration problems. Conflicts often center on competing interests among different levels of government, between different agencies and jurisdictions, and between various industries and interest groups. Water resource issues are often the most controversial to consider—economically, socially, environmentally, and personally.

Framework for Analysis

In examining watershed planning and restoration in a regional context, two general approaches have been identified: (1) the regulatory or bureaucratic approach, and (2) the negotiated or grassroots approach. Planning efforts can incorporate elements of each of these approaches, although organizations usually stress one approach over the other. The regulatory approach to watershed planning is one in which an organization's programs are directed toward control of activities by mandated requirements. The negotiated approach brings together key parties concerned with or affected by the environmental issues, in order to reach a consensus on some action for problem resolution or environmental restoration.

The regulatory approach can often be territorial, parochial, and ineffective due to interagency competition. The regulatory function of government often ignores political and equity concerns. Regulatory approaches are most valuable when there is clear consensus on a regional need, as exemplified by the discussion of the Puget Sound Water Quality Authority.

The negotiated approach to watershed planning can result in piecemeal and ineffective planning efforts. The negotiated approach may be difficult to employ when dealing with multidisciplinary issues associated with water quality, land use, and watershed and estuarine management. If the negotiated approach is used as part of a strategically applied planning process, it can be successful as illustrated in the discussion of the Marine Resources Council of East Florida.

Both approaches influence and are affected by economic or market factors. Environmental regulations often place demands on the market (e.g., property taxes to control runoff, or impact fees for development). In other instances, withdrawal of federal support for planning and transfer of these responsibilities to the state and local levels has encouraged formation of grassroots environmental organizations. Consideration of multiple impacts in watershed planning requires recognition of conflicting market interests. These interests vary depending on perception of the problems, the solutions, and the costs.

Often key areas of concern that link watershed planning to estuarine restoration are triggered by some market-related crisis, such as closure of shellfish beds or regional requirements for upgraded sewage treatment. Quite often, only the perception of a market crisis has prompted public suspicion that some ecological calamity has occurred. The market crisis in shellfishing has often been linked initially to ineffective sewage treatment, although more extensive problems resulting from watershed runoff are involved.

After initial recognition of a water quality problem, there are many possible solutions to be considered in the formulation of a watershed restoration plan. Implementing the solutions requires different kinds of participation from the various parties in the watershed, including voluntary actions and acceptance of new land use controls.

The following two examples describe watershed planning efforts that resulted from potential crises related to water quality. In the Florida example, a negotiated approach was used to seek resolution of the problem. In contrast, a regulatory scheme was adopted in the Puget Sound case for addressing the problems of water quality degradation.

The Marine Resources Council

The Marine Resources Council of East Florida (MRC) is a regional grassroots organization concerned with enhancement, restoration, and management of the Indian River Lagoon and its watershed. The Indian River Lagoon is located along a 250-km stretch of the Florida Atlantic coastline. As defined by MRC, the lagoon system includes the Indian River, Banana River, Hobe Sound, and the southern portion of Mosquito Lagoon. Together with the watershed, the Indian River Lagoon system includes over 5,900 km^2 and is located within six counties.

In the early stages of recognizing and defining the problems of the lagoon, a multidisciplinary conference was held on the resources of the Indian River Lagoon system (Taylor and Whittier 1983). The conference provided the impetus for the MRC to be organized. The MRC has continued to address the issues concerning resources of the lagoon system in a strategic manner, by involving citizen participation, public education on issues, and political activism.

The goals of the MRC include utilization, enhancement, and maintenance of the marine resources in the region of the Indian River Lagoon (Barile et al. 1987). The MRC acts in a coordinating role, providing a network of information needed for informed participation and decision-making by individuals, government agencies, and elected officials. Objectives are related to helping establish resource goals, and providing help in citizen consensus building, public awareness, and education.

The MRC is a coalition with members representing individuals and organizations from at least 33 broad-based special interest groups, including commercial and recreational fishermen, scientists, educators, environmentalists, active citizens, farmers, realtors, home builders, boaters, port officials, government personnel, and elected officials (Barile et al. 1987). The MRC has a board of directors elected from among the membership.

Much of the program development for MRC is accomplished by volunteer advisory committees. The MRC has a small staff consisting of an executive director and an administrative assistant. Project related work is often accomplished by community volunteers and/or paid graduate students from Florida Institute of Technology, which also rents facilities to the Council. The MRC has informal alliances with the regulatory agencies of the State of Florida. The Council often does special studies for these agencies. Depending on the funding source, MRC has different reporting requirements. Activities of the MRC rely for continued funding on strong political connections to the legislators and the heads and technical staff of agencies.

Issue definition has been accomplished by various public awareness and education projects including an "American Assembly" process as described in Barile et al. (1987). This process includes participants representing a broad cross section of views and interests. Two assemblies have been conducted resulting in consensus on a series of issues related to the Indian River Lagoon and a set of action items for future direction. Actions have included direct requests to the Governor for the formation of an Interagency Management Task Force to prepare a report detailing intergovernmental coordination for development of a management plan (Panico and Barile 1986). Legislation has been sponsored by a number of state lawmakers, who have also served on the assemblies. The legislature has directed the MRC to oversee and coordinate research activities with monies directed through the Florida Sea Grant College Program.

Plans for future activities include more intensive coordination with local and county governments during preparation of coastal zone elements of their comprehensive plans. Public education projects are ongoing and are considered very important.

The Puget Sound Water Quality Authority

The Puget Sound Water Quality Authority was first formed in 1983 to identify and evaluate pollution-related threats to marine life and human health, and to investigate the need for coordination among agencies responsible for protecting water quality in the Puget Sound basin of the state of Washington. The basin includes the entire land area south of the Canadian border that drains into Puget Sound and the Strait of Juan de Fuca, and encompasses 12 county jurisdictions. The initial examination of Puget Sound water quality issues led to a recommendation for the development of a long-range, coordinated management plan for the Puget Sound basin. In 1985 the state legislature restructured the Authority and commissioned it to engage in a comprehensive planning process through 1991.

The Authority's primary objective is to facilitate implementation of a water quality management plan. The goal of the Puget Sound Water Quality Management Plan (PSWQA 1987a) focuses on prevention of further degradation of water quality, and on preservation and restoration of resources. The Authority is comprised of seven members and two ex- officio members, all appointed by the Governor. An advisory committee and a scientific review panel representing local governments, the business and industrial community, state and federal agencies, Native American tribes, scientists, and public interest groups were chosen by the Authority and its staff to assist with evaluation of issues and development of the plan. Following the formation of the Authority and its advisory committees, public meetings were held throughout the region to gain citizen input to identify the most important issues. The enabling legislation for the Authority (state statute RCW 90.70) specifically required the plan to address nonpoint source pollution management, industrial pre-treatment of toxic wastes, dredge spoil disposal, and the protection, preservation, and restoration of wetlands, wildlife habitat, and shellfish beds (PSWQA 1987a). In addition, the legislation required the plan to set priorities for management and clean-up activities, with cost estimates and implementation criteria that assure coordinated federal and state efforts, and local government-initiated planning.

The Authority has no specific implementation or enforcement powers, but functions as a negotiator/advisor between affected parties and local governments and the Washington Department of Ecology. The Puget Sound Water Quality Act requires state agencies and local governments to implement the plan. The Authority's responsibility to assure implementation is accomplished by monitoring and reviewing local government compliance with the policies of the Puget Sound Water Quality Act. The Authority also contributes to and/or intervenes in administrative or judicial proceedings to assure consistency with the plan. Recently, the Authority proposed a draft nonpoint source administrative rule (PSWQA 1987b) for state adoption to carry out the priority watershed and nonpoint source action programs of the plan.

State and local governments identified in the plan must report biennially to the Authority on their compliance with the plan. The Authority in turn reports to the legislature and the governor. State and local implementing agencies are also reviewed by the Authority to determine budgetary needs for carrying out local programs. Noncompliance with the plan could result in state retention of funds for local-level planning and the transfer of local program development and implementation to the Washington Department of Ecology. The planning role of the Authority is scheduled to undergo a "sunset" review in 1991. At that time, the Washington Department of Ecology could become the responsible agency for continued planning and implementation of water quality management plans.

Discussion

It is important to recognize that the organizations described above were formed in part to address anticipated economic impacts of growth on water- dependent uses, including the harvest of marine food resources, recreational opportunities, tourism, and water-oriented lifestyles. Yet a purely economic or market-based approach for resolving these issues was not feasible due to the many noneconomic factors involved. These included environmental, aesthetic, and social values which emerged as a result of heightened public awareness.

In both cases, the large number of governmental entities, private interest groups, and industries that affect water quality had limited individual authority to attempt solutions targeted at restoration of water quality degradation. In Florida and Washington a new entity or "third party" was introduced to provide a more comprehensive and integrated approach to problems and solutions.

Traditionally, public decision making involves interaction between three key groups: (1) the legislature, or similar decision-making bodies, (2) the bureaucracy, or regulatory agencies, and (3) interest groups, or active citizens (The Conservation Foundation 1984). These groups do interact, but generally operate independently, in a competitive and conflicting fashion. In the two examples, the new entities facilitate communication between the three traditionally disparate groups.

A grassroots coalition began to facilitate communication in the Indian River Lagoon watershed. Activists and organizers sought to involve all pertinent parties and key interest groups who had some stake in resolving the crisis. This strategy typifies the negotiated approach to watershed planning. Strengths of the MRC effort include: (1) the promotion of trust among interest groups and individuals, (2) the establishment of access to politicians and the media, and (3) the development of power through information networking.

In the Puget Sound basin, a state-mandated planning agency has a coordination role between the legislature, regulatory agencies, and interest groups. The Puget Sound Water Quality Authority follows a modified regulatory approach, which enables it to perform critical analysis and planning for improving water quality, without enforcing actual statutes. There are a number of strengths in this particular approach. First, the Authority enjoys official political sanction, having been created by the state legislature. Second, the Authority is well funded by the state. Third, although the Authority acts as a planning agency, it is able to exercise substantial influence with its formal ties to regulatory agencies. Finally, a prescribed schedule enables the Authority to follow an orderly process for achieving its identified goals.

An important weakness of the coalition and negotiation approach used in Florida is the real possibility that efforts at reaching a consensus among the key parties may not be successful. Compromises may produce ineffective solutions. Stalemates may occur which deadlock the entire process. Significant parties may agree to a decision, but do little to help realize successful implementation.

Several weaknesses also are possible in the regulatory approach used by the Authority in the Puget Sound region. The Authority adopted a process requiring some planning action at the local level. Funding is being provided for local communities to prepare subbasin watershed plans. However, without local commitment to implementation, such plans merely maintain the status quo and have little consequence for existing or future activities.

Although many limitations can be identified in both of the approaches presented, they each have the potential of moving beyond the typical piecemeal and patchwork pattern evident in most public problem-solving processes. Indeed, there is potential in both examples for developing a more integrated and comprehensive process for restoring water quality.

Intergovernmental Solutions

The coordination role is an interjurisdictional mechanism that has met with initial success in both situations examined. This role has helped to establish communication between independent jurisdictions and governmental entities. However, both examples deviate from the ideal role of a coordinator (Figure 1). The Authority would like to establish more broad- based citizen activism and bring politically active interest groups into the planning process.

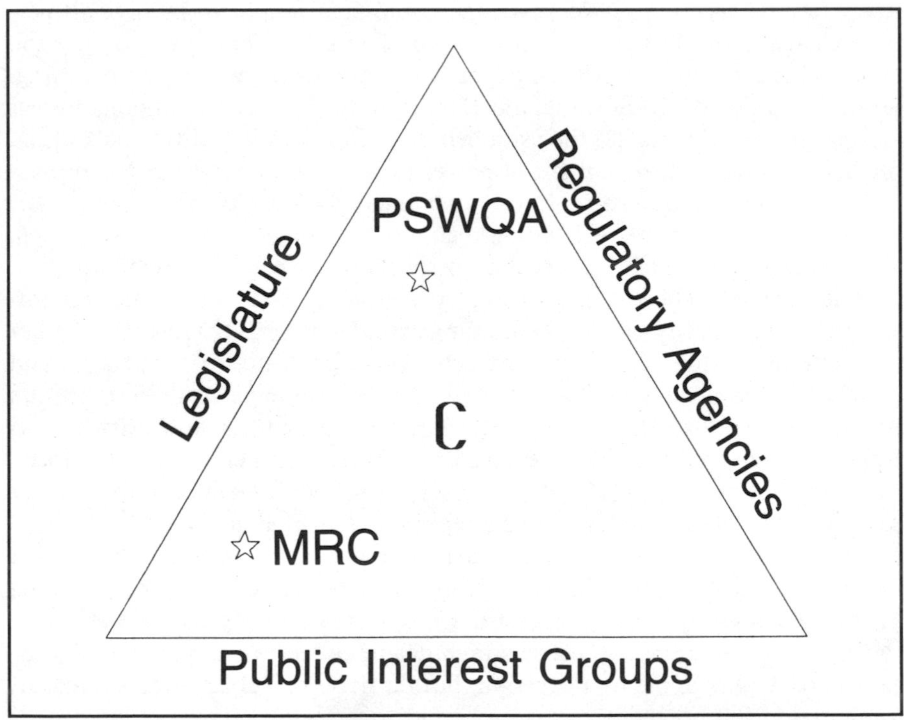

FIGURE 1. *Interjurisdictional coordination role in public decision-making. The symbol "C" represents the ideal location for effective coordination between legislative, regulatory, and public interest groups. The Puget Sound Water Quality Authority (PSWQA) and the Marine Resources Council of East Florida (MRC) positions indicate their respective relationship with decision-making entities.*

The MRC has tried to exert political influence in an effort to become the established coordinating body for the Indian River Lagoon, including the development of formal ties to the state's regulatory agencies and legislature. In both examples, the strategies used have been helpful in initiating watershed planning activities and in building public awareness. Future roles of these organizations will require working with interjurisdictional solutions to implement the land use controls needed for water quality restoration and resource management.

Environmental regulations, especially in the context of impact assessment, have produced little change from business-as-usual and have done little to affect comprehensive watershed planning. The costs of environmental impact assessment have not produced equitable benefits. Impact assessment alone cannot force the inclusion of watershed planning goals into local decision making (The Conservation Foundation 1984).

Intergovernmental agreements have been used for applications such as providing urban services, settling resource use conflicts, coordinating environmental review, and establishing responsibility for restoration and enhancement projects. Structured liaisons between government agencies can be effective in addressing the problems encountered in watershed planning.

Political realities do not make sweeping changes of institutional structures very likely. Thus, avenues for reducing the complications of fragmentation must focus on working with structures already in place. To implement land use controls effectively, intergovernmental agreements are more politically feasible. As governments, agencies, and interest groups begin to work together to promote common solutions to common problems, then broader, more effective regulatory mandates can be developed and adopted.

Administrative consolidation could possibly follow, but may not be desirable or possible. For example, consolidation may mean that mechanisms for coordination between various interest groups may not be available. The coordinating entity may easily take on all of the characteristics of a bureaucratic agency, losing its function as the critical tie between the legislature, bureaucracy, and interest groups.

Summary

The institutional coordination role that the Authority has been given and the MRC seeks to attain is an excellent strategy for achieving comprehensive and integrated interjurisdictional programs when backed by political commitment. Without such commitment to a particular course of action, there is potentially little incentive for cooperation between jurisdictions in order to achieve regional restoration goals.

With all elements in place (i.e., enthusiastically involved interest groups, effective regulatory entities, and a politically committed legislature), successful watershed planning and water quality restoration and management becomes achievable. A coordinating entity with ties to all elements can effectively provide the communication link that is necessary for implementing local land use controls and other voluntary and statutory mechanisms for watershed management.

The Authority and the MRC are both coordinating the development of watershed subbasin plans. Both organizations realize that public education is a key to success since it supplements mandatory regulation with public peer pressure. In the future, these organizations will be coordinating the implementation of land use controls.

References

Barile, D. D., C. A. Panico, M. B. Corrigan, and M. Dombrowski. 1987. Estuarine management. *Coastal zone '87/proceedings of the fifth symposium on coastal and ocean management,* 1:237–249. New York: American Society of Civil Engineers.

The Conservation Foundation. 1984. *America's water/current trends and emerging issues.* Washington, DC: The Conservation Foundation.

Panico, C. A., and D. D. Barile. 1986. *Interim report/management of the Indian River Lagoon.* Melbourne, FL: Marine Resources Council of East Florida.

PSWQA. See Puget Sound Water Quality Authority.

Puget Sound Water Quality Authority. 1987a. *1987 Puget Sound water quality management plan.* Seattle, WA: Puget Sound Water Quality Authority.

Puget Sound Water Quality Authority. 1987b. *Proposed rule/local planning and management of nonpoint source pollution,* Chapter 400–12 WAC, 12 November 1987. Seattle, WA: Puget Sound Water Quality Authority.

RCW 90.70. Revised Code of Washington State Statutes.

Taylor, W. K., and H. O. Whittier, eds. 1983. Future of the Indian River system. *Florida Scientist* 46(3/4):124–431.

Kittie E. Ford is an environmental planner, Huckell/Weinman Associates, Inc., Seattle, WA 98115. Karen A. Glatzel is a research associate, Department of Urban Design and Planning, University of Washington, Seattle, WA 98195. Rocky E. Piro is a research assistant and an instructor of interjurisdictional planning, Department of Urban Design and Planning, University of Washington, Seattle, WA 98195.

Resolving Coastal Restoration Conflicts: Environmental Protection Guarantees

David Shonman

Abstract: *Habitat restoration projects which are part of proposed developments have been controversial. Opposition has been based on the inadequacy of restoration plans to accomplish agreed-upon goals. To deal with these problems, project elements are recommended which deal with short-term and long-term environmental protection guarantees, verifiable project criteria, long-term monitoring, correction of conditions which fail to meet agreed-upon criteria, and long-term funding to ensure that all recommended elements can be carried out. These components are especially critical for restoration plans which involve fragile habitats and/or rare, threatened, or endangered species.*

Key words: *restoration, development, monitoring, enforcement, funding.*

Introduction

While habitat restoration might seem a laudable goal to most of us, there are many for whom it is a subject of intense controversy. This is often the case when restoration of disturbed habitat is made a condition of approval for a development permit. To a developer who must restore environmentally disturbed land, restoration might be just another expensive and time-consuming hoop to jump through in order to complete a project. To some environmental activists, restoration might represent a short-sighted deal between a developer and a government agency which allows development of pristine habitat in return for a promise to restore damaged habitat, a promise which often goes unfulfilled.

The arguments which surround these projects are usually highly charged with emotion. Often, much of the opposition stems from dislike of the development itself; the restoration element is seen only as a sugar-coating, tossed in to help increase public acceptance of the development proposal. At other times, opponents may object to using restoration as a bargaining chip: a mitigation which permits the development of undisturbed habitat in return for the promised restoration of damaged land. When used in this way, restoration is sure to be controversial.

At the root of this controversy is the nature of habitat restoration. Like other forms of mitigation, restoration can be used to lessen or counter-balance the negative environmental impacts which might result from development. But restoration differs from many other mitigation options in an important respect: restoration is a long-term, experimental, sometimes expensive process which can often lack clearly definable goals and whose success can never be fully guaranteed. (The lack of guarantee is often at the heart of many restoration controversies.) Restoration clearly contrasts with the more straightforward forms of mitigation (e.g. reduction of the development's scope or dedication of an access- or conservation-easement) which are easy to verify and have benefits that can be felt almost immediately.

Biologists involved in the restoration process have been surprised by the intensity of opposition to their project. To a biologist, restoration is strictly a biological activity. But restoration is more than just biological; it is political, legal, economic, and in some respects, philosophical. A successful restoration project must address all these concerns.

These controversies serve to remind us that restoration/development projects must be accountable to the community's needs. The following recommendations deal with some of the more controversial aspects of these projects. If included in restoration/development agreements, these recommended elements should contribute to the long-term success of these restoration projects.

Restoration of Disturbed Habitat Should Not Be Equated with Protection of Undisturbed Habitat

In view of its experimental nature, the use of habitat restoration as a way of mitigating a development's detrimental impacts must be carefully evaluated. While restoration does have the potential to increase the ecological value of already disturbed land, it may not match the value of protecting prime undisturbed habitat. Whenever possible, protection of existing habitat should be given precedence over restoration of disturbed habitat.

Restoration Must Be Sensitive to the Local Biogeographic Area and Include the Appropriate Biota

In the Monterey Bay coastal dunes, for example, some restoration projects have introduced exotic species along with native plants. Some exotics, like statice (*Limonium* spp.), introduced by the California Department of Transportation in the dunes alongside of Highway 1 in the City of Marina, are a minor, but unnecessary, nuisance. Others, like Hottentot-fig ice plant (*Carpobrotus edulis*) and European dune grass (*Ammophila arenaria*), are extremely invasive, and can out-compete and eventually replace native dune plants, including those which serve as habitats for native animals; unfortunately, these exotics have been introduced into coastal dunes in such areas as Moss Landing, Sand City, and Monterey.

Even the use of native California plants does not automatically guarantee project success if those particular species are not a natural part of the community found at the site. In another unfortunate example from the Highway 1 freeway project through the City of Marina, landscapers used the common bright orange California poppy (*Eschscholzia californica*), apparently without realizing that in the Marina dunes, the proper plant should be the coastal variety of California poppy, *E. californica maritima*, which has a predominantly yellow flower.

A restoration which is sensitive to the local biota must be based on more than just a species list. For example, a recent botanical study of the Marina Dunes in Monterey Bay reported that the plants found in those portions of the dunes formed during the end of the last ice age ("Flandrian dunes") represent a unique assemblage (Pavlik and Zoger 1987). All of the native species recorded here are also found in other parts of the Monterey Bay coastal dune system, but the proportional species composition and patterns of distribution at this site are not duplicated anywhere else in California. To restore the dunes at this site, a successful project should be designed on the basis of a complete biotic survey, one which characterizes the area with enough accuracy and detail to allow the restorers to recreate the habitat as much as possible.

Restoration Should Set Short- and Long-term Goals Which are Specific and Easily Verifiable

Success in achieving goals which are too vague cannot be measured. For example, a mitigation which requires that a developer simply "restore the habitat" is too general to be enforced. Restoration/development projects should set specific goals to be achieved during a specified time period. These goals may include projected species composition and densities of native plants, allowable densities of exotic invasives (preferably as low as possible), and methods for protecting the restored area from damage caused by trampling and off-road vehicles. These goals will serve as the criteria against which the project's success will be measured.

Restoration Must Include a Provision for Monitoring

This is a critical aspect of any mitigation-implementation agreement because it is the only way to verify if the developer is meeting the agreed-upon terms.

While monitoring might seem to be too intrusive, a property owner will usually derive great benefit (financial and otherwise) from the development of property, especially along the coast. Most often, the only benefit the environment receives will come from the promised mitigations. Since these promises are part of the basis of approval of the development permit, they must be checked. As Reagan said to Gorbachev in December 1987: "*Doveryai no proveryai* (Trust but verify)."

Obviously, the success of any restoration monitoring will depend, in part, on the clarity and specificity of the restoration goals, as described above. These will give direction to the monitor and describe what conditions must be observed and measured. Some conditions will be easy to verify, such as the existence of an intact fence or the lack of invasion by aggressive exotics. However, if the agreement requires that the owner maintain a viable population of Menzies' wallflower (*Erysimum menziesii*), then the monitor must be able to determine what constitutes a viable population. Should the agreement describe a minimum threshold number of these plants which must be maintained, or just describe an area which must be set aside and protected? How should the monitor deal with *natural* problems (e.g., adverse weather, deer predation, etc.)? These may constitute threats to the restoration project, but they are clearly out of the control of the property owner/developer/restorer. The above are but a few of the kinds of questions which can be expected to arise. But they must be answered in order to establish criteria which are clear and not open to misinterpretation—this will best benefit the monitor, the property owner, and the habitat.

Restoration Must Include an Established Mechanism for Correcting Problems Reported by the Monitor

Problem-correction responsibilities might range from simply repairing a fence to eradicating regrowth of invasive exotics or replanting native species in a restored area which failed to meet the established criteria. These procedures must be carried out in a manner conforming to the original agreement. If, for example, methods for removal of invasive exotics and planting of appropriate native species were described in the original restoration/development agreement, then these same methods should be employed if necessary to correct any problems reported by the monitor.

For example, if the needed repairs involve replanting Menzies wallflower, then the revegetation methods might include collecting seeds only from local dunes, then outplanting seedlings during late fall to early winter without using artificial irrigation. If ice plant must be stopped before it can encroach on a newly restored area, then it might be necessary to remove the ice plant by hand, wherever it is found growing within 2 m of the restoration area, rather than to spray it with an herbicide.

This portion of the restoration/development agreement should clearly designate who has the responsibility for correcting problems. If the responsible party is the property owner, then the terms of the restoration/development agreement should be recorded as a deed restriction, which will automatically pass to the new owner whenever the property changes hands. This will clearly establish where the legal responsibility rests, and will be helpful in those situations where a cooperative owner sells the property to someone who has less environmental sensitivity.

Restoration Must Include a Long-term Funding Mechanism

The establishment of a long-term funding mechanism is perhaps one of the most important elements in assuring the success of a restoration/development program. This is necessary to guarantee that the critical program elements described above can be carried out. Long-term funding could be obtained through an assessment of a "user fee" from those who benefit from the development. For example, a motel which will soon be built in the Marina Dunes will be paying an $18,000 lump sum as well as $0.35 per occupied room per day into a dune protection fund.

If there are many properties in the same geographic area which are involved in similar restoration/development projects, then another possible funding mechanism is to form a Resource Protection District which can receive taxes and, in return, provide such services as monitoring, restoration expertise, greenhouse support, etc. This may be modeled after the concept of a dune preservation district described by Nordstrom and Psuty (1980).

A third method is to seek funds from government resource protection agencies and private foundations and land trusts. During the past several years, these organizations have become increasingly aware that long-term habitat maintenance is as important as habitat acquisition.

Restoration/Development Agreements Must Be Enforceable

Mitigation implementation agreements are often violated, sometimes accidentally, sometimes purposely. Whatever the cause, there must be an effective way to redress these violations in a competent and timely manner (once they are reported by the monitor). In some cases, the responsible party may not readily be willing to pay for the necessary corrections. To avoid this, there are several mechanisms available which can relieve the immediate problem:

Performance Bonds

This is a commonly used method of assuring that a contractor's work will meet a specific level of quality. Often, a specified amount of money will be held in an interest-bearing account until the finished project has been inspected and given final approval. If made a part of a restoration/development agreement, these bonds can be applied to situations where the responsible party fails to perform an agreed-upon task necessary to the long-term success of the restoration project. In this case, the performance bond is similar to a renter's security deposit.

Penalties/Fines

In some cases, specific fines can be made a part of a restoration/development agreement. For example, as part of a recent development near Tucson, Arizona, one of the conditions of permit approval was that the developer would protect saguaro cactuses (*Carnegeia gigantea*) in the development area. During the construction phase, large signs were hung on each cactus warning the bulldozer operators how much money would be charged for any damage (Phillips 1986). For development/conservation agreements, the most effective use of this method would be to assess a penalty in addition to the cost of correcting any problem.

Liens

Another available mechanism is through the placement of a lien against the responsible party's property in order to pay for necessary actions. This can be carried out by the appropriate government agency (e.g., local jurisdiction).

In all these cases where an enforcement mechanism is necessary, the restoration/development agreement must be clearly written and the responsibilities clearly set. In order to reduce environmental damage to an absolute minimum, corrective actions must be performed in a timely and competent manner; anything less must constitute a violation of the agreement.

The preceding recommendations were originally developed to deal with coastal dune restoration when it becomes a requirement for development approval. These recommendations are especially critical for restoration plans which involve fragile habitats and/or rare, threatened, or endangered species. However, they are general enough to apply to a wide variety of restoration projects. There can be no way to fully guarantee, in advance, the success of a habitat restoration project, but these recommendations will help increase the chances of success by creating a framework which sets project goals and establishes the financial responsibility for achieving them.

Given the impacts of development projects on the environment, any enterprises which seek to restore damaged habitat must be held in the highest regard. However, those involved in habitat restoration have the responsibility to use the best available techniques, backed by sufficient financial resources, in order to make a good-faith effort to meet all promised goals. Anything less diminishes the quality of the environment and violates the public trust.

References

Nordstrom, K. F. and N. P. Psuty. 1980. Dune district management: A framework for shorefront protection and land use control. *Coastal Zone Management Journal* 7(1):1–23.

Pavlik, B., and A. Zoger. 1987. Marina Dunes rare plant survey. Unpublished report for Marina Coastal Zone Planning Task Force.

Phillips, P. 1986. Fitting in: A desert resort respects its environment. *Urban Land* (June):36–37.

David Shonman is a coastal biologist, former Chairman of the Marina Coastal Zone Planning Task Force, and co-founder of the Monterey Bay Dunes Coalition, P.O. Box 2116, Carmel, CA 93921.

Restoration of Endangered Species: A Strategy for Conservation

Donald A. Falk

Abstract: *Over 5,000 types of higher plants are currently at risk of extinction in the United States, representing a serious threat to continental biodiversity. Traditional strategies emphasizing acquisition of land and legal protection of species must be complemented by offsite propagation, research, and reintroduction efforts in an integrated conservation model. Moreover, restoration of communities and ecosystems should be an integral component of species survival strategies. By joining conservation and restoration efforts at multiple levels of biological hierarchy, more flexible and successful strategies may be developed for protecting planetary diversity.*

Key words: *endangered species, integrated conservation, conservation strategies, restoration.*

Introduction

Over 5,000 types of plants—at least fifteen percent of the native flora of the United States—are currently at risk of extinction (CPC 1989). Species loss on this scale destroys the elements of diversity, which are in turn the building blocks for more complex communities and ecosystems (Falk and McMahan 1988; Falk in press [a]). There is an essential continuity of conservation practice at various levels of hierarchy: individual organisms, populations, ecotypes, communities, habitat types, and so on. Any effective effort will take place at multiple biological levels, with specific strategies evolved to address the particular problems of each. Recognition of the interdependence of efforts at these many levels is ultimately the key to understanding the *integrated conservation strategy* approach (Falk 1987; Falk in press [a]; Falk in press [b]).

There is continuity across another dimension as well: the spectrum of conservation, management, and restoration practice. These fields of action are not neat packages with precisely defined boundaries. Rather, they are more like points along a spectrum, with seemingly infinite intermediates involving elements of all three (Falk 1987; Diamond 1985). In fact, it is probably more useful to seek out the common ground among the variety of conservation and restoration methods, rather than belabor their technical or philosophical differences (Jordan 1986; Jordan et al. 1987). In a real sense restoration, habitat management, and on- and off-site conservation are all facets of a single unified discipline, the benign or altruistic human intervention into biological processes toward restorative ends. If we view conservation and restoration as separate, or (worse) mutually exclusive, we miss the point and the greatest opportunity for action.

Restorative strategies for endangered species are now being developed and tested across the United States. By applying specifically selected combinations of land preservation and management, offsite research and conservation, and habitat restoration techniques, a greater overall degree of protection for multiple levels of biological diversity is possible than can be attained by focussing on one level of hierarchy, or by using a single approach.

The Conservation Spectrum

The Royal catchfly (*Silene regia*) is now known from only a few remaining sites, mostly prairie remnants in the midwestern United States. Like many prairie plants it is not naturally endemic in distribution, and was probably found over a broad continuous range. One of the easternmost remaining sites is in Adams County in southwestern Ohio, on a 0.2 ha site known as the Bigelow Cemetery. Its designation as a cemetery when the area was first settled in the nineteenth century protected it from decades of ax, plow, and bulldozer; as a consequence, it is one of the few undisturbed sites in the area. And yet, an area 0.2 ha in size, with mowing and intensive management, is clearly not purely a preserved habitat parcel, nor just a prairie relic, nor just a refugium—it is that oddest and most contradictory of land uses, a "managed natural area." Such parcels defy categorization as "natural," "managed," or "wild," and, in fact, illustrate the confluence of all of these in many sites. One can only hope that the Bigelow Cemetery's nickname, the "Prairie Cemetery," is not prophetic of the fate of prairie habitat in the United States.

This small example illustrates a point on the *conservation spectrum*, the first of the two central concepts in this paper. There is an oversimplified view that "conservation" means the protection of pristine wilderness, that "natural areas" are places that do not show the human hand in any way, and that the only true conservation action is to acquire large tracts of virgin land and leave them alone. Moreover, according to this view, offsite research and restoration have nothing of biological value to contribute to the protection of biological diversity. These attitudes may once have been appropriate; indeed, they were responsible for the protection of millions of hectares in national parks and forests earlier in this century, and continue to be most crucial strategic elements in many parts of the country (such as the passage of the 1980 Alaska National Interest Land Conservation Act). But most conservation in the United States now is practiced in the interstices of development; this has profound implications for conservation strategy (Falk 1987; Jordan 1986; Jordan et al. 1987). Only in the 1930s and '40s did the concept of *stewardship* begin to coalesce; this was the beginning of the new unified discipline. Stewardship implies, once basic steps have been taken to protect a *site* (by legal measures, acquisition, and so on), that there is a further responsibility to *manage* it for a desired end. For instance, a site may be managed to maximize species diversity, or to enhance the population of a particular species. Stewardship, by definition, therefore implies *active* intervention into natural processes and is thus an expression of a new approach to conservation. This in turn opens up the full spectrum of activities related by acceptance of a degree of human intervention in the maintenance of natural environments. It is ironic that in stewardship, as in restoration, the very tools of contemporary practice—the chainsaw, bulldozer, and herbicide—are in different hands the tools of destruction.

One theoretical end of this spectrum is completely *laissez-faire,* hands-off protection of vast tracts of untouched wilderness. On the other extreme one might find cryogenic storage of germplasm samples of a population in a seed bank. Both of these extremes have an air of unreality about them—the former because such wilderness no longer exists in the United States, the latter because reducing species preservation to a computerized gene bank lacks many of the characteristics we call "nature."

In fact, *most* conservation management lies somewhere between these two, and the categorical distinctions between conservation, management, and restoration are, in fact, increasingly artificial. By the time an agency is engaged in an aggressive biological management program on a site fenced to exclude feral herbivores—with periodic thinning to reduce understory growth and site revegetation using material propagated offsite—protection, management, and restoration have become virtually indistinguishable. Such a combination of techniques is perhaps becoming the *rule,* not the exception, in conservation land management in the United States today. If this is true, then "restoration" suddenly takes on immense importance, for it is the most synthetic of disciplines, itself uniting agriculture, horticulture, forestry, population biology, community ecology, genetics, and a fair dose of civil engineering. Few fields can claim to be so inherently interdisciplinary.

Integrated Conservation Strategies

From the idea of the conservation/restoration spectrum, it is but a short leap to the most exciting emerging possibility, in what may be called *integrated conservation strategies.* It is in such synthetic approaches that management, offsite research and propagation, and restoration become intertwined to create a powerful, new, unified discipline.

The integrated conservation strategy concept is grounded on the principle that, for any given situation, there is an optimal mix of conservation, management, and restoration that will be maximally effective. The conservation spectrum becomes a palette, a repertoire of techniques we can apply in whatever order and proportion will achieve the best results. This approach is valid at any level of biological hierarchy (as described above), as well as for a range of endangerment causes or conservation needs. One situation will emphasize land acquisition, another listing under the U.S. Endangered Species Act, another offsite research, yet another restoration or revegetation of a degraded site. But most will involve *several* of these techniques used in harmony; the practitioner thus begins to resemble a conductor or choreographer as much as a scientist.

An excellent current example is a project involving restoration of the Malheur wire-lettuce (*Stephanomeria malheurensis*) of central Oregon. The species was described by University of California–Davis geneticist Leslie Gottlieb in 1978, using specimens from the type locality near Burns, Oregon. *S. malheurensis* is of particular scientific interest because it appeared to be of relatively recent hybrid origin, possibly representing an incipient speciation. Gottlieb took a considerable seed sample as well as voucher specimens when he described the taxon. In 1982 disaster struck in the form of fire followed by an invasion of the site by cheatgrass (*Bromus tectorum*). The population, which was already under heavy grazing pressure, failed to regenerate, and within two years it appeared that the species had been extirpated from its only known site. Six years after it was described, *S. malheurensis* seemed extinct in the wild (U.S. Fish and Wildlife Service 1982). Gottlieb's seed collection at Davis was now for all practical purposes the only remaining living germplasm of the species. In 1986, at the instigation of the Bureau of Land Management (BLM) botanist Cheryl McCaffrey, a plan was developed to reintroduce *S. malheurensis* on the site in a cooperative project involving the BLM, the Berry Botanic Garden of Portland, Oregon, the Oregon Native Plant Society, the Center for Plant Conservation, and the regional office of the U.S. Fish and Wildlife Service (USFWS). In the spring of 1987, germinated seedlings were transplanted to the site, which had been fenced and prepared by volunteers. BLM added an element of controlled experimentation to the reintroduction by preparing herbivore exclosures with four different dominant vegetative covers. The seedlings were established during 1987, and, at last report, the population seemed to be reestablishing on the site. With luck and a lot of work and dedication by staff and volunteers, *S. malheurensis* will once again grow wild in its natural range.

Conservation and Restoration: Looking Ahead

In a field as new as restoration, we must inevitably rely on belief in the future of our work as much as citing current or past examples of success. Nonetheless, there is an increasing body of experience with restoration applied to endangered species conservation, demonstrating that the integrated approach will be feasible and effective. A few such examples are cited below; no attempt has been made to assemble a complete list of current projects. The intent here is only to provide a sample reference for those interested in possible applications to conservation of endangered species.

A restoration project involving an extirpated taxon (*Arctostaphylos uva-ursi* var. *leobreweri*), is being carried out by the Tilden Regional Botanic Garden (Berkeley, California) in cooperation with the San Mateo Parks Department and state authorities (Edwards, personal communication). The taxon was endemic to San Bruno Mountain south of San Francisco until the community was destroyed by fire and invasive exotic species in the 1960s and 1970s (Reid and Walsh 1987). Original material from the site had been collected and restored. The reintroduction will take place near the original site and will be maintained cooperatively until the population is reestablished.

Other restoration, conservation, and research projects for endangered species have been developed for Menzies wallflower (*Erysimum menzeisii*) (Ferreira and Smith 1987), Arizona agave (*Agave arizonica*) (Delamater and Hodgson 1987), Knowlton's pincushion cactus (*Pediocactus knowltonii*) (Olwell et al. 1987; Milne 1986), Barrett's penstemon (*Penstemon barrettiae*) (Kierstead 1986), Texas snowbells (*Styrax texana*) (Falk and McMahan 1988; Cox 1987; U.S. Fish and Wildlife Service 1987), and others.

Conclusion

The field of conservation is currently undergoing an intellectual metamorphosis, characterized by evolutionary change in several dimensions. One facet of this growth is the increasing concern that conservation and restoration should be grounded on biologically sound models, supported by experimental data wherever possible (CPC 1986, 1988; Soulé 1986). Another vector of change is the growing emphasis on multidisciplinary approaches integrating restoration and research.

Where endangered species are concerned, the most critical need is often to provide suitable habitat for a population to survive; whether the site is "natural" or restored may ultimately be academic. Correspondingly, restoration needs the elements of species diversity—and genetic diversity within those species—as its raw materials to create or recreate complex biotic environments. These connections illustrate the essential unity of conservation and restoration practice. It is hardly an exaggeration to say that the future of protection of biological diversity lies in the wise recognition of this synergy, and the recognition that conservation and restoration are just different ways of doing the same dance.

References

CPC (Center for Plant Conservation). 1986. *Recommendations for the collection and* ex situ *management of germplasm resources for rare wild plants.* Jamaica Plain, MA: Center for Plant Conservation.

———. 1988. *A conference on the genetics of rare plant conservation: Considerations for offsite conservation and management.* Jamaica Plain, MA: Center for Plant Conservation.

———. 1989. Database listing of threatened and endangered plants. Jamaica Plain, MA: Center for Plant Conservation.

Cox, P. 1987. Chasing the wild Texas snowbells. *Plant Conservation* 2(4):1.

Delamater, R. and W. Hodgson. 1987. *Agave arizonica*: An endangered species, a hybrid, or does it matter? In *Conservation and management of rare and endangered plants: Proceedings of California Native Plant Society conference,* ed. T. S. Elias, 305–310. Sacramento, CA: California Native Plant Society.

Diamond, J. 1985. How and why eroded ecosystems should be restored. *Nature* 313:629–630.

Edwards, S. 1988. Personal communication. Director, Tilden Regional Botanic Garden. Berkeley, CA.

Elias, T., ed. 1987. *Conservation and management of endangered plants.* Sacramento, CA: California Native Plant Society.

Falk, D. 1987. Endangered species conservation *ex situ*: The national view. In *Conservation and management of endangered plants,* ed. T. Elias. Sacramento, CA: California Native Plant Society.

Falk, D. In press (a). The theory of integrated conservation strategies for biological diversity. In *Ecosystem management and conservation of rare and endangered species,* ed. D. Leopold. Syracuse, NY: Natural Areas Association.

Falk, D. In press (b). Integrated conservation strategies in theory and practice. *Annals of the Missouri Botanical Garden.*

Falk, D. and L. R. McMahan. 1988. Endangered plant conservation: Managing for diversity. *Natural Areas Journal* 8(2):91–99.

Ferreira, J., and S. Smith. 1987. Methods of increasing native populations of *Erysimum menziesii.* In *Conservation and management of endangered plants,* ed. T. Elias. Sacramento, CA: California Native Plant Society.

Jordan, W. R. III. 1986. Restoration and re-entry of nature. *Orion* 5(2):14–25.

Jordan, W. R. III, M. E. Gilpin, and J. D. Aber, eds. 1987. *Restoration ecology: A synthetic approach to ecological research.* New York: Cambridge University Press.

Kierstead, J. 1986. Barrett's penstemon: A story. *Plant Conservation* 1(2):1, 8.

Milne, J. 1986. Conserving some of the world's tiniest cacti. *Plant Conservation* 2(1):1.

Olwell, P., A. Cully, P. Knight, and S. Brack. 1987. *Pediocactus knowltonii* recovery efforts. In *Conservation and management of endangered plants,* ed. T. Elias, 519–522. Sacramento, CA: California Native Plant Society.

Reid, T. S., and R. C. Walsh. 1987. Habitat reclamation for endangered species on San Bruno Mountain. In *Conservation and management of endangered plants,* ed. T. Elias, 493–500. Sacramento, CA: California Native Plant Society.

Soulé, M. 1986. *Conservation biology: The science of scarcity and diversity.* Sunderland, MA: Sinaur Associates.

U.S. Fish and Wildlife Service. 1982. Malheur wire-lettuce listed with critical habitat. *Endangered Species Technical Bulletin* 7(12):1.

———. 1987. Texas snowbells (*Styrax texana*) recovery plan. Albuquerque, NM: U.S. Fish and Wildlife Service, Southwest Regional Office.

Wallace, S. R., and L. R. McMahan. 1988. A place in the sun for plants. *Garden* 12(1):20–23.

Donald A. Falk is Executive Director, Center for Plant Conservation, Jamaica Plain, MA 02130.

State Coastal Zone Resource Restoration and the Common Law Public Trust Doctrine

Deborah C. Trefts

ABSTRACT: *The coastal ecosystems of the United States continually sustain damage caused by humans and nature. Management efforts generally focus on preventing and mitigating future harm, while damage which has been sustained in the past is often left in disrepair, and inadequate provision is made for redressing present and future injury after it occurs. This paper contends that the common law public trust doctrine could supplement existing approaches to restoration if courts and legislatures were to broaden the doctrine's geographic scope, limit somewhat their protection of the traditional public trust uses, and extend the concept beyond mitigation to coastal resource restoration.*

KEY WORDS: *restoration, public trust doctrine, tidelands, shorelands, coastal management.*

Introduction: The Concept of Coastal Resource Restoration

It is well recognized in the United States that coastal ecosystems function as critical components of our natural environment. During the past two decades the federal government and most coastal states have been managing our use of the coastal zone, largely through programs established pursuant to the Coastal Zone Management Act of 1972 (16 U.S.C. 1451 [1982 and Supp. IV 1986]). These management efforts have focused predominantly on "mitigating" or preventing future harm to coastal resources, including barrier islands, submerged lands and vegetation, wetlands, sand dunes, and beaches. Meanwhile, coastal resources which have already been degraded, either by humans or nature, are commonly left in disrepair. In some areas of the United States the adverse effects of this neglect have been dramatic. In others they have been much less apparent, temporarily, at least.

Serious attention has been given in recent years to the development of specific techniques for successfully restoring or rehabilitating damaged natural resources and areas. The term "restoration" is used to refer to activities undertaken to re-create or upgrade resources, including land, which have been damaged or destroyed, so that their biological purpose and potential can once again be realized (Bradshaw and Chadwick 1980). Here it means in-kind, on-site resource reestablishment, and therefore, technically can be classified as a type of "mitigation" (see Ashe 1982). It does not, however, signify attempts to address and compensate for unavoidable losses which accompany necessary, project-specific development activity, as through in-lieu mitigation payments.

Restoration of degraded natural resources is not a new concept. The term is used in several federal statutes governing natural resources, most of which are also applicable to coastal environments; in the regulations promulgated pursuant to these laws; and in case law resulting from these laws and regulations. Many states have enacted similar laws. Examples of these federal statutes are the Federal Aid in Wildlife Restoration Act of 1937, the Federal Water Pollution Control Act of 1948 (FWPCA) and subsequent amendments, the National Environmental Policy Act of 1969 (NEPA), the Coastal Zone Management Act of 1972 (CZMA), the Outer Continental Shelf Lands Act Amendments of 1978 (OCSLA), and the Comprehensive Environmental Response, Compensation, and Liability Act of 1980 (CERCLA), as amended by the Superfund Amendments and Reauthorization Act of 1986 (Superfund).

However, despite statutory, regulatory, and case law references to and requirements of resource rehabilitation, and despite the increased recognition of the concept by resource managers, restoration has, to a large degree, been loosely interpreted, infrequently implemented, and weakly enforced, to the detriment of our coastal environment. This paper provides a summary discussion of a common law doctrine which appears to be a viable means of addressing these problems. It does not suggest, however, that restoration activities can be undertaken in every instance. On the contrary, rigorous, case-by-case evaluation is necessary to ensure not only that restoration would be technically, economically, and administratively feasible, but also that experts agree on such fundamental issues as restoration goals and techniques. Additionally, it is necessary to determine whether a higher return might be secured were sites put to a new and different use, or were other resources, sites, and/or types of mitigation activities undertaken.

Restoration and the Public Trust: A Case Study

Coastal resource restoration is a major focus of the court in *Commonwealth of Puerto Rico v. SS Zoe Colocotroni* (*Puerto Rico* 1980). In this case the court ordered restoration of marine resources damaged in 1973 by the discharge of 1.5 million gallons of crude oil from the tanker, *SS Zoe Colocotroni,* in order to lighten the load and extricate the ship after she went aground 4.8 km south of a village in Puerto Rico. The resulting slick inundated a reef, and damaged 8 ha of marine organisms and 9 ha of mangrove swamp. The bay was so heavily impacted by the crude oil that five years later there was still considerable evidence of its occurrence.

In the U.S. District Court (Puerto Rico) opinion, the judge asserted that the Commonwealth has title to and therefore owns "both the living and nonliving resources located in the navigable waters of the Commonwealth and those on the bottom and its subsoil, as well as those located within the . . . maritime-terrestrial zone." He stated that *"[t]he Commonwealth holds title in trust to the public property and domain,* and is charged with the protection of the people's interest in the same." Consequently, Puerto Rico could "bring legal action in court to protect its property and to recover damages to the same." The court then awarded damages based on the cost of restoring the area heavily impacted by the oil to the condition it was in before the spill and of making Puerto Rico "whole." Recognizing that "no market value, in the sense of loss of market profits, [could] be ascribed to the biological components of the . . . ecosystem," it used the "market cost" of replacing the impaired organisms in deciding the amount of harm actually sustained.

The U.S. Court of Appeals, First Circuit, affirmed the lower court opinion, except for the amount of damages awarded. It cited the oil spill liability provisions in the Clean Water Act (CWA) Amendments of 1977 (33 U.S.C. 1321 [f] [4] & [5]), and the OCSLA Amendments of 1978 (43 U.S.C. 1813 [a] [2] [C] & [D]). The former authorize the recovery by the federal government and the states of the costs or expenses incurred for restoration or natural resource replacement, and conclude that the President or a state representative shall act as trustee of the natural resources to recover these costs. The latter amendments contain similar provisions. The court noted that Puerto Rico had not proposed to buy marine animals from a biological supply laboratory to place in the coastal sediments, and reasoned that replacement should be a "component in a practicable plan for actual restoration." The judge remanded the case to give Puerto Rico's Environmental Quality Board (EQB) a chance to demonstrate "that some lesser steps are feasible that would have a beneficial effect on the . . . ecosystem without excessive destruction of the existing natural resources or disproportionate cost." He left it to the EQB to decide whether or not to consider alternative-site restoration as a measure of damages. On February 23, 1981, the U.S. Supreme Court denied the *SS Zoe Colocotroni*'s petitions for writs of certiorari.

Restoration of Coastal Resources via the Public Trust Doctrine

In *Puerto Rico* (1980), the concept of coastal resource restoration is linked with that of the common law public trust doctrine in natural resources through judicial enforcement of federal statutory law. Few judicial opinions have combined the two concepts so integrally (*Just* 1972; *Allied Towing* 1979). In considering the issue of water quality degradation, other opinions have asserted the right of states to recover damages for injury to and diminution of their trust property, but in so doing have not mentioned, much less focused upon, actual resource restoration (*Maryland v. Amerada Hess* 1972; *State Department of Environmental Protection v. New Jersey Central Power and Light* 1973). Nevertheless, the public trust doctrine could be used to supplement both the existing regulatory and nonregulatory approaches to coastal resource restoration. Before pursuing this notion further, a brief discussion of the public trust doctrine is necessary.

The Public Trust Doctrine in Natural Resources

Much has been written about the history of the public trust doctrine (e.g., Ausness 1986), hence, this section will merely highlight the main premises of the concept to provide background for further discussion. Essentially, the American public trust doctrine is a collection of common law principles which developed "from the early American courts' perception of the English common law" (Hannig 1983). While these principles have occasionally been incorporated into state constitutions and legislation, the concept has mainly evolved through the courts, which have supplemented it with property theories containing similar premises (Rodgers 1977).

The public trust doctrine recognizes that the government, as trustee, holds in trust for the benefit of its people some natural resources to be protected and perpetuated. These resources are generally associated with waterways, as the doctrine apparently evolved from the need of European kings to ensure public access to them. When the United States won its independence from England, the citizens of each state gained the rights to the waterways, with the exception of any paramount rights given over by the states to the federal government upon their admission to the Union (*Martin v. Lessee of Waddell* 1842). Despite retaining a property interest in the trust lands – for the public's benefit – states are restrained in their power over these lands, as the judiciary can scrutinize closely actions which diminish these properties. In Massachusetts and California, both of which have experienced development pressures critically affecting their water resources, public trust law tends to be more clearly defined than it is elsewhere (Heath 1983).

Historically, the doctrine was quite narrow, both with respect to its geographical scope – the tidelands and navigable inland waters of the state – and the uses it encompassed – navigation, commerce, and fishing. The "tidelands" or "foreshore" are just that portion of the shore covered and uncovered by the ebb and flow of ordinary tides (Black's 1979). The term "navigable waters" has come to mean "those waters which afford a channel for useful commerce" (Black's 1979). (In California, however, the courts have "progressively redefined 'navigable waters' in order to increase public trust uses of the state's water resources. The modern California standard is the 'recreational boating test' " [Hannig 1983], which is met if the waters "are capable of being navigated by oar or motor-propelled small craft" [*People ex rel. Baker v. Mack* 1971].)

The foundations for the contemporary public trust doctrine are attributed to U.S. Supreme Court Justice Field's opinion in *Illinois Central* (1892) which entrusted the state of Illinois with specific public trust responsibilities. The court upheld the 1873 revocation by the state of Illinois of a grant it had made to Illinois Central Railroad of the right and title to the submerged lands of Lake Michigan. Justice Field "recognized uses incidental to or in aid of [the] traditional categories," such as wharves, docks, and piers, as improving the interest of the people in navigation and commerce, but not the unrestricted alienation of the submerged lands. "So long as the minimum standards of *Illinois Central* are met, it remains a matter of state common law as to what waters are subject to the public trust doctrine. Further, the states have the ability to expand the public's rights beyond the historically recognized commerce, navigation, and fishing rights" (Hannig 1983).

Through the years the geographic scope of the doctrine has been broadened in some jurisdictions to include submerged lands (the land lying to the seaward of the tidelands (as in *Illinois Central*), dry sands, and park lands. Likewise, the uses protected by the doctrine have expanded. "Navigation has come to include recreational as well as commercial boating or shipping. Commerce has come to mean water *related* commerce beyond the water-borne activity itself. And fisheries encompasses 'the right to fish, hunt, bathe, swim. . . .' Nor have courts felt constrained to limit the evolution of the doctrine to only the logical extensions of these three historically based [uses]" (Morrison and Dollahite 1985). In some jurisdictions the concept "sees all natural resources in the state as impressed with a trust for usage and conservation as a state resource" (*Chain O Lakes Protective Association v. Moses* 1972). "In administering the trust the state is not burdened with an outmoded classification favoring one mode of utilization over another. . . . There is a growing public recognition that one of the most important public uses of the tidelands—a use encompassed within the tidelands trust—is the preservation of those lands in their natural state, so that they may serve as ecological units for scientific study, as open space, and as environments which provide food and habitat for birds and marine life, and which favorably affect the scenery and climate of the area. It is not necessary to . . . define precisely all the public uses which encumber tidelands" (*Marks v. Whitney* 1971). Often state legislatures leave it to the courts to define public trust uses and to establish the best use, or the preferred use if there is a conflict, in case-by-case analysis.

While the public trust can be used to protect and perpetuate particular coastal resources via the common law, state statutes, and state constitutions, there are also federal statutes applicable to coastal resources which incorporate the trust concept. They are the CWA Amendments to the FWPCA; NEPA, as amended; OCSLA, as amended in 1978; and CERCLA, as amended by Superfund. (As noted, each of these acts also contains references to natural resource restoration.)

Restoration as an Element of the Public Trust Doctrine

While the public trust concept has permeated virtually every type of legal format, its meaning is that which the courts give it. For the doctrine to expand further to fully encompass coastal resource restoration, courts and legislatures throughout the United States must extend it:

1. beyond its narrow geographic scope;
2. beyond its traditional purposes, while also limiting them; and
3. from "mitigation" to "restoration."

A discussion of each of these points follows.

Broadening the Geographic Scope of the Public Trust

Historically, the exact geographic scope of the public trust doctrine has been unclear. Consequently, in the United States it varies from state to state. With respect to the oceans and to navigable tidal lakes and streams, resources between the "high-water mark" and the "low-water mark" (the tidelands) have been considered traditionally as part of the states' trust property. Since there are many different high- and low-tide lines, it is up to each state to determine which will serve as its boundary markers. States can choose the water marks existing when they were admitted to the Union (Hannig 1983).

There are, however, exceptions to this general rule. In some states on the east coast, the shore may be owned privately as far seaward as the low-water line, though rights are reserved for the general public for purposes such as navigation, commerce, and fishing (Kiefer 1987). And in other states, the geographic scope has been broadened to include other elements of the greater coastal ecosystem. For instance, in Wisconsin it has been expanded to include wetlands and swamps (*Just* 1972). In Oregon the public trust has been connected with the custom theory such that the public has an established right, for recreational purposes, to the continued use of and access to the dry sands between the mean high-tide line and the visible vegetation line along the entire Pacific coast of the state (*State ex rel. Thornton v. Hay* 1969). Furthermore, courts have generally extended the doctrine to nontidal lakes and streams.

Given the flexible and dynamic nature of the public trust doctrine, it could be expanded further geographically to encompass more resources and areas—both landward and seaward. In the United States it is commonly recognized that coastal areas are extremely valuable and biologically productive yet fragile ecosystems, of which the tidelands are only a component (*Just* 1972). In fact, preservation, protection, development, and, where possible, restoration or enhancement of our nation's coastal zone resources for present and succeeding generations is the first of the four policies of the federal CZMA, as amended (6 U.S.C. 1452[1] [1982 & Supp. IV 1986]). Given their special nature and immense ecological importance, our coastal ecosystems merit the same public trust protection that is now generally bestowed only upon the tidelands. Environmental studies have shown repeatedly that impacts to the ecosystem seaward and shoreward of the tidelands can have profound effects on the tidelands themselves – thus harming the trust property. While the integral relationship of the various subsystems of the coastal ecosystem was not known when the public trust doctrine began to evolve, it is now an accepted fact. Hence, in the interest of truly protecting the public trust, the narrow geographic reach of the doctrine should be broadened considerably to encompass far more, if not all, of the coastal or maritime-terrestrial zone. If nothing else, it should encompass a buffer zone around the tidelands to lessen the potentially adverse impacts of human activity.

Expanding and Limiting the Trust's Traditional Purposes

While some public trust cases have considered water quality degradation, most have dealt with the issues of title to resources, public access, public rights, and protection of in-place uses of the trust corpus (Rodgers 1977; Johnson 1981). It is evident from the public trust case law, as well as from the types of activities traditionally protected, that the doctrine has emphasized the *use* of trust resources, rather than their preservation. The basis of the trust concept is protection and promotion of, as well as access to, the historical public uses of navigation, commerce, fishing, and more recently, recreation. "Even before *Illinois Central* was decided, it was recognized in California that the state had the authority as administrator of the trust on behalf of the public to dispose absolutely of title to private persons if the purpose of the conveyance was to promote navigation and commerce" (*City of Berkeley v. Superior Court* 1980).

Nevertheless, the Supreme Court of California has ruled that "[a]ll uses of water, including the public trust uses, must now conform to the standard of reasonable use" (*National Audubon* 1983). In Oregon, the Court of Appeals, in reversing a state agency decision to grant a permit to fill 13 ha of an estuary in order to extend an airport runway, asserted that "[t]he tacit principle underlying the public trust doctrine is that water resources should be devoted to uses which are consistent with their nature and should be protected from inimical uses" (*Morse* 1978). An example of a consistent use is found in a recent Washington Supreme Court decision which declared that a state statute allowing owners of residential property bordering state-owned tidelands to put up recreational docks without having to pay the state did not violate the public trust doctrine, but instead promoted public interests by recognizing that docks are a beneficial use of the tidelands adjacent to private homes because they enable homeowners and their guests to have access to navigable water for recreation. The statute also gave up little right of control over public ownership and did not convey title to any state-owned tidelands or shorelands (*Caminiti v. Boyle* 1987).

Inherent in the trust is the notion of fiduciary responsibility to future generations, and thus, a presumption favoring continuing public use. This presumption forms the procedural rule underlying trust cases, while an opportunity for the public to use and enjoy the trust property is included among the substantive implications of these cases (Brooks 1980). Thus, trust *uses* must not only be protected, but also maintained and preserved. A related principle of the public trust doctrine as it has developed in the United States is that of no significant deterioration, or of nondegradation, of public rights in public resources. This principle is evident in cases involving resources impressed with public easements, in which legislative actions to qualify, refine, and enhance the trust corpus have been protected against claims of taking without just compensation (*Just* 1972).

There is, however, a basic conflict between traditional public trust uses and newer uses, such as the protection and preservation of the environment. "Liberal expansion of public trust uses, without strict regard to traditional uses of fishing, navigation and commerce, has usually involved balancing of competing uses. In such cases, the implementation of the public trust doctrine, when multiple interests (private, public and governmental) conflict, has required decisions as to the best, paramount or most democratic use" (Althaus 1978). "[N]o single interest in the use of navigable waters, though afforded the protection of the public trust doctrine, is absolute. Some public uses must yield if other public uses are to exist at all. The uses must be balanced and accommodated on a case-by-case basis" (*Wisconsin v. Village of Lake Delton* 1979). Consequently, some uses may be impaired so as to promote other uses (Ausness 1986). Thus, the use which is ultimately chosen is not necessarily the one which is most environmentally benign. For instance, "[u]se of tidelands for oil drilling and construction of restaurants, bars, motels, swimming pools, convention centers, and apartment buildings has been held consistent with the trust when such uses further trust purposes" (Corfield 1987).

There is, therefore, a need to limit somewhat the protection afforded by courts and legislatures to traditional trust uses when they adversely impact other trust uses, so that newer uses, such as preservation, are protected sufficiently. They are at least as important to states today as the historical uses were when the doctrine was conceived. Given its flexibility, our common law system could bring them more prominently within the public trust doctrine's reach without stretching that reach unreasonably.

Extending the Public Trust from Mitigation to Restoration

"The duty of the trustee to preserve the resource and protect against loss, dissipation or diminution and to act with diligence, fairness and faithfulness in doing so is well established in trust law" (Yannacone et al. 1972). Maintenance and stewardship of resources is a notion central to the public trust doctrine. In general, in administering trust property there is always the duty to "use care and skill" to preserve it (Scott and Fratcher 1987). Just as a trustee has the power and the duty to spend trust money to keep buildings in repair, a trustee of the public trust should have a duty to keep natural resources in good repair. Often simple public management of trust resources is not sufficient to protect the environment.

The public trust concept has been applied to require a reasonable effort to mitigate resource harm where the destruction of a public trust resource was justified because an overriding public purpose was served (Johnson 1981). That a resource has been degraded does not mean it can be alienated to private control or otherwise removed from the corpus of the trust. "An individual may abandon his private property, but a public trustee cannot abandon public property" (*State v. Cleveland and Pittsburgh R. R.* 1916), unless abandonment is consistent with the purposes of the trust (*National Audubon* 1983). The trust property must thus be maintained for the purposes of the trust; it cannot be wasted or misappropriated."[S]tatutes purporting to authorize an abandonment of . . . public use will be carefully scanned to ascertain whether or not such was the legislative intention, and that intent must be clearly expressed or necessarily implied. It will not be implied if any other inference is reasonably possible. And if any interpretation of the statute is reasonably possible which would not involve a destruction of the public use or an intention to terminate it in violation of the trust, the courts will give the statute such interpretation" (*People v. California Fish* 1913).

Since trust uses must be maintained, it follows that the resources and areas supporting and enabling these uses must also be maintained. Damage done to them should be rectified.

The severe restriction upon the power of the state as trustee to modify water resources is predicated not only upon the importance of the public use of such waters and lands, but upon the exhaustible and irreplaceable nature of the resources and their fundamental importance to our society and to our environment. These resources, after all, can only be spent once. Therefore, the law has historically and consistently . . . required the highest degree of protection from the public trustee (*Morse* 1978).

While coastal resource restoration has only rarely been a focus of public trust litigation, despite its providing one of the highest degrees of protection, it is the most assured means of maintaining the trust property for future use. Mitigation of damage is necessary to protect resources which are threatened, and restoration, when technically, economically, and otherwise feasible, is necessary for those resources which have already been harmed, if their full biological potential is again to be realized.

Three of the four federal natural resource statutes mentioned earlier as including trust language attach a requirement of restoration directly to that trust responsibility. The two concepts are integrally linked within the liability provisions of the CWA, the OCSLA, and Superfund. In *Puerto Rico* (1980) and in *Allied Towing* (1979) the court specifically cited the trust/restoration language in the CWA and the OCSLA. While the public trust doctrine could be broadened to encompass coastal resource restoration based upon its own merits alone, these statutory authorities give such expansion all the more credence.

Conclusion

The common law public trust doctrine in natural resources would not be a panacea, even if courts and legislatures in coastal states were to extend it to encompass coastal resource restoration as a way of maintaining and preserving uses and purposes protected by the doctrine now and in the future. However, it appears to be the most promising means of supplementing existing methods of achieving restoration—the most common of which is the regulatory approach, whereby agencies attach conditions to permits allowing a coastal site or resource to be impacted by a development activity—which are, on the whole, used infrequently and inadequately. The more often the public trust doctrine is put forth as a basis for legally enforcing coastal restoration activities, the more accepted it will become as an effective tool enabling coastal resource managers to justify, require, and enforce restoration of damaged coastal resources.

Acknowledgments

Special thanks go to The Jesse Smith Noyes Foundation and to Dr. Richard O. Brooks of Vermont Law School for their generous support of earlier versions of this paper.

References

Allied Towing, in the matter of the complaint of Allied Towing Corporation, 478 F. Supp. 398 (1979).

Althaus, H. F. 1978. *Public trust rights.* Washington, DC: U.S. Government Printing Office.

Ashe, D. M. 1982. Fish and wildlife mitigation: Description and analysis of estuarine applications. 10 *Coastal Zone Management Journal* 1(2):1–50.

Ausness, R. 1986. Water rights, the public trust doctrine, and the protection of instream uses. *University of Illinois Law Review* 2:407–437.

Black's Law Dictionary (rev. 5th ed. 1979).

Bradshaw, A., and M. Chadwick. 1980. *The restoration of land.* Berkeley, CA: University of California Press.

Brooks, R. O. 1980. The future public trust in natural resources. Presentation at Iowa State University.

Caminiti v. Boyle. 1987. 107 Wash.2d 662.

Chain O'Lakes Protective Association. 1972. 193 N.W.2d 708, 710.

City of Berkeley v. Superior Court. 1980. 26 Cal.3d 515, 523.

Corfield, M. A. 1987. Sand rights: Using California's public trust doctrine to protect against coastal erosion. 24 *San Diego Law Review* 3:727–750.

Hannig, T. J. 1983. The public trust doctrine expansion and integration: a proposed balancing test. 23 *Santa Clara Law Review* 211–236.

Heath, M. A. 1983. A tidelands trust for Georgia. 17 *Georgia Law Review* 3:851–884.

Illinois Central Railroad Co. v. Illinois, 146 U.S. 387, 13 S. Ct. 110, 36 L. Ed. 1918 (1892).

Johnson, R. 1981. Public trust protection for stream flows and lake levels. In *The public trust doctrine in natural resources law and management,* conference proceedings, ed. H. C. Dunning, 108–131. Davis, CA: The Regents of the University of California.

Just v. Marinette County, 56 Wis. 2d 7, 201 N. W. 2d 761 (1972).

Kiefer, M. J. 1987. The public trust doctrine: state limitations on private waterfront development. 16 *Real Estate Law Journal* 1:146–171.

Lazarus, R. J. 1986. Changing conceptions of property and sovereignty in natural resources: questioning the public trust doctrine. 71 *Iowa Law Review* 3:631–716.

Marks v. Whitney. 1971 at 491 P.2d 374,380.

Martin v. Lessee of Waddell. 1842. 41 U.S. (6 Pet.) 366.

Maryland v. Amerada Hess. 1972. 350 F.Supp 1060.

Morrison, M. D., and Dollahite, M. K. 1985. The public trust doctrine: Insuring the needs of Texas bays and estuaries. 37 *Baylor Law Review* 365–424.

Morse v. Oregon, 581 P. 2d 520 (1978).

National Audubon Society v. Superior Court of Alpine County, 658 P. 2d 709 (1983).

People ex rel. Baker v. Mark. 1971. 91 California Reporter 448, 454.

People v. California Fish. 1913 at 166 Cal. 576, 597.

Puerto Rico (Commonwealth of Puerto Rico v. SS Zoe Colocotroni), 628 F. 2d 652 (1980), affirming in part, vacating in part and remanding 456 F. Supp.1327 (1978), cert. den. 450 U.S. 912, 67 L. Ed. 2d 336, 101 S. Ct. 1350 (1981).

Rodgers, W. H., Jr. 1977. *Handbook on environmental law.* St. Paul, MN: West Publishing Co.

Scott, A. W. and W. F. Fratcher, 1987. *The law of trusts,* IIA, 4th ed. Boston, MA: Little, Brown & Co.

State Department of Environmental Protection v. Jersey Central Power and Light. 1973. 308 A.2d 671.

State ex rel. Thornton v. Hay. 1969. 462 p.2d 671, 673.

State v. Cleveland and Pittsburgh R.R. 1916. 113 N.E. 677.

Wisconsin v. Village of Lake Dalton. 1979 at 286 N.W. 2d 622, 632.

Yannacone, V. J., Jr, B. S. Cohen and S. G. Davison. 1972. *Environmental rights and remedies.* Rochester, NY: The Lawyer's Cooperative Publishing Co.

Deborah C. Trefts is a Coastal Project Coordinator for the Secretary of Environmental Affairs, State of California, Sacramento. The views expressed here are those of the author.

Some Factors Affecting Management Strategies for Restoring the Earth

John Cairns, Jr.

Abstract: *Five issues deserve attention when major restoration projects are undertaken: (1) preventing damaged ecosystems from exporting deleterious materials or pest species to adjacent ecosystems, (2) controlling invading exotic species, (3) choosing (an) appropriate model(s) for the restoration process, (4) developing less prescriptive legislation for the mandated restoration projects, (5) providing pedigreed sources of colonizing species that will not jeopardize unique or fragile ecosystems.*

Key words: *restoration ecology, restoration management strategies, restoration models.*

Introduction

This brief presentation calls attention to five areas of restoration ecology that should be publicly debated before new major restoration projects are undertaken. These are by no means the only factors that deserve more attention (see Cairns 1985). It is, however, difficult to think of a major restoration project where these five areas do not deserve attention.

Five Ecosystem Management Concerns

1. Preventing Damaged Ecosystems from Exporting Deleterious Materials or Pest Species to Adjacent Ecosystems

The potential of damaged ecosystems to export deleterious materials and inappropriate species to surrounding ecosystems has been seriously underestimated. Surface-mined lands not properly restored can export heavy metals to surrounding ecosystems (particularly aquatic ones), lead to contamination of groundwater, and the like. Although the response may be to attempt restoration to original condition immediately, under certain circumstances ecological stabilization eliminating these deleterious effects upon adjacent ecosystems might not occur with sufficient rapidity to justify this course of action. This would be particularly true if the adjacent ecosystems are particularly fragile, vulnerable, or unique. If the initial management goal is reducing or eliminating damage to adjacent ecosystems, the management strategy would be quite different than if the initial goal were reestablishing the original condition. For preventing runoff of heavy metals and other hazardous materials into surrounding ecosystems or groundwater, a rapid reestablishment of vegetative cover is desirable. Ideally, species should be chosen that would not become so well established that they would exclude the possibility of choosing other options eventually, such as restoring to original condition. Species assemblages that require continual management in order to survive on the restoration site would be a good initial choice if a quite different course of action is the ultimate restoration goal. Thus, merely removing the management support system would eliminate the recently colonized species. Determining the effectiveness of the initial management strategy will require monitoring and assessment not only on the restoration site but on the surrounding ecosystems as well. Once the plan is operating effectively, one can then make a decision about ultimate restoration practices leading to a permanent, self-sustaining ecosystem that is either a close approximation of the original condition or of some alternative ecosystem. Since groundwater contamination has attracted considerable public and political attention in recent years, obtaining funds for restoration that will protect groundwater quality, as well as providing the other benefits just described, would be easier to obtain than they would have been a few years ago.

2. Controlling Invading Exotic Species

Many ecologists who have assumed that indigenous communities, whether natural or restored, can relatively easily exclude invading exotics will find the experience of Dr. David Wingate on Nonesuch Island in Bermuda interesting. Over a number of years, Wingate has been engaged in a remarkably successful attempt to restore the native flora and fauna of Bermuda on this island. This is not to say that naturalist W. Charles Beebe would recognize it in its present condition, but it is certainly much closer to the condition he knew than it would have been before Wingate began his commendable efforts. When I had the pleasure of touring Nonesuch Island with Wingate about a year ago, he was constantly uprooting small exotic plants, many of which would have enthralled the indoor plant hobbyist in the United States. Many of these resulted from seeds brought over by birds that fed on the mainland islands and came back each evening to roost on Nonesuch, thus providing a continual introduction of exotic species. If the damaged ecosystem happens to be an ecological island surrounded by dissimilar ecosystems, particularly altered ones, it is quite possible that a comparable effort will be required even after the original biota or a substantial portion thereof has been reestablished.

3. Choosing a Model for the Restoration Process

Restoration to original condition may not be a viable option for those responsible for restoring a particular ecosystem for the following reasons: (a) There may be no adequate data base on both the structure and function of the damaged ecosystem before damage occurred. (b) Even in the unlikely event that both structural and functional characteristics of the ecosystem are well documented, the predictive capabilities of ecological science are often not sufficiently precise to offer a reasonable guarantee that, if certain measures are undertaken successfully, the outcome will be certain, namely restoration to original condition. (c) Even if conditions (a) and (b) are met, adequate species for recolonizing the area may not be available without damage to a comparable area elsewhere. For example, if a relatively uncommon ecosystem is damaged by an oil spill, and there are only two other areas with appropriate species, it may not be possible to transport a sufficient number of these species to successfully recolonize the damaged area without damaging the two remaining ecosystems in this category. (d) If the pre-damaged ecosystem is a small patch of a relatively common ecosystem type, then the decision may well be to restore not to original condition but rather to a very scarce type of ecosystem. This could be done to preserve a rare and endangered species or to expand the species pool of certain types of species for recolonizing other damaged areas.

The instinctive reaction, at least for those concerned with environmental degradation, is to restore damaged ecosystems to their original condition as rapidly as possible. However, when examined from a regional, continental, or global standpoint, this may not be the best management strategy (Cairns 1988). Some types of ecosystems are being destroyed at a rate much greater than others and under circumstances in which restoration to original condition in the foreseeable future is highly improbable. For example, wetlands filled in for urban development, highly productive agricultural lands, or areas damaged by off-road vehicles have powerful lobbies to ensure continuing use. Suppose one wanted to provide new nesting habitat for the piping plover in the hope of reversing its declining numbers. It is not entirely clear that the loss of nesting sites through shoreline development and intensive use of some crucial nesting areas by off-road vehicles and fishermen are the primary cause for the decline. However, if they were, alternative sites might be provided elsewhere as part of a restoration process. Thus, even though the alternative ecosystem condition might not have been originally suitable as a nesting ground for the piping plover, enough is known about the piping plover's nesting requirements to make it a reasonably good gamble to restore an area for this purpose.

4. Governing the Restoration Process with Legislation

Of all the dilemmas in the restoration process, it is possible that developing adequate legislation is the worst. We all wish to prevent ecosystem damage and to ensure that when it occurs those responsible make full restitution expeditiously. Abundant evidence of poor stewardship in the past makes highly prescriptive legislation almost mandatory. We want to ensure that all of the appropriate measures are taken and we wish to spell this out in great detail at every opportunity. Unfortunately, the science of restoration ecology (see Jordan et al. 1987) is not nearly sufficiently well advanced to do this and is unlikely to be so in the foreseeable future. Furthermore, experimental evidence that is so necessary for this predictive capability must be developed in the process of restoring already damaged ecosystems, because it would be unthinkable to deliberately damage more to acquire this evidence. The basic problem is that, whatever restoration measures we take, the outcome is highly uncertain. For example, what should the criteria be for cessation of restoration practices (i.e., at what point in the restoration process can natural processes alone suffice)? Full restoration for a complex ecosystem is likely to take many years, often exceeding normal human life span. Even the simplest ecosystems should be able to survive normal, cyclic events of drought, floods, temperature shifts, etc., which are not likely to occur in a short time span. For the more complex ecosystems, the fact that certain successional processes are underway replicating reasonably well those thought to be the early successional stages in the normal development of the ecosystem does not assure that the outcome will be the same. The precise sequence of climatic and other events leading to the formation of the ecosystem will be highly unlikely to recur in the same pattern. Finally, funding for such massive experiments is not likely to come from the usual foundations and other sources because they are unable to meet existing demands, much less a new one of this magnitude. If experiments undertaken on existing sites are to be successful, they must not all be carried out in the same pattern or on the basis of the same series of strategies, because this will not furnish the type of information needed for developing a sound predictive capability. As a consequence, legislation cannot be prescriptive, at least in every case, or we will not generate the quantity of timely information necessary for skillful restoration. Even the guilty party, namely the individual or institution causing the ecosystem damage, might have a solid legal case for avoiding seemingly endless research. Therefore, if the legislation is to be sufficiently flexible, it will almost certainly require a mixture of private (the ecosystem damager's) and public funds. Furthermore, if the research is truly that, some projects will fail and provide useful information but little satisfaction to those who want that particular ecosystem restored. Ecosystems that are likely to furnish the most useful information can be chosen on a global, national, or even regional basis. An appropriate committee could set priorities that would help determine which ecosystem restoration project should be experimental and which should be highly prescriptive. Local residents may have quite different opinions about this. This will be particularly difficult when even the criteria for failure (that the ecosystem restoration is not going as predicted) are not well worked out.

5. Providing Sources of Colonizing Species

Most environmentalists favor restoring damaged ecosystems to original condition if possible and to an ecologically more respectable condition if not. However, not enough consideration has been given to the source of colonizing species. In cases where the reproductive rate is far in excess of that needed for a sustained population, there is probably no problem. However, for many ecosystems the replacement rate may be just barely adequate to sustain existing populations or, in more severe cases, not even that. Consequently, when obtaining colonizing species for unique ecosystems, fragile ecosystems, or similar ecosystems at some distance from the damaged site, considerable judgment will be required in order to avoid damage to the presently undamaged ecosystems. One possible solution to this problem is to establish ecosystem preserves where certain types of species can be produced in adequate quantity for recolonizing damaged areas in such a way that removal of seed stock from thriving and healthy natural systems will not affect them. This adds an intermediate step in the process, but it will show which species can be handled in this way and which cannot. In situations where the production of colonizing organisms is impossible or extremely difficult in the ecosystem preserve, it seems reasonable to assume that comparable difficulties would ensue in the damaged ecosystem. In this case, an alternative ecosystem type should be chosen.

References

Cairns, J., Jr. 1985. Facing some awkward questions concerning rehabilitation management practices on mined lands. In *Wetlands and water management of mined lands,* eds. R. P. Brooks, D. E. Samuel, J. B. Hill, 9–17. University Park, PA: The Pennsylvania State University.

Cairns, J., Jr., ed. 1988. *Rehabilitating damaged ecosystems.* 2 vols. Boca Raton, FL: CRC Press.

Jordan, W. R., III, M. E. Gilpin, J. D. Aber, eds. 1987. *Restoration ecology.* Cambridge, England: Cambridge University Press.

John Cairns, Jr., University Center for Environmental and Hazardous Materials Studies and Department of Biology, Virginia Polytechnic Institute and State University, Blacksburg, VA 24061.

Restoring Florida's Everglades: A Strategic Planning Approach

Barbara C. Brumback

Abstract: *This paper describes the strategic planning approach that is being used to restore Florida's Everglades, beginning with the government of Florida's 1983 executive order initiating the Save Our Everglades Program and continuing through the establishment of two key resource planning and management committees.*

Key words: *Florida, Everglades, strategic planning, resource planning and management.*

Introduction

The Everglades are unique in the world. Originally covering most of southeast Florida, they are a subtropical system of marshes, sloughs, tree islands, and cypress forests. The Indian name for the Everglades is Pa-hay-okee, meaning grassy water, and the Everglades themselves are essentially a wide, shallow, slow-moving river of sawgrass (*Cladium jamaicense*) (Douglas 1947).

Approximately 65–70% of the Everglades' vegetation is sawgrass, sometimes mixed with other grasses, sedges and the like, but the trees are on the banks of alligator holes or on islands of higher ground (Fernald and Patton 1984). As the sawgrass has died, in the 4,000 years that it has grown in south Florida, it has formed layer upon layer of peat.

The story of the natural functioning of the Everglades system is, essentially, a story of fire and water. Fire maintains the Everglades as a river of grass, burning out the hardwoods that take hold during the winter dry season. The steady, north-to-south sheet-flow of water over the marshlands in the rainy season is what makes the Everglades a river of grass. The health of the Everglades system, and the many plants and animals it supports, depends on the natural water fluctuations between summer's wet season and the dry season of winter (Fernald and Patton 1984).

Lake Okeechobee, at the north of the Everglades, is the third largest freshwater lake totally within the United States' boundaries. The lake is large but shallow, with a maximum depth of about 4.3 m. It is fed primarily by the Kissimmee River to the north and by rainfall. As late as the beginning of this century, Lake Okeechobee lay in the center of the state like an enormous platter, filled with water, tilting slightly to the south. In the rainy season, the lake overflowed its southern shore and water coursed across the Everglades in a wide swath, making its way to the tip of the state and the estuaries of the Florida Bay and the Ten Thousand Islands. The Kissimmee River, one of the lake's main sources of water, meandered 158 km from the Kissimmee chain of lakes to the north. The river flowed in a series of oxbows through a 1.6–3.2 km wide marshy floodplain.

The Modification of the Everglades: A Story of Water and Money

When Florida became a state in 1845, it was small, poor, and wet. In 1855, less than 150,000 people lived in the state. Over two-thirds of Florida was covered with swamp, marsh, or other wetlands, most of which belonged to the federal government. After the Seminole Indian Wars ended in the 1840s, new residents arrived, with settlers from Ohio, Pennsylvania, and other northern states moving into south Florida. For them, the swampy conditions made south Florida a hostile place, unsuited to agriculture as they knew it, so they began to correct this "problem."

The first state legislature declared the Everglades "wholly valueless," and asked Congress for help in draining this wasteland in order to put it to productive use (Executive Office of the Governor, undated). In 1850, the federal government gave the state 8.1 million hectares of land under the Swamp and Overflowed Lands Act, most of which was under water at least part of the year. This land grant gave the state most of south Florida. But the land, which was "unfit for cultivation due to its swampy and overflowed condition," came with the requirement that it be drained and sold, with the proceeds being used to drain more land. The Swamp and Overflowed Lands Act set the stage for the drama of water and money that has characterized the state's short history in managing the Everglades.

Looking for growth and development, but lacking money, the state turned to the railroads. Florida pledged its land—its only asset—to guarantee railroad construction bonds. The Civil War bankrupted railroads in the state, and in 1865 Florida was hopelessly in debt to railroad bond holders (Fernald and Patton 1984).

Courting growth and development, Florida sought investors. Hamilton Disston, heir to a Philadelphia fortune, had visited Florida in 1877 and dreamed of creating an agricultural and urban empire by draining the Everglades. In 1881, in what is said to be the country's biggest land deal, Florida sold Disston 1.6 million hectares at 62 cents a hectare and promised him ownership of half of all the additional land he could drain. The drainage of the Everglades began in earnest, and the story of water and money continued.

By the early 1900s, the state joined private efforts to drain south Florida, and Governor Napoleon Bonaparte Broward (1905–1909) was elected on a campaign promise to drain the Everglades. Basically, drainage was and is accomplished through a network of canals that carry water off the land and discharge it into the Atlantic Ocean and the Gulf of Mexico. The foundation for south Florida's drainage system was laid by the 1920s.

Early drainage in the Everglades encouraged growth and development, especially agricultural development in the peat soil areas to the south of Lake Okeechobee. But the drainage network was not adequate to protect the growing region from flood, especially flooding caused by hurricanes. The region was hit by hurricanes in 1922, 1924, and 1926, but the damage and death caused by these storms paled in comparison to the hurricane of 1928. The 1928 hurricane lifted the water out of Lake Okeechobee and broke through its 1.5 m dike, causing 2,000 deaths, one of the worst hurricane catastrophes in history (Fernald and Patton 1984).

The federal government then began a major flood control program in south Florida, including the construction of the 137 km long, earthen Hoover Dike around Lake Okeechobee. Hoover Dike rises about 6.1 m above the level of Lake Okeechobee and transforms the lake from a huge platter into an enormous bowl of water. In the 1960s, mainly to provide flood control in the upper Kissimmee River basin, the Army Corps of Engineers transformed the 158 km wandering river into a straight, 77.2 km white-walled canal, called the C-38, or as it has come to be known informally among environmentalists, "the wicked ditch."

The Effects of Draining, Ditching, Diking, and Developing

South Florida today has about 2,252 km of canals and levees and 143 water control structures, giving Florida the world's largest plumbing system. Draining and ditching opened up more land to development. Far from being a small state, Florida is now the fourth largest in population, with another new resident moving in every 90 seconds. Projections show that 80% of the new residents will continue to move to coastal counties, and one third of them will move to Dade (Miami), Broward (Fort Lauderdale), and Palm Beach (West Palm Beach/Boca Raton) counties. Urban and agricultural activity in south Florida has increasingly moved west, putting such activities in greater competition with the remnants of the Everglades.

Diking Lake Okeechobee, digging canals, and building east/west roads blocked the natural north/south sheet-flow of water that the Everglades depends on. For years, the managed water system did not follow the natural wet and dry cycles of the Everglades, including Everglades National Park. In abnormally wet years, canals cannot drain the water off the land fast enough, and water has to be released from Lake Okeechobee to protect people and property from potential hurricane flooding. Water is released into water conservation areas, which are levied portions of over 2,590 km^2 miles of Everglades. These areas then turn into vast shallow lakes, with disastrous effects on wildlife. When the water conservation areas are full and agricultural and urban areas are at stake, water is released into the Everglades National Park (Fernald and Patton 1984).

The effect on native wildlife has been disastrous. Between the 1870s—before ditching, diking and draining, and developing—and the 1970s, the population of wading birds in south Florida dropped 95%, from about 2.5 million to 128,400 (Robertson and Kushland 1974).

There are four major effects on the environment caused by the reclamation of the Everglades:

1. Overdrainage has brought about a loss of natural habitat so that wetlands are often invaded by other species, often exotics.
2. Stabilized water levels mean that wetlands are wetter than they would be under natural conditions and seasonal wetlands are drier, resulting in degradation or destruction of habitat.
3. Breaking up the natural sheet-flow of water across the Everglades means that the Everglades National Park is wetter than it should be in the rainy season and drier than it should be in the dry season.
4. Water quality problems have been caused by urban and agricultural runoff and discharge into wetlands, the Kissimmee River, and Lake Okeechobee and its tributaries (Fernald and Patton 1984). Many fear that Lake Okeechobee is dying, and massive algae blooms and fish kills in the past two years point to severe water quality problems (Martin 1987).

The Save Our Everglades Initiative—Water and Money for Restoration

The goal of the Save Our Everglades initiative was to assure "that the Everglades of the year 2000 looks and functions more like it did in 1900 than it does today." A strategic planning approach was chosen because it was recognized that a single, comprehensive solution to restoration was impossible. The key to this strategic planning approach is identifying strategic issues and devising methods for resolving them. The breadth of the initiative's goals—the restoration of specific natural functions and resource values—and the complexity of achieving them in a high-growth area required the use of an intergovernmental approach that identified key "actors" and issues.

The seven specific issues identified in the Save Our Everglades program, and the progress toward achieving them are described below.

1. Reestablishing the Values of the Kissimmee River

Channelizing the river resulted in the drainage of about 809,200 ha of wetlands and a reduction of groundwater levels. While the restoration of the river had been studied since 1973, actual restoration work began in 1984. It is the largest dechannelization ever attempted. In a demonstration project, three weirs were built across the canal (that reflooded 19.3 linear km of the original floodplain) at a cost of $1.3 million. Its effects are being monitored and physical and numerical modeling is being done to study the effects of dechannelization. The study is expected to be completed in 1989. In the meantime, efforts are underway to acquire land in the old Kissimmee floodplain.

2. Protecting Lake Okeechobee

High levels of nutrients, largely from agricultural uses, have harmed water quality in the lake. Measures to reduce the amount of nutrients focus on reducing the amount of nutrient-rich water pumped into the lake from nearby agricultural areas. Studies are underway on methods to further reduce these nutrients. Cleaning up Lake Okeechobee is expected to cost tens of millions of dollars.

3. Restoring the Area Adjacent to the Northern Water Conservation Areas

This area was overdrained and has become infested with upland species and subject to peat fires. Privately owned land in the area is being acquired, and water management structures are being put in place to allow restoration of Everglades habitat and appropriate seasonal reflooding, at an estimated cost of $7.5 million.

4. Managing the Deer Herd to Survive High Water

In 1982, especially in the water conservation areas, many deer died due to flooding that cut them off from food supplies and migratory routes. Establishing a more natural seasonal flow of water through the Everglades and modifying even minor agency actions (such as allowing deer additional fodder by not mowing leaves) has resulted in a stabilization of the deer herd. In the rainy season of 1986, with 33 cm of rain in one month (138% above average), no deer died due to flooding.

5. Protecting Wildlife from Highway Traffic

East-west roads blocked the natural sheet-flow through the Everglades. Construction of I-75 will use an existing roadbed (Alligator Alley), but will be designed to improve the overland flow of water through the Everglades and Big Cypress Swamp by including bridges and culverts and to provide improved protection for wildlife through 36 animal underpasses. Severance payments to private landowners will make it possible to add over 40,460 ha to the Big Cypress Preserve. The estimated cost of these improvements is over $100 million.

6. Restoring the Values of Everglades National Park (ENP)

ENP is the second largest national park in the coterminous United States (Yellowstone is larger), and one of eight areas that the United Nations has designated as an international biosphere reserve. Traditionally, water was managed in south Florida to meet human needs, and for years, ENP suffered from either too much or too little water. Environmental needs for water have been recognized, and water managers are experimenting with ways to restore a more natural flow of water into ENP. Levees have been breached, canals filled, and a new water delivery plan for the park is being implemented.

7. Protecting the Florida Panther from Extinction

Panther once ranged throughout the southeast. About 50 are struggling to survive in the Everglades and Big Cypress. Five panther were killed on the state road that traverses the Big Cypress Preserve in the past five years. Actions to protect the panther include acquisition of habitat and assurance of prey by further restricting hunting and construction of highways (Executive Office of the Governor 1986).

The Save Our Everglades initiative, put in place by executive order of the governor, differs from other environmental protection approaches attempted in the state. It is not an overall land use planning and regulatory effort but an exercise in problem solving that includes such key actors as national nonprofit conservation groups, federal government agencies (such as the Park Service, U.S. Fish and Wildlife Service, the Army Corps of Engineers, and the Department of Transportation), state agencies (including the state park and conservation, environmental protection, planning, game and fish, and transportation agencies) and regional agencies (most importantly, the regional water management district).

The first step in the strategic planning process is to agree that there is a problem and to establish a mission. In the Save Our Everglades program, that mission is for the Everglades to look and function more like it did in 1900 by the year 2000. The next step is to determine how to achieve that mission. The seven points of the Save Our Everglades program above outline these strategies. The third step is to draw up a blueprint for action, including budgets.

Finally, an evaluation process is needed, so progress toward the mission can be monitored. Many of the strategies outlined have studies attached so that progress toward the mission can be checked. The mission statement itself includes benchmark years against which progress can be measured. The Save Our Everglades program has published an annual "report card" evaluating the achievement in each of the seven areas.

Resource Planning and Management Committees: More Specific Strategic Planning

The Save Our Everglades program is a broad strategic planning approach to restoration. It is not the only example of how this approach has been used to resolve conflicts over land and water in restoring an area of the Everglades subjected to intense development pressure. To understand this more specific iteration of strategic planning in the Everglades, the state's special area planning program needs to be explained.

By 1972, Florida began to have doubts about its long, unquestioned love affair with growth. In that year, among other efforts, the state passed the Area of Critical State Concern program (Chapter 380, Florida Statutes). Areas of Critical State Concern are specific geographic areas of statewide natural or historic importance. The areas are designated by the governor and cabinet, both of whom are elected statewide (DeGrove 1984). Once designated, local governments in an Area of Critical State Concern must draft land use plans and regulations that deal with identified issues and meet certain state standards. The state has continuous oversight over development in the designated area, and all local development orders in an Area of Critical State Concern are subject to state review and appeal. Since passage of this law, three critical areas have been designated, one of which, the Big Cypress Swamp, is in the Everglades.

Changes to the Area of Critical State Concern legislation in 1979 transformed the strict designation process and added greater flexibility (see 380.045, Florida Statutes). The result of these changes has been that the critical area approach has blossomed into an effective tool for managing growth and balancing natural resource concerns in areas of statewide significance. The key change to the process was the institution of resource planning and management committees, a mandated strategic planning and conflict resolution mechanism. Before an area can be designated, a resource planning and management committee must be established, and, if local governments adopt the measures recommended by the committee, actual designation can be avoided (DeGrove 1988). Of the 12 resource planning and management areas put in place, two—East Everglades and the lower Kissimmee River basin—concern the restoration of the Everglades and are, in essence, a means of furthering the mission of the Save Our Everglades initiative.

The resource planning and management committee that will undertake the study is appointed by the governor. The law sets out committee membership, which must include local elected officials and planning officials (whose participation is voluntary) and relevant regional and state agencies. The governor has the option of adding other governmental or private sector officials. This flexibility has led to the inclusion of environmental, development, and industry representatives that reflect the concerns in the region. Often, despite sharing concerns about the area's growth, the resource planning and management process is the first time that the affected interests come together.

The legislation also specifies how the committee will work: the first six months are spent in data gathering and identifying the issues, while the second six months are spent in designing means of resolving those issues (DeGrove 1988).

The strategies outlined can, and typically do, involve state and local government actions. Local government participation is voluntary, but if the local governments do not adopt the strategies outlined by the committee, they face designation as an Area of Critical State Concern and the continued state oversight that goes with it.

East Everglades: A Case Study of Water and Money in Action

The East Everglades Resource Planning and Management Area lies in western Dade County, with its western boundary abutting the Everglades National Park. For years, the county and the water management district used limited flood control and other restrictions on services to establish an informal urban growth boundary. Agricultural uses increasingly moved west, and during dry periods of the 1970s, residential uses began to spring up beyond the perimeter of flood protection.

Tropical Storm Dennis flooded the East Everglades and amalgamated agricultural interests and residents who demanded better flood protection. By this time, the park was finally receiving the amount of water it needed from the regional water management district, but without regard to the timing of its delivery. Unless the timing of wet and dry cycles in the park mimicked the natural sheet-flow pattern, species such as wood storks were unable to breed and feed.

When Governor Bob Graham established the resource planning and management committee for the East Everglades, he gave the committee specific charges. Those charges outlined the goals (to establish the natural sheet-flow pattern of water through the park); strategies (to develop a flood protection strategy for the East Everglades); and implementation methods (to include land acquisition) for the committee's deliberations.

The outcome of the process was, in part, a revised management plan for the area that includes both agricultural and residential interests in a more restricted area, located near an existing canal system with better drainage. Existing agricultural practices would continue, but in a more limited area under interim guidelines. Land acquisition would be mainly in areas closest to development to create a swath of protected lands, rather than in more pristine natural systems under less development pressure (Light 1987; Abrams 1987).

The restoration of the Everglades is far from complete, but some good lessons have been learned from the processes used to bring about that restoration. First, restoration of such a vast and complex system requires an iterative approach. All of the information is not in, and as more is learned about the natural functioning of the system, the approaches must be modified. Thus, monitoring of the environmental system is essential. Similarly, organizations must review their traditional means of operation in the light of this information.

Second, for a strategic planning approach to work, all affected players must be involved in the issue identification and resolution process. The EPA did not participate in the East Everglades Resource Planning and Management Committee, and is now issuing an intent to deny a permit for rock plowing in an area governed by the interim rule.

Finally, someone must be keeper of the process. Despite a change in state administration, the Save Our Everglades program remains intact and in force under a newly elected governor. The resource planning and management committee process has not fared quite as well. There is neither permanent funding nor permanent staff for resource planning and management committees, which has the potential of undermining their full effectiveness.

In conclusion, Florida's brief history in relation to the Everglades has been a story of water and money. Initially, in seeking to reclaim the Everglades for productive human use, the state sought and received funding to drain the Everglades. More recently, with some four million people living in south Florida, the state is taking the lead in committing water and money to restore the natural functions of this unique and complex environmental system.

References

Abrams, K. 1987. Overview of East Everglades/Everglades National Park 380 committee. Presentation before the Urban Land Institute, Federal Permit Working Group Steering Committee, November 6, 1987, Washington, DC.

Douglas, M. S. 1947. *The Everglades: River of grass.* New York: Rinehart & Co., Inc.

DeGrove, J. M. 1984. *Land, growth, and politics.* Chicago, IL: The American Planning Association.

DeGrove, J. M. 1988. Critical area program in Florida: Creative balancing of growth and the environment. *Journal of Urban and Contemporary Law,* pending publication.

Executive Office of the Governor. 1986. *Save our Everglades: Third anniversary report card.* Tallahassee, FL: Executive Office of the Governor.

Executive Office of the Governor. Undated. Presentation by Governor Bob Graham. Tallahassee, FL: Executive Office of the Governor.

Fernald, E. A., and D. J. Patton, eds. 1984. *Water resources atlas of Florida.* Tallahassee, FL: Institute of Science and Public Affairs.

Light, S. 1987. Restoring water deliveries to the southern Everglades. Presentation before the Urban Land Institute, Washington, DC: Federal Permit Working Group Steering Committee, November 6, 1987.

Martin, S. 1987. The surface water improvement act: A new chance for Florida. *Florida Environmental and Urban Issues* 15(1):4–8.

Robertson, W., and J. Kushland. 1974. The south Florida avi-fauna. In *Environments in South Florida: Present and past,* ed. P. Gleason. Miami, FL: Miami Geological Society.

State Comprehensive Plan Committee. 1987. *Final report: Keys to Florida's future: Winning in a competitive world.* Tallahassee, FL: Department of Community Affairs.

U.S. Department of Commerce, Bureau of the Census. 1980. *1980 census of population.* Washington, DC: U.S. Government Printing Office.

Barbara C. Brumback is Senior Research Associate, Florida Atlantic University/ Florida International University Joint Center for Environmental and Urban Problems, Ft. Lauderdale, FL 33301.

Toward a New Federalism for Environmental Restoration: The Case of Air Quality Conservation Through Intergovernmental Action—From Community to Global

Irwin Mussen

Abstract: *This paper deals with air quality protection through networks and interorganizational arrangements, including joint planning, involving governmental institutions and actions. Some of these networks currently exist (e.g., for joint planning and/or program administration); others are now emerging. The argument presented here rests on the premise that the public sector contains the optimal potential for regulation and other public actions for environmental protection, conservation, and restoration. This does not diminish the important roles of firms and individuals, environmental advocacy groups, other interested organizations, or the need for public/private cooperation.*

The discussion focuses, geographically, mainly on the regional scale and specifically on the San Francisco Bay Area as an example of one hub of intergovernmental relationships. Achievement of health-based air quality standards in a "non-attainment air basin" such as the Bay Area would constitute a critical aspect of environmental restoration. Many of the principles discussed may apply to other geographic scales, to other regions, to other environmental issues, and their interrelationships.

Key words: *air quality, Bay Area Air Quality Management District.*

Air Quality

Environmental problems often tend to be more complex than they first appear, interwoven with other environmental subsystems and frequently related to human activities that may not traditionally be defined as environmental. There is a direct parallel in governmental intervention in environmental affairs; a public agency with responsibilities in one environmental area will often find that it must call upon the involvement of other agencies with other areas of responsibility.

Air pollution is obviously not a single environmental problem. The ambient air—in addition to being essential for most forms of life—is a vehicle, or medium of distribution, for health-damaging and environment-degrading contaminants. Under certain conditions, the atmosphere also acts as a caldron for adverse chemical reactions.

For instance, carbon monoxide (CO), inhalable particulates (including airborne heavy metals), and some other contaminants can be concentrated at hot-spots consisting of a single road intersection or a single urban land parcel. Yet CO, as well as ozone and other air pollutants, may also be distributed in clouds extending over a multi-city subregion. Hydrocarbons and nitrogen oxides may be emitted in one local area, yet produce impacts elsewhere as they react in sunlight to form ozone downwind at the other end of a large air basin. Ozone can also at times be transported between regional air basins.

Governmental Intervention

The concept of governmental intervention at the regional level is based on scientific evidence about the relationships between contaminant distribution and meteorology and topography. However, the creation of air-basin-wide special governmental agencies, such as the Bay Area Air Quality Management District (BAAQMD), could lull cities and counties into a false sense of complacency. Local government may be inclined to neglect the responsibilities in air quality protection and/or roles that they can exercise more efficiently than can regional air quality agencies. For instance, in California it is the cities and counties that are mandated to adopt general plans, including land use, circulation, housing , and other required elements that have significant impacts on automobile use. Motor vehicle usage is the major source of air contaminants such as CO, precursors of ozone, particulates, lead, formaldehyde, and benzene. It is a federal agency, the Environmental Protection Agency (EPA), and—in California—a state agency, the Air Resources Board, that regulate vehicle emissions standards. Those agencies also play a strong role in the setting of emissions standards from stationary sources, such as industries, and in approving regional air quality management plans. Thus a regional agency with responsibility for air quality restoration and conservation needs to relate to—i.e., to depend on and cooperate with—the national and state governments above, and the city and county governments within: a fine illustration of the concept of federalism.

The Local General Plan

In California today, air pollution is widely recognized as a major issue. Yet air quality is not the subject of a mandated general plan element, or even suggested—as are water and soil quality—as a subsection of the conservation element required under California general planning law. An air quality element of the city or county general plan —or, perhaps better from the point of view of integration of environmental concerns, a full air quality section of the conservation element—would compel local planners to recognize the relationships between air quality and all plans and projects. By law it would force other plan elements, which are already required and whose contents are prescribed under State General Plan Guidelines, to be consistent with air quality requirements. Therefore the BAAQMD Board of Directors has recently urged the 103 local units of governments, within its nine-county jurisdiction, to incorporate air quality elements in their general plans. The board has also advocated the adoption of transportation control measures by its constituent cities and counties; these measures may be seen as important implementing actions of the circulation element, as well as the air quality element, of a local general plan. The district offers its staff to assist as needed since it is unlikely that local planning agencies will include air quality specialists as staff members. It has also distributed a *Guidelines* for local governmental use in formulating air quality elements and in air quality sections of environmental impact assessment documents (BAAQMD 1985).

The Regional Air Quality Plan

In the San Francisco Bay Area, the regional air quality management district works not only with its constituent city and county units of government, but also with other regional agencies. Regional air quality plan is a joint responsibility of the Metropolitan Transportation Agency and the Association of Bay Area Governments, as well as of BAAQMD (ABAG et al. 1982). The Bay Area Air Quality Plan, in turn, must be accepted by the State of California Air Resources Board and become part of the State Implementation Plan (SIP). The SIP is then submitted to the EPA for approval, as set out by the Clean Air Act. These horizontal and vertical intergovernmental relationships constitute a system of federalism. (The boards of the regional agencies themselves constitute interlevel networks since the directors are all elected county supervisors, city mayors or council persons, or appointed representatives of locally elected officials.)

Other Regional Relationships

A recent addition to the intergovernmental web of relationships is seen in the work on the Integrated Environmental Management Plan (IEMP) for the Santa Clara Valley. The focus is on air, water, and land as media for toxics contamination. The critical need for such an integrated approach becomes clear when it is realized that often the least expensive, most expedient method for reducing toxic contamination in water and soil is air stripping—i.e., transferring the toxics to the air. The IEMP, established under an EPA grant, includes an Intergovernmental Coordinating Committee with members representing towns and cities, the county, the regional water quality control board, the subregional water district, and state offices of health, water resources and environmental affairs—as well as BAAQMD. Nongovernmental members represent a university, environmental advocacy groups, and industry. A Public Advisory Committee has an even broader base of representatives of governmental and nongovernmental organizations.

In still other programs, BAAQMD is cooperating with at least two other adjoining air quality districts in analyzing ozone transport between air basins. New relationships between the state's Department of Health Services and BAAQMD have been established regarding the new methodology of health risk assessment. As new issues emerge and/or increase in salience—e.g., airborne toxics, acid rain, intermedia transfer of contaminants, etc.—there will be a continuing need to modify existing networks and to create new relationships for effective public action.

Governmental Structure

There are long-standing questions about how best to structure intergovernmental coordination. One model exists at the federal level; all federal environmental concerns are supposedly within the EPA's realm. Yet the Clean Air Act Amendments of 1977 recognized that the Department of Transportation also needs to be involved because of the significant relationship of air quality to transportation. The Department of Energy is concerned with alternative fuels—which have critical air quality implications. The Department of the Interior is concerned with visibility impairment and other air quality issues at national parks and other so-called Class One areas; on the other side, it is involved with permitting offshore oil drilling which raises substantial air quality questions. The Department of Commerce is, in various ways, interested in economic development; industrial location—and not only of old-fashioned smokestack industry—often involves air quality issues. There are, of course, also connections between air quality issues and the responsibilities of the U.S. Department of Health and Human Services. While the relationship of air quality to health is obvious, some others may be less so. For instance, there are issues of social welfare and equity since the poor are more likely to be impacted by air contaminants emitted by nearby industry and/or waste dumps and by motor vehicle fumes along major arterials; these are where the least expensive housing in urban areas is likely to be located. This leads directly to the role of the Department of Housing and Urban Development—at least to the extent that it plays a role in determining how much and where housing subsidies and programs are applied. HUD's policies and programs may strongly affect the affordability of housing readily accessible to workplaces and, therefore, the mode of transportation available and length of the journey to work, and consequent use of automobiles and production of the air contaminants. On another front, the U.S. Department of Education could play a vital role in fostering environmental awareness. Even supposed integration of environmental issues into a single governmental agency—in this case on the federal level—does not obviate the need for intergovernmental networking.

The list of examples of linkages, existing and potential, is virtually endless. Just to illustrate with a rather far-out example: in first and second century Israel/Palestine it was the religious organizational structure that determined environmental protection rules. The rabbinical writings in the Talmud prescribe specific buffer-zone distances from residential areas and down-wind locations for odor-generating uses such as tanneries, dyeing works, bakeries, and stables (Mamane 1987).

In the United States today, many of the needs for linkage at the federal department level may be mirrored at the state, regional, and local government levels. Yet at the state level in California, while there is a Secretary of Environmental Affairs with coordination responsibilities—and he or she has always been the same official who also directs the State Air Resources Board—there is not a true structural integration even among traditional environmental protection functions, such as water and air quality management.

Regional Structure for Environmental Protection

At the regional level, governmental organizations concerned with environmental protection are quite separate in the Bay Area. Water quality is controlled by a regional office of the State Water Quality Control Board governed by state appointees; air quality is controlled by a regional special government agency with a board of directors made up of city and county elected officials. Transportation planning is assigned to one agency while general regional planning is the responsibility of another. (These functions are usually integrated into a single regional Council of Governments/Metropolitan [transportation] Planning Organization in other major urban regions of this state.) Conservation of the Bay itself is handled by an agency separate from that which regulates coastal (ocean coast) conservation.

A web of cooperative working relationships has been forged among relevant Bay Area agencies, and more far-reaching approaches have been proposed in the past for the Bay Area. More than a decade ago, state legislation was introduced to consolidate most or all Bay Area regional governmental agencies into a single multi-purpose regional agency. The measure lost at least twice. A separate move was made to consolidate air and water quality control in regions throughout California; it, too, was defeated. The argumentation included functional concerns, as well as those dealing with historical and political realities, which prevailed in maintenance of the separation of powers here. An underlying question in both cases was whether it was preferable to retain the focus, clarity, and zeal that, hopefully, characterize single-purpose agencies, or sacrifice these objectives in order to attempt to gain the potential efficiency and ease of interrelated action that might be achieved through consolidation.

Conclusion/Recommendations

A few relatively modest steps should be taken that might have value in strengthening regional capability to coordinate protection of the environment. The details apply to the state of California and to specific needs in the San Francisco Bay Area. The underlying concepts may be applicable, if sensitively adapted, to other states and regions as well.

State law and guidelines concerning local general plans should be substantially expanded to cover all significant environmental aspects, including air quality. The relationships among elements or sections should be analyzed and solutions to problems programmed—whether the links are with factors also defined as environmental, social, economic, or other. Protection of each environmental medium, and the entire interrelated ecosystem, should be given a central, proactive position in the planning process, not left to an environmental impact assessment process.

State law should mandate—not just enable on a voluntary basis—the formulation and adoption of regional *plans, dealing at least with land use, transportation, housing affordability, and all regional environmental and related planning concerns.* These regional plans should be given a status and power at least equivalent to the current power of city and county plans. (The paradox is that, in California, an air quality plan is required on the regional, but not on the local level: land-use and related planning is required on the local, but not on the regional level.)

Regional plans should be coordinated by an agency with broad regional responsibilities, e.g. the regional Council of Governments (COGs), in cooperation with relevant special-purpose regional and local bodies. (In the Bay Area there had been an Environmental Management Plan which at least attempted to integrate concerns for air quality with those for water quality, water supply, and solid waste [ABAG 1978]; however, it ran into strong political criticism which resulted in a weakening of recommendations to tie land use and development controls to environmental protection.) It would be naive to assume that a proposal of a state requirement for stronger, integrated regional planning would have smooth political sailing. Though there is the inescapable need, discussed above, to deal with issues that are inextricably intertwined, opposition is sure to come from interests wedded to the cause of local autonomy in planning and development. However, a strengthened regional planning function may be much more digestible politically than complete regional government or the merging of established special purpose regional agencies.

In order to foster meaningful regional plans, the state should revise its formula for the institution of regional planning bodies and/or the COGs themselves. At the very least, all cities and counties should be required to participate, to remain in them and contribute their fair share toward financing them—instead of the existing system of voluntary participation through joint-exercise-of-powers agreements whereby disgruntled cities and counties can simply back away from their regional responsibilities.

On a much more general plane, surely applicable as well outside of the Bay Area and California, *all public agencies with environmental functions should establish and/or strengthen clear mechanisms to facilitate coordination.* This means, at very least, assigning one or more high-level officials—or whole divisions—within each organization to the specific role of intergovernmental liaison. The responsibility of each organization to establish working relationships with other organizations should be explicitly spelled out in its charter. Interagency research, planning, and program implementation should be jointly administered and financed—including cost and revenue sharing where needed.

The need exists, and will probably persist, at most levels of public service, to articulate a process for maintenance of existing and continuous forging of new networks—ad hoc or permanent—for creative joint action programs. These programs should not only restore and protect an essential environmental feature, such as the air, but also relate that restoration objective to society's other compelling environmental, health, social, and economic goals.

References

Association of Bay Area Governments (ABAG). 1978. *San Francisco Bay Area environmental management plan.* Berkeley, CA.

Association of Bay Area Governments (ABAG), Bay Area Air Quality Management District (BAAQMD), and Metropolitan Transportation Commission (MTC). 1982. *Bay Area air quality plan.* Berkeley, CA.

Bay Area Air Quality Management District (BAAQMD). 1985. *Air quality and urban development: Guidelines for assessing impacts of projects and plans.* San Francisco, CA.

Mamane, Y. 1987. Air pollution control in Israel during the first and second century. *Atmospheric Environment* (Great Britain) 21(8):1861–1863.

Irwin Mussen is the Senior Planner of the Bay Area Air Quality Management District, 939 Ellis St., San Francisco, CA 94109, and a consultant in urban, regional and environmental conservation and development. The views expressed in this paper are not necessarily the opinions of the Board of Directors or staff of the Bay Area Air Quality Management District or any other agency.—IM

Wetland Issue Conflict Management

Radford S. Hall

ABSTRACT: *Five cases of conflict associated with permit applications to fill saltwater and/or freshwater wetlands and mitigate expected wetland losses were studied. The mitigation plans involved saltwater and freshwater wetland restoration. Strong disagreement existed among the various participants in the process as to the measurement of compensatory values, restoration design, and the likelihood of successful implementation.*

Evaluation efforts focused on the implementability of a designed conflict management strategy and the ability of that strategy to resolve the wetland restoration/environmental issues. The conflict management strategy was applied to two of the cases. Through a "quasi-experimental" research design, qualitative case descriptions were prepared and analyzed using data obtained through direct observation, review of the permit process documentation, and a questionnaire.

KEY WORDS: *conflict management, wetlands, mitigation, restoration, permits.*

Radford S. Hall is Chief, Permits Section, Corps of Engineers, San Francisco, CA 95105.

SOME PRINCIPLES OF CONFLICT RESOLUTION AND CITIZEN DESIGN WORKSHOPS

JOSEPH E. PETRILLO

ABSTRACT: *Awareness of the tools available to resolve conflict over the use of a natural resource is essential if our natural heritage is to be saved from destruction. In negotiating with representatives of the many interests often contending over a threatened resource, one must have a facility in using the tools available, persistence, and a nonjudgmental attitude.*

This paper explores some of those tools with an emphasis on the role resource restoration can play in reducing land use conflicts.

KEY WORDS: *conflict resolution, negotiation, community, participation.*

Joseph E. Petrillo is an attorney with Baker and McKenzie, 2 Embarcadero Center, 24th Floor, San Francisco, CA 94111, specializing in coastal issues. He was formerly Executive Officer, California State Coastal Conservancy, 1977–1985.

"ECONOMOLOGY": THE MERGING OF ECONOMICS AND ECOLOGY

WAYNE TYSON

ABSTRACT: *We, and a good part of the rest of the world, believe that the restoration of the earth will come about through some combination of technological fixes: that righteousness can be bought. We are born-again true believers in numbers, and in facts represented only by numbers.*

Building a society that can restore the earth requires that we more closely question our assumptions. Economics (house-management) and ecology (house-study) are now kept separate, and we assume that it must always be so. Economics is too perverted in its present form (obsession with numbers alone, the management concept having long ago been subordinated to mere numbers instead of numbers being restricted to serving the values of management). Ecology may be too "pure" in its "hands-off" detachment from real events. These two disciplines are spontaneously coming together, driven by desperate circumstances. This reunification may not be happening quickly enough to be a pivotal force for restoring the earth. Therefore, let us begin to merge the concepts of economics and sociology. With this merger, ecology is taken out of the realm of mere study and into the realm of management, and management is restored to its rightful meaning within economics. Numbers with value are implied.

A term is needed to define this value approach to earth-house management. May I suggest "economology," which would mean the study and value-management of our whole earth-house, to distinguish it from the more narrow, but still valid concepts of economics and ecology? Rather than subordinating value to technique, in economology, technique continuously arises from the study of value as a dynamic process. The "dismal science" of economics could go on being obsessed with numbers and price alone, or, if economoloy becomes widely practiced, it would perhaps be made less dismal. Ecology should go on meaning the study of our earth-house, but it, too, might benefit from inclusion within the broader principle of economology.

To restore the earth, we must come to a deep understanding that merely to consider both economics and ecology separately will not do. We must unify the tools associated with numbers and money in an equal if not superior effort to understand both the meaning of numbers and of the values beyond them.

KEY WORDS: *ecological restoration and management, ecology, economics, "Economology," systems theory, lexicology, revegetation.*

Wayne Tyson is Manager, Land Restoration Associates, Box 3120 Hillcrest Station, San Diego, CA 92103.

Cost-effective Resource Management Planning

Elgar Hill

Abstract: *Planning is the basis for more orderly and consistent decision making. It is necessary that the formulation of goals, matters of priority and responsibility, and taking advantage of opportunities be carefully addressed.*

A management plan for the lands associated with a large gold mine has just been completed. Goals include (1) ecosystem protection, (2) prediction of problems, (3) superior land stewardship, and (4) a cost-effective approach to land management. The plan's emphasis is on understanding natural processes and intervening only when necessary to avoid conflicts or to improve habitat. It is believed that the same approach can be used on other projects, whether large or small, and that beneficial results can be achieved at lower cost.

Key words: *land use planning, land management, habitat restoration, fire hazard reduction, shoreline management, sensitive plant maintenance.*

Elgar Hill is a planning consultant and architect for Environmental Analysis and Planning, P.O. Box 690, Penngrove, CA 94951.

Protection of Environmental Resources Through a Comprehensive Watershed Ordinance and Long-term Planning in the City of Austin, Texas

Jim Eldred

ABSTRACT: *Austin, Texas, like many cities, has its richest natural resources outside its corporate city limits in its extra-territorial jurisdiction (ETJ). These resources include most of the city's water supply watersheds, much of the Edward's aquifer recharge zone, wetlands, remnant grasslands, sensitive canyon heads and associated headwaters, mature woodlands, and rare, threatened, or endangered species habitat. In the recent past, Austin experienced a rapid rate of subdivision and commercial development activity in much of its extra-territoritial jurisdiction and in its more environmentally sensitive watersheds. That kind of growth led to a rapid alteration of many of the area's valuable natural features.*

Cities in Texas are restricted in what they can do to regulate development and protect these valuable natural resources outside their city limits. They can regulate subdivisions and site development there only through ordinances or regulations aimed at water pollution abatement. This paper introduces what is being done in Austin in resource protection and environmental planning to help protect water and other valuable natural resources.

KEY WORDS: *watershed ordinances, erosion control, setbacks, long-term planning, water quality zones, Austin plan.*

Jim W. Eldred is an Environmental Specialist, Department of Environmental Protection, City of Austin, TX 78767.

Selected Bibliography

This selected bibliography contains books, articles, and monographs of more general interest than the specific references cited in the preceding articles. It was prepared by Arlene Magarian from the notes of John Berger.

Adams, T. E., and B. L. Kay. 1982. *Seeding for erosion control in coastal and central California.* Cooperative Extension Leaflet 21304. Berkeley, CA: University of California, Division of Agricultural Sciences.

Adams, T. E. 1983. *Seeding for erosion control in mountain areas of California.* Cooperative Extension Leaflet 21346. Berkeley, CA: University of California, Division of Agricultural Sciences.

Adams, T. E., P. B. Sands, W. H. Weitkamp, N. K. McDougal, and J. Barolome. 1987. Enemies of white oak regeneration in California. In *Proceedings of the Symposium on Multiple-Use Management of California's Hardwood Resources,* Gen. Tech. Report PSW-100, 459–462. Berkeley, CA: U.S. Department of Agriculture, Pacific Southwest Forest and Range Experiment Station.

Alexander, C. 1977. *A pattern language.* New York: Oxford University Press.

———. 1979. *The timeless way of building.* New York: Oxford University Press.

Amimoto, P. 1978. *Erosion and sediment control handbook.* Sacramento, CA: California Resources Agency, Department of Conservation.

Arvola, T. 1978. *California forestry handbook.* Sacramento, CA: California Department of Forestry.

Ashby, W. C., et al. 1978. *Our reclamation future: The missing bet on trees.* Chicago, IL: Institute for Environmental Quality.

Association of Bay Area Governments. 1981. *Erosion and sediment control training handbook.* Berkeley, CA: Association of Bay Area Governments.

Beazeley, A. 1900. *Reclamation of land from tidal waters: A handbook for engineers, landed proprietors and others interested in works of reclamation.* London: C. Lockwood & Son.

Berger, J. J. 1985. *Restoring the earth: How Americans are working to restore our damaged environment.* New York: Alfred Knopf, Inc./Doubleday & Co., Inc. [1987].

Bradshaw, A. D., and M. J. Chadwick. 1980. *The restoration of land: The ecology and reclamation of derelict and degraded land.* Berkeley and Los Angeles, CA: University of California Press.

Brewer, R., ed. 1983. *Proceedings of the eighth North American prairie conference.* Kalamazoo, MI: Department of Biology, Western Michigan University.

British Columbia Ministry of Environment. 1980. *Stream enhancement guide.* Vancouver, B.C.: Minister of Environment, Fish and Wildlife.

Broughton, F. 1985. The reclamation of derelict land for agriculture: Technical, economic and land-use planning issues. *Landscape and Urban Planning* 12:1.

Brown, D., et al. 1986. *Reclamation and revegetation of problem soils and disturbed lands.* Park Ridge, NJ: Noyes Data Corporation.

Cairns, J., Jr. 1980. *The recovery process in damaged ecosystems.* Ann Arbor, MI: Ann Arbor Science Publications.

______, ed. 1988. *Rehabilitating damaged ecosystems.* Boca Raton, FL: CRC Press, Inc.

Cairns, J., Jr., K. L. Dickson, and E. E. Herricks. 1977. *Recovery and restoration of damaged ecosystems.* Charlottesville, VA: University Press of Virginia.

Chadwick, M. J., and G. T. Goodman. 1975. *The ecology of resource degradation and renewal.* New York: John Wiley & Sons.

Chan, F. J. 1985. *Vegetation establishment.* San Francisco, CA: International Erosion Control Association.

Chisholm, A., and R. Dumsday, eds. 1988. *Land degradation: Problems and policies.* New York: Cambridge University Press.

Clambey, G. K. and R. H. Pemble, eds. 1986. *The prairie—past, present, and future: Proceedings of the ninth North American prairie conference.* Fargo, ND: Tri-College University.

Clark, W. C. and R. E. Munn, eds. 1986. *Sustainable development of the biosphere.* New York: Cambridge University Press.

Clary, R. F., and R. D. Slayback. 1985. Revegetation in the Mojave Desert using native woody plants. In *Native Plant Revegetation Symposium,* 42–47. San Diego, CA: California Native Plant Society.

Colorado State University Environmental Resources Center Information. 1974. *Proceedings of workshops on revegetation of high-altitude disturbed lands.* Series no. 10, 21, and 28. Fort Collins, CO: Environmental Resources Center, Colorado State University.

Conservation Foundation. *The Conservation Foundation Letter.* Washington, DC.

Cooke, G. D., E. B. Welch, S. A. Peterson, and P. R. Newroth. 1986. *Lake and reservoir restoration.* Stoneham, MA: Butterworth Publishers.

Council of State Governments. Center for the Environment and Natural Resources. 1988. *Wetlands: User's manual to a national database of state wetland protection programs and contacts.* Lexington, KY: Council of State Governments.

Critchfield, R. 1983. *Villages.* Garden City, NY: Anchor Press/Doubleday.

Dasmann, R. F. 1985. Achieving the sustainable use of species and ecosystems. *Landscape and Urban Planning* 12:3.

Davis, A., and G. Stanford, eds. 1988. *The prairie: Roots of our culture, foundation of our economy: Proceedings of the tenth North American prairie conference.* Dallas, TX: The Native Prairies Association of Texas.

Detweiler, R., J. Sutherland, M. S. Weatherman, eds. 1973. *Environmental decay in its historical context.* Glenview, IL: Scott Foresman.

Diamant, R., J. G. Eugster, and C. J. Duerksen. 1984. *A citizen's guide to river conservation.* Washington, DC: Conservation Foundation.

Duff, D. L. 1988. *Indexed bibliography on stream habitat improvement.* U.S. Department of Agriculture, Forest Service, Intermountain Region. Logan, UT: Forest Service.

DuPlat-Taylor, F. M., and M. G. Ustavus. 1981. *Reclamation of land from the sea.* London: Constable and Company, Ltd.

Eagles, P. F. J. 1984. *The planning and management of environmentally sensitive areas.* A Longman Scientific & Technical Publication. Washington, DC: Heldref Publications.

Faludi, A. 1985. A decision-centred view of environmental planning. *Landscape and Urban Planning* 12:3.

Frey, W. 1985. Plant propagation trials of the nipomo dune flora. In *Proceedings of conference XVI, International Erosion Control Association,* 41–45. San Francisco, CA: International Erosion Control Association.

Fukuoka, N. 1978. *One straw revolution.* Emmaus, PA: Rodale Press, Inc.

Gilden, E., R. S. Adams, and C. Hawkes. 1973. *Planting California forest land.* Davis, CA: University of California, Agricultural Extension, California Division of Forestry, U.S. Department of Agriculture, Forest Service.

Glenn-Lewin, D. C., and R. Q. Landers, Jr., eds. 1978. *Fifth midwest prairie proceedings.* Ames, IA: Extension Courses and Conferences, Iowa State University.

Global Education Associates, eds. 1982. Land and world order. In *The whole earth papers* No. 17. East Orange, NJ: Global Education Associates.

Gore, J. A. 1985. *Restoration of rivers and streams.* Stoneham, MA: Butterworth Publishers.

Graves, W. L., B. L. Kay and T. Ham. 1980. Rose clover controls erosion in Southern California. *California Agriculture* 34(4):4–5.

Gray, D. H., and A. T. Leiser. 1982. *Biotechnical slope protection and erosion control.* New York: Van Nostrand Reinhold.

Greenbie, B. B. 1976. *Design for diversity.* New York: Elsevier Science Publishing Co., Inc.

Hagevik, G. 1972. *Relationship of land use and transportation planning to air quality management.* New Brunswick, NJ: Center for Urban Policy Research and Conferences Department, University Extension, Rutgers University.

Harvey & Stanley Associates, Inc. 1983. *Revegetation manual for the Alameda County Flood Control and Water Conservation District revegetation program.* Hayward, CA: Alameda County Public Works Agency.

Hashagen, K., C. Toole, B. Wyatt, K. Somarstrom, and S. Taylor, eds. 1984. *Report of the second California salmon and steelhead restoration conference.* University of California Sea Grant MAP-21. Eureka, CA: California Sea Grant Marine Advisory Program.

Heede, B. H. 1976. *Gully development and control: the status of our knowledge.* Research Paper RM-169. Ft. Collins, CO: Rocky Mountain Forest and Range Experiment Station, U.S. Forest Service.

Holdgate, M. W., and M. J. Woodman. 1978. *The breakdown and restoration of ecosystems.* New York: Plenum Press.

Hossner, L. R. 1987. *Reclamation of surface-mined lands.* Boca Raton, FL: CRC Press.

Hough, M. 1985. *City form and natural process: Towards a new urban vernacular.* Florence, KY: Van Nostrand Reinhold.

Hulbert, L. C. 1973. Management of Konza Prairie to approximate pre-white-man fire influences. In *Third midwest prairie conference proceedings,* ed. L. C. Hulbert, 14–16. Manhattan, KS: Division of Biology, Kansas State University.

Hulbert, L. C. 1978. Controlling experimental bluestem prairie fires. In *Fifth Midwest Prairie Proceedings,* ed. D. C. Glenn-Lewin and R. Q. Landers, Jr., 169–171. Ames, IA: Extension Courses and Conferences, Iowa State University.

Hutnik, R. J., and G. Davis. 1973. *Ecology and reclamation of devastated land.* New York: Gordon and Breach.

Jacobs, J. 1961. *The death and life of great American cities.* New York: Vintage Books/Random House.

Jacobs, P. 1985. A sustainable society through sustainable development: Towards a regional development strategy for Northern Quebec. *Landscape and Urban Planning* 12:3.

______. 1985. Preface: Achieving sustainable development. *Landscape and Urban Planning* 12:3.

Jacobsen, J., and M. Webb. 1982. *U.S. carrying capacity: An introduction.* Washington DC: Carrying Capacity.

Johnson, R. L. 1981. Oak seeding—it can work. *Southern Journal of Applied Forestry* 5(1):28–33.

Jordon, W. R., III, M. E. Gilpin, and J. D. Aber, eds. 1988. *Restoration ecology: A synthetic approach to ecological research.* New York: Cambridge University Press.

Jorgensen, S. E., and W. J. Mitsch, eds. 1983. Application of ecological modelling in environmental management, Part B. *Developments in environmental modelling,* Vol. 4B. New York: Elsevier Science Publishing Co., Inc.

Jorgensen, S. E., ed. 1983. Application of ecological modelling in environmental management, Part A. *Developments in environmental modelling,* Vol. 4A. New York: Elsevier Science Publishing Co., Inc.

______, ed. 1984. Modelling the fate and effect of toxic substances in the environment. *Developments in environmental modelling,* Vol. 6. New York: Elsevier Science Publishing Co., Inc.

______, ed. 1986. Fundamentals of ecological modelling. *Developments in environmental modelling,* Vol. 9 New York: Elsevier Science Publishing Co., Inc.

______, ed. 1986. *Ecological modelling: international journal on ecological modelling and engineering and systems ecology.* Vols. 31–34. Copenhagen: Vaerlose. New York: Elsevier Science Publishing Co., Inc.

Kaiser, E. J. et al. 1973. *Promoting environmental quality through urban planning and controls.* Washington, DC: U.S. Environmental Protection Agency. Chapel Hill, NC: Center for Urban and Regional Studies, University of North Carolina.

Kay, B. L. 1978. Mulches for erosion control and plant establishment on disturbed sites. University of California, Davis, Agricultural Experiment Station, Cooperative Extension. *Agronomy Progress Report* 87:1–19.

Kay, B. L., et al. 1981. Zorro: Annual fescue for emergency revegetation. *California Agriculture* 35, (1 & 2):15–17.

Kay, B. L., R. M. Love, and R. D. Slayback. 1981. Discussion: Revegetation with native grasses: A disappointing history. *Fremontia* 9(3):11–14.

Kay, B. L., and W. L. Graves. 1983. History of revegetation studies in the California deserts. In *Environmental effects of off-road vehicles: Impacts and management in arid regions,* ed. R. H. Webb and H. G. Wilshire. New York: Springer-Verlag.

———. 1983. Revegetation and stabilization techniques for disturbed desert vegetation. In *Environmental effects of off-road vehicles: Impacts and management in arid regions,* ed. R. H. Webb and H. G. Wilshire. New York: Springer-Verlag.

Kay, B. L., and R. D. Slayback. 1985. California interagency guide for erosion control plantings. In *Erosion control—a challenge for our times: The proceedings of conference 16 of the International Erosion Control Association.* Davis, CA: Department of Agronomy and Range Science, University of California, Davis.

Ketchum, B. H., ed. 1983. *Estuaries and enclosed seas.* New York: Elsevier Science Publishing Co., Inc.

Knox, R. C. et al. 1986. *Aquifer restoration: State of the art.* Dark Ridge, NJ: Noyes Publications.

Kraebel, C. J., and A. F. Pillsbury. [1934] 1980. *Handbook of erosion control in mountain meadows.* U.S. Department of Agriculture, Forest Service. Washington, DC: Government Printing Office.

Kucera, C. L., ed. 1983. *Proceedings of the seventh North American prairie conference.* Springfield, MO: Southwest Missouri State University.

Lauenroth, W. K., G. V. Skogerboe, and M. Flug. 1983. Analysis of ecological systems: State-of-the-art in ecological modelling. *Developments in environmental modelling,* Vol. 5. New York: Elsevier Science Publishing Co., Inc.

Law, D. L. 1985. *Mined-land rehabilitation.* Florence, KY: Van Nostrand Reinhold.

Leiser, A. T. et al. 1974. *Revegetation of disturbed soils in the Tahoe Basin.* Davis, CA: Department Agronomy and Range Science, University of California, Davis.

Lewis, R. R., III 1982. *Creation and restoration of coastal plant communities.* Boca Raton, FL: CRC Press, Inc.

Lifshits, L. 1976. Rejuvenated soil. *Soviet Life* 3:44–47.

Lutz, P., and P. Sautmire. 1973. *Ecological renewal: Introductory report of the Normandy Reservoir salvage project.* Knoxville, TN: University of Tennessee Press.

Lyle, S. 1986. *Surface mine reclamation.* New York: Elsevier Science Publishing Co., Inc.

Margolin, M. 1986. *The earth manual.* Second Edition. Boston, MA: Houghton Mifflin.

McCreary, S. T. 1985. Needed: a national agenda for restoration of degraded wetland ecosystems. In *Proceedings: National wetlands assessment symposium,* 271–277. Portland, ME: Association of State Wetland Managers.

Miller, G., and G. Miller, Jr. 1972. *Replenish the earth: A primer in human ecology.* Belmont, CA: Wadsworth Publishing Company.

Mitsch, W. J., R. W. Bosserman, and J. M. Klopatek, eds. Energy and ecological modelling. *Developments in environmental modelling,* Vol. 1. New York: Elsevier Science Publishing Co., Inc.

National Academy of Sciences. 1974. *Rehabilitation potential of western coal lands.* Cambridge, MA: Ballinger Publishing Company.

National Water Well Association. 1983. *Proceedings of the third national symposium and exposition on acquifer restoration and ground water monitoring.* Worthington, OH: National Water Well Association.

———. 1984. *Proceedings of the fourth national symposium and exposition on acquifer restoration and ground water monitoring.* Worthington, OH: National Water Well Association.

———. 1985. *Proceedings of the fifth national symposium and exposition on acquifer restoration and ground water monitoring.* Worthington, OH: National Water Well Association.

———. 1986. *Proceedings of the sixth national symposium and exposition on acquifer restoration and ground water monitoring.* Dublin, OH: National Water Well Association.

———. 1987. *National October action conference on aquifer restoration, ground water monitoring, and geological methods.* Dublin, OH: National Water Well Association.

Nichols, O. G. and B. A. Carbon, et al. 1985. Rehabilitation after bauxite mining in south-western Australia. *Landscape and Urban Planning* 12:1.

Nielson, D. M., ed. 1982. *Proceedings of the second national symposium on aquifer restoration and ground water monitoring.* Columbus, OH: National Water Well Association.

Olkowski, H., et al. 1979. *The integral urban house: Self-reliant living in the city.* San Francisco, CA: Sierra Club Books.

Paulson, A., ed. *The National Wildlfower Research Center's wildflower handbook.* Austin, TX: Texas Monthly Press.

Payne, N. F. , and F. Copes. 1986. *Wildlife and fisheries habitat improvement handbook.* Wildlife and Fisheries Administrative Report. Washington, DC: U.S. Department of Agriculture, Forest Service.

Plummer, A. P., D. R. Christensen, and S. B. Monsen. 1968. *Restoring big-game range in Utah.* Publication no. 6803, Utah Division of Fish and Game. Salt Lake City, UT: Utah Division of Wildlife Resources.

Postel, S., and Heise, L. 1976. Reforesting the earth. *Futurist* 10(1):35.

———. 1988. *Reforesting the earth.* Washington, DC: Worldwatch Institute.

Randall, A., et al. 1978. Reclaiming coal surface mines in central Appalachia: A case study of the benefits and costs. *Land Economics* 54:472–489.

Reeves, F. B., D. Wagner, T. Moorman, and J. Kiel. 1979. The role of endomycorrhizae in revegetation practices in the semiarid West. *American Journal of Botany* 66(I):6–13.

Rieger, J. P. 1984. *Proceedings of the first native plant revegetation symposium.* San Diego, CA: John P. Rieger and Associates.

Rieger, J. P., and B. K. Williams. 1987. *Proceedings of the second native plant revegetation symposium.* Madison, WI: Society for Ecological Restoration and Management.

Robinette, G. O., ed. 1985. *How to make cities liveable.* Florence, KY: Van Nostrand Reinhold.

Schiechtl, H. 1980. *Bioengineering for land reclamation and conservation.* Edmonton, Alberta: University of Alberta Press.

Schlottmann, A. J. 1976. Economic impacts of surface mine reclamation. *Land Economics* 52:265–277.

Schramm, P., ed. 1970. *Proceedings of a symposium on prairie restoration.* Special Publication no. 3. Galesburg, IL: Knox College Biological Field Station.

Schueler, D. G. 1980. *Preserving the Pascagoula.* Jackson, MS: University Press of Mississippi.

Seehorn, M. E. 1980. *Proceedings of the trout stream habitat improvement workshop.* U.S. Department of Agriculture. Ashville, NC: Forest Service and Trout Unlimited.

———. 1985. *Fish habitat improvement handbook.* Forest Service Technical Publication R8-TP7. Atlanta, GA: U.S. Department of Agriculture, Forest Service, Southern Region.

Shaller, F. W., and P. Sutton, eds. 1978. *Reclamation of drastically disturbed lands.* Madison, WI: American Society of Agronomy.

Shubert, G. H. and R. S. Adams. 1971. *Reforestation practices for conifers in California.* Sacramento, CA: California Department of Forestry.

Shukla, J. B., T. G. Hallam, and V. Capasso, eds. 1987. Mathematical modelling of environmental and ecological systems. *Developments in environmental modelling, 11.* New York: Elsevier Science Publishing Co., Inc.

Singer, S. 1980. *Guidelines for gully control.* Aptos, CA: Santa Cruz County RCD.

Soulé, M. E. 1986. *Conservation biology: The science of scarcity and diversity.* Sunderland, MA: Sinauer Associates, Inc.

———. 1987. *Viable populations for conservation.* New York: Cambridge University Press.

Straskraba, M., and A. H. Gnauck. 1985. Freshwater ecosystems: Modelling and simulation. *Developments in environmental modelling, 8.* New York: Elsevier Science Publishing Co., Inc.

Stuckey, R. L., and K. J. Reese, eds. *The prairie peninsula—in the 'shadow' of transeau: Proceedings of the sixth North American prairie conference.* Ohio Biological Survey Notes no. 15. Columbus, OH: Ohio Biological Survey.

Sullivan, B. J., A. C. McGraw, and D. O. Johnson. 1988. *Environmental aspects of rights-of-way for natural gas transmission pipelines: An updated bibliog-*

raphy. Argon National Laboratory, Report GRI-88/0319. Chicago, IL: Gas Research Institute.

Summers, D., and N. Abbott. 1976. *The great level: A history of drainage and land reclamation in the fens.* North Pomfret, VT: David and Charles.

Thames, J. L., ed. 1977. *Reclamation and use of disturbed land in the Southwest.* Tucson, AZ: The University of Arizona Press.

Timmerhaus, K. 1981. Energy resource recovery in arid lands. In *Earth surface processes.* Albuquerque, NM: University of New Mexico Press.

Todd, J., and N. J. Todd, eds. 1980. *The village as solar ecology.* East Falmouth, MA: The New Alchemy Institute.

Todd, K. 1985. *Site, space, and structure.* Florence, KY: Van Nostrand Reinhold.

Toole, C., B. Wyatt, and S. Taylor. 1985. *Report of the third California salmon and steelhead restoration conference.* University of California Sea Grant MAP-85-4. Eureka, CA: California Sea Grant Marine Advisory Program.

Toole, C., B. Wyatt, S. Somarstrom, and K. Hashagen, eds. 1983. *Report of the first California salmon and steelhead restoration conference.* University of California Sea Grant MAP-18. Eureka, CA: California Sea Grant Marine Advisory Program.

Tourbier, J., and R. Pierson. 1976. *Biological control of water pollution.* Philadelphia, PA: University of Pennsylvania Press.

U.S. Congress, Senate. Committee on Energy and Natural Resources. Subcommittee on Public Lands and Resources. 1977. *Surface Mining Control and Reclamation Act of 1977.* Hearings, 95th Congress. 1st Session on S.7.

U.S. Congress, Senate. Committee on Interior and Insular Affairs. Subcommittee on Energy Research and Water Resources. 1976. *Reclamation Projects.* Hearing, 94th Congress.

U.S. Congress. Conference Committees. 1977. *Surface Mining Control and Reclamation Act of 1977.* Conference Report to Accompany H.R. 2.

U.S. Congress. House. Committee on Interior and Insular Affairs. Subcommittee on Energy Research and the Environment. 1977. *Reclamation Practices and Environmental Problems of Surface Mining.* Hearings, 95th Congress, 1st Session on H.R. 2.

U.S. Department of Commerce. National Technical Information Service. 1986. *Highways and wetlands: Compensating wetland losses.* FHWA-IP-86-22/LBR. Springfield, VA: National Technical Information Service.

U.S. Department of Transportation. Federal Highway Administration. 1979. *Restoration of fish habitat in relocated streams.* FHWA-IP-79-3. Washington DC: Government Printing Office.

U.S. Environmental Protection Agency. Office of Air Programs. 1971. *A guide for reducing air pollution through urban planning.* Washington, DC: Government Printing Office.

Uppal, H. L. 1963. *Reclamation of land rendered barren by salinity and alkalinity.* Punjab, India: Irrigation and Power Research Institute.

Van Kekerix, L., and B. Kay. 1986. *Regeneration of disturbed land in California:*

An element of mined-land reclamation. Sacramento, CA: California Department of Conservation, Division of Mines and Geology.

Vance, M. 1987. *Reclamation of land: A bibliography.* Monticello, IL: Vance Bibliographies.

Vogel, W. G. 1981. *A guide for revegetating coal minesoils in the eastern United States.* U.S. Department of Agriculture, Forest Service, Northeastern Forest Experiment Station. Washington, DC: Government Printing Office.

Wali, M. K., ed. 1975. *Practices and problems of land reclamation in western North America.* Grand Forks, ND: University of North Dakota Press.

———, ed. 1975. *Prairie: A multiple view.* Grand Forks, ND: University of North Dakota Press.

Wali, M. K., and E. M. Watkin, eds. 1987. *Reclamation and revegetation research 6.* New York: Elsevier Science Publishing Co., Inc.

Walker, M. M. 1986. *Botanical clubs and native plant societies of the United States.* Framingham, MA: New England Wild Flower Society, Inc.

Weaver, W. 1985. *Technical specifications for hand-labor erosion control methods.* Unpublished report. Arcata, CA: Redwood National Park.

Webb, R. H., and H. G. Wilshire, eds. 1983. *Environmental effects of off-road vehicles: Impacts and management in arid regions.* New York: Springer-Verlag.

Weddle, A. E., ed. 1987. *Landscape and urban planning 14.* New York: Elsevier Science Publishing Co., Inc.

Westman, Walt. 1978. Measuring the inertia and resilience of ecosystems. *BioScience* 28:705–710.

Whitelaw, G., P. Hubbard, and G. Mulamoottil. 1988. *Restoration of swampland drained for agriculture: Guidelines and recommendations.* Wroxeter, Ontario: Maitland Valley Conservation Authority.

Wright, R. A., ed. 1978. *The reclamation of disturbed arid lands.* Albuquerque, NM: University of New Mexico Press.

Zimmerman, J. H., ed. 1972. *Proceedings of the second midwest prairie conference.* Madison, WI: James H. Zimmerman.

RESTORING THE EARTH: THE ORGANIZATION

HISTORICAL BACKGROUND

The beginnings of modern restoration—the repair of damaged resources, ecosystems, and habitats—can be traced to early twentieth century landscape architect Jens Jensen at the University of Chicago, to the activities of Aldo Leopold and others at the University of Wisconsin in the 1930s, and to various Depression-era Civilian Conservation Corp projects and legislation, including the Federal Aid in Wildlife Restoration Act of 1937. Restoration was little known and seldom practiced until recently, however, outside of a few government agencies. The current burgeoning interest derives, in part, from John J. Berger's book, *Restoring the Earth: How Americans Are Working to Renew Our Damaged Environment* (Knopf 1985; Doubleday 1987), and the activities of Restoring the Earth (RTE), the organization he founded in 1985 to continue the work described and publicized by the book.

Restoring the Earth, a selection of two book clubs, describes fifteen exemplary cases of significant restorations. Each chapter focuses on concerned, dedicated individuals, their motivations, and how they constructed the coalitions necessary to tackle large-scale environmental damage. The book's overall message is the hopeful one that we can move beyond defending a steadily decreasing undamaged resource base to reclaim lost ground, that we need not acquiesce to the destruction of the environment.

ACCOMPLISHMENTS

RTE's first major event, Restoring the Earth Conference • 88, was held January 13–16 at the University of California, Berkeley. Initially conceived as a half-day workshop at which restorationists could share information, a four-day national meeting resulted with more than 1,000 participants, 200 speakers, and extensive media coverage. Millions of Americans learned about restoration through major stories in the *New York Times,* the *Los Angeles Times, Newsweek,* the *Christian Science Monitor, U.S. News and World Report,* and National Public Radio. A TV program on the conference, produced by the University of California, Berkeley, TV Office, was broadcast on cable stations throughout California.

David Brower, Chairman and Founder of Earth Island Institute, has described the conference as "the beginning of a whole new direction in the environmental movement." Since the conference, RTE has received many requests for information on restoration from journalists and people interested in starting restoration projects in their own communities. A summary of conference media coverage and selected comments of conference participants (both available from RTE) further attest to the meeting's impact.

RTE's second major project to date is a proposed multi-part documentary TV series on environmental restoration. A five-part, prime-time version of the series has recently been proposed and is under submission. In the course of approaching a large number of potential corporate underwriters, PBS stations, cable networks, producers, and others in the TV industry, we have made substantial progress toward eventual production of the series. When produced, it will inspire millions of people to become involved with restoration. (A brief brochure and one-page summary of the series are available.)

In 1988 RTE completed two consulting assignments. The first was a study for the College of Natural Resources at the University of California, Berkeley, that surveyed the college's present and potential constituencies to determine how the college could better serve them. We have also set up meetings on a pro bono basis for a California State Assemblyman and his staff to discuss legislation that could be introduced to commit the state to a policy of restoration.

Further details on these and other RTE activities and accomplishments are described in our publication, "Recent Accomplishments," which is available from RTE.

Philosophy

RTE's approach is based on the following premises:

- Restoration requires the broadest possible coalition of interests, involving both technical specialists and concerned citizens in a common effort.
- Restoration, conservation, preservation, pollution abatement, and environmentally sensitive planning and development are compatible, mutually reinforcing responses to our current environmental emergency.
- Without restoration, the relatively pristine areas still remaining will deteriorate. For example, an unrestored toxic waste site will inevitably contaminate an underlying unpolluted aquifer; run-off from a clear-cut forest will affect the integrity and health of downhill creeks.
- If resources such as forests, are not restored, the incentive for destroying the few remaining pristine resources will be irresistible.
- While restoration projects perforce reflect local needs and concerns, their initiators have much to learn from similar experiences elsewhere and in understanding the relationship of their work to regional, national, and global issues.

- Because the core of restoration's appeal is the hope it engenders, the restoration movement must hold out stirring visions as well as practical, realistic means and goals.
- Economic growth and development ultimately depend on a healthy natural environment. By creating jobs, renewing resources, and increasing the wealth and health of communities, restoration can provide important economic benefits.
- Restoration should not be used to justify destructive development. It is necessary to guard against such potential abuse of the concept.

Address and Phone

Restoring the Earth
1713 C Martin Luther King Jr. Way
Berkeley, California 94709
(415) 843-2645

Conference Papers and Tapes

Those papers that could not be included in this volume in their entirety were included as abstracts and may be ordered at cost from the office of the conference convenor.

The entire conference is available on audio tape from:

Conference Recording Service
1308 Gilman Street
Berkeley, California 94702
(415) 527-3600

About the Editor

John J. Berger

Restoring the Earth's executive director, John J. Berger, is a graduate of Stanford University with a master's degree in energy and resources from the University of California, Berkeley. He is the author of a book on nuclear energy and is currently a Ph.D. candidate in Ecology at the University of California, Davis. His dissertation deals with ecological and economic aspects of restoration. He has written on restoration and other science topics for *Audubon, Omni, Sierra,* the *Los Angeles Times,* the *Boston Globe,* and has published with Knopf, Random House, Doubleday, and Ramparts Press. His most recent book, *Restoring the Earth: How Americans Are Working to Renew Our Damaged Environment* (Alfred A. Knopf, 1985; Doubleday, 1987), is in its third edition after acquisition by two book clubs. Mr. Berger previously founded and directed two nonprofit organizations: Alternative Features Service, Inc. (engaged in public education activities); and the ongoing Nuclear Information and Resource Service of Washington, D.C. Founded in 1978, this nonprofit group provides information and services on alternative energy to hundreds of citizen energy groups and to the media, while monitoring developments in nuclear power regulation and siting.

Index

Also Available from Island Press

The Challenge of Global Warming
Edited by Dean Edwin Abrahamson

The Complete Guide to Environmental Careers
by The CEIP Fund

Down by the River: The Impact of Federal Water Projects and Policies on Biological Diversity
By Constance E. Hunt with Verne Huser

Forests and Forestry in China
By Dennis Richardson

Natural Resources for the 21st Century
Edited by R. Neil Sampson and Dwight Hair

Research Priorities for Conservation Biology
ISLAND PRESS CRITICAL ISSUES SERIES
Edited by Michael E. Soulé and Kathryn A. Kohm

Rivers at Risk: The Concerned Citizen's Guide to Hydropower
By John D. Echeverria, Pope Barrow, and Richard Roos-Collins

Rush to Burn: Solving America's Garbage Crisis?
From *Newsday*

Saving the Tropical Forests
By Judith Gradwohl and Russell Greenberg

Shading Our Cities: Resource Guide for Urban and Community Forests
Edited by Gary Moll and Sara Ebenreck

War on Waste: Can America Win Its Battle with Garbage?
By Louis Blumberg and Robert Gottlieb

Wildlife on the Florida Keys: A Natural History
By James D. Lazell, Jr.

For additional information about Island Press publishing services and a catalog of current and forthcoming titles, contact Island Press, P.O. Box 7, Covelo, California 95428.

Island Press Board of Directors